SHARECROPPING
AND
SHARECROPPERS

SHARECROPPING AND SHARECROPPERS

Edited by
T.J. Byres

FRANK CASS

First published 1983 in Great Britain by
FRANK CASS AND COMPANY LIMITED
Gainsborough House, 11 Gainsborough Road,
London E11 1RS

and in the United States of America by
FRANK CASS AND COMPANY LIMITED
c/o Biblio Distribution Center
81 Adams Drive, P.O. Box 327, Totowa, N.J. 07511

British Library Cataloguing in Publication Data

Sharecropping and sharecroppers – (The library
of peasant studies; 6)
1. Métayer system – Congresses
I. Byres, T.J. II Series
333.5′63 HD1478

ISBN 0-7146-3223-6

This group of studies first appeared in a Special Issue on 'Sharecropping and Sharecroppers' of *The Journal of Peasant Studies*, Vol. 10, Nos. 2 & 3, published by Frank Cass & Co. Ltd.

Printed in Great Britain by
John Wright and Sons (Printing) Ltd,
at The Stonebridge Press, Bristol

CONTENTS

Editor's Note

Of all the social relationships that exist in the countryside in contemporary poor countries, and which have existed in the past in 'developed' countries, that of share tenancy is among the most significant and the most fascinating. It is, and has been, geographically widespread, varied in its manifestations, and historically tenacious. Sharecropping has been singled out frequently in land reform programmes as a candidate for elimination. Yet it persists, often in disguised form. It raises difficult theoretical issues, which have attracted the attention of some of the outstanding economists – from Adam Smith, through John Stuart Mill, Karl Marx, and Alfred Marshall – and which remain contentious. Sharecroppers, moreover, have sometimes been involved in important political movements in the countryside. This, too, has given rise to considerable debate. In this double special number of the *Journal of Peasant Studies*, these varied issues are given extensive and rigorous treatment within a predominantly political economy framework. Sharecropping and sharecroppers are examined both in general terms, in a number of theoretical contributions, and in a rich variety of regional contexts, in which their specific manifestations emerge. No 'editor's introduction' is provided. The opening paper is intended, in part, as an Introduction, which attempts to situate the articles by providing overall historical perspectives.

T. J. BYRES

A. Sharecropping in Historical Perspective: A General Treatment

Historical Perspectives on Sharecropping

T. J. Byres*

1. INTRODUCTION

Sharecropping is as old as recorded history. The manner of its first appearance and its precise early origins are, unfortunately, lost in the mists of antiquity, and can only be the subject of speculation based upon the most fragmentary of information. We know remarkably little either about its extent in ancient societies or about the exact nature and implications of its operation there. Yet, that this means of surplus appropriation has an ancient lineage is beyond doubt. When today we grapple with its complexities, puzzle over its inner logic, consider why it should be more widespread or more deeply rooted in some regions than in others, and contemplate the possible circumstances of its demise, it behoves us to recall that we are dealing with a relationship that stretches far back in time. Our endeavours in this respect may have a wider significance than we realise. They may, in unsuspected ways, be capable of casting light upon the past, but they may, too, be severely limited by the absence of any sense of history. Such historical awareness is not always present in contemporary writing on sharecropping. In the present collection, something, at least, of the necessary historical content may be detected. It is to be hoped that this is an initiative which, however provisional, will be built upon in future research.

This cannot be the place for a detailed exploration of sharecropping's earliest manifestations or of its spread in the vast diversity of places in which it has appeared. The essential, painstaking work upon which one might base such a study, that was at all adequate, has not been done by the historians: although even a sifting of the available fragments, in the relevant secondary literature, might itself constitute a volume on its own. All that is possible are a few tentative observations which may establish a little of the necessary historical perspective.

2. SHARECROPPING'S ANCIENT LINEAGE

a. Ancient Greece: the hektēmoroi and after

Classical scholars point to the existence in Attica, before the time of Solon, of a 'class of peasants' [*Hammond and Scullard,* 1970: 492], the *hektēmoroi* (literally, 'sixth-parters' or 'sixth-part men'), who, in the standard accounts,[1]

**School of Oriental and African Studies, University of London.* *I am indebted to the following, who were generous in the help they gave me, providing me with information and directing me to the relevant sources: G.E.M. de Ste. Croix, Charles Curwen and Peter Nolan (on, respectively, ancient Greece, ancient China and modern China). Where I have used the information given by them in personal communications I have indicated that in the relevant notes. They are, of course, in no way responsible for any misguided interpretation I may have put upon the information in question. I have also learned greatly from the contributors to this special issue.*

are represented as having been involved in a kind of sharecropping relationship. It is here that our pursuit of the lost origins of sharecropping begins. At once we encounter issues which recur in the modern literature on sharecropping. We are not, it seems, on alien territory.

The *hektēmoroi* make their first recorded appearance at the time of Solon's reforms (594-593 BC), and seem to have disappeared thereafter [*Ehrenberg,* 1973: 65], but we are cautioned that 'exactly what they were and what Solon did for them is not clearly remembered and is much disputed' [*Hammond and Scullard,* 1970: 492]. The circumstances of their emergence, their extent, and their exact nature are shrouded in mystery, but, long after Solon, Aristotle (384-322 BC), in his treatment of Solon, refers to their existence, and 'evidently thought of them as leasehold tenants, paying one sixth of their crop in rent'.[2] It is this tradition that appears to have survived as the standard version, and to justify their being seen as one of the earliest known instances of sharecropping. In one rendering:

> Most probably they were in origin free small-holders who in difficult times had borrowed corn from wealthier neighbours. These, in default of repayment, progressively took over their land . . . but allowed them to continue to farm it . . . on condition they handed over one sixth of the produce. If they failed to do this, they and theirs were apparently liable to be sold into slavery [*Hammond and Scullard,* 1970: 490].

Such is the view that takes the *hektēmoroi* within our ambit.

For those familiar with sharecropping in more recent times, such an account possesses considerable *a priori* plausibility. Indebtedness has been a common device both for transforming 'free peasants' into sharecroppers and for maintaining and reinforcing the 'web of dependency'[3] of which sharecropping is a part (cf. Bhaduri, [1973] and see the papers by Cooper, Keyder, Halim, Keegan, Stolcke and Hall in this volume). Where, however, familiarity with the wide array of established sharecropping practice would give one pause for thought is in the apparent level of the 'share' paid by the *hektēmoroi.* One sixth is very low by comparison with standard sharecropping agreements encountered pervasively elsewhere. A 50:50 share is the archetypal, recurring division between sharecropper and landlord, but with far higher agreed shares going to the landlord in numerous cases. There are, it is true, recorded examples of landlords taking less than 50 per cent, where, for example, the soil is poor. But that is unusual, and a landlord share of as low as a sixth (i.e. 16.7 per cent) would be most unlikely in circumstances in which, as here, the 'landlords' had the whip hand. We shall encounter the matter of the share in greater detail below.

It has, indeed, been suggested by some classical scholars that the *hektēmoroi* were 'sixth-parters', not in the sense suggested by Aristotle (of paying one sixth and retaining five sixths) but, rather, inasmuch as they handed over five sixths and retained only a sixth (see Woodhouse, [1938: 125-6] and Rostovtzeff [1926: 218-9], both cited in Tuma, [1965: 22 and 252 note 12]). This has been firmly rejected by more recent classicists: Ehrenberg, for example, insisting that 'it is certain that the hectemorii paid one-sixth and not five-sixths' [1973:

401 note 21].[4] One observes, moreover, that just as one sixth (or 16.7 per cent) appears very low for a sharecropping relationship, so five sixths (or 83 per cent) is very high, and would surely have placed in severe doubt the capacity of these cultivators to survive (i.e. 'to reproduce their conditions of labour, in addition to their subsistence' [*Marx,* 1962: 771]). There are limits to the amount of surplus that can be appropriated.

Accepting that the *hektēmoroi* were 'sixth-parters' in the sense that they handed over one sixth, de Ste. Croix has disputed their possible identification as sharecroppers.[5] This he does on the grounds, firstly, that one sixth, as we have suggested, is a very low proportion of the crop to pay in rent for a sharecropper, and that 'even from the later Greek and Roman world, something like one half would be much more normal'; and, secondly, that, indeed, the 'sixth' which gave the *hektēmoroi* their name represented not a sixth of the total crop in rent, but a sixth of the annually taken loan as interest (either a sixth at harvest time or a sixth for every month of the loan). There are, even, other possibilities. For example, might one speculate that although the figure of one sixth is enshrined in the name *hektēmoroi,* the cultivator might have paid in practice a higher proportion? Such divergence between apparent reality and actual practice is by no means unknown in more recent experience. The question of how, precisely, one categorises the *hektēmoroi* remains an open one. But the *hektēmoroi,* however one identifies them, seem to have had cause to be grateful to Solon, since he 'cancelled their debts and restored their land to them' [*Hammond and Scullard,* 1970: 492]. The first historical example, perhaps, of a land reform with the eradication of sharecropping as its central aim.

Thereafter, 'there is extraordinarily little evidence for tenancies of any kind in Attica until around 400 BC'[6]. But, even if we have to wait until the fourth century BC (or somewhat later)[7] for the first true appearance of sharecropping relationships in ancient Greece, sharecropping's ancient lineage is in no doubt. Our brief foray into classical antiquity demonstrates with great clarity, moreover, the relevance of some knowledge of 'sharecropping as an institution'[8] to the historical puzzles of that epoch.

b. Ancient China

Turning from ancient Greece to ancient China, one finds a not dissimilar situation.[9] There is the possibility that sharecropping arrangements existed as early as the Spring and Autumn Period (722-481 BC) – roughly coeval with the pre-Solon *hektēmoroi.* There is a reference to a sharecropping kind of relationship in a compilation of essays called the *Guanzi,* with the essay in question possibly relating to the Spring and Autumn Period. The relevant passage reads:

> If land is equally allocated to the people for their separate use, they will understand the importance of the seasons, and that when time presses they will be faced with hunger and cold and will therefore rise early and go to sleep late. Old and young will not neglect their work, and will not shirk because they are tired: the people will not fear hardship. The

> disadvantage of inequality is that the benefits of the land cannot be fully realised, nor can the people's labour be fully used. If they are not told about the seasons they will not know: if they are not directed they will not work.
>
> If the goods [product] are divided, the people will know how much is theirs and how much is due in tax [*zheng*]. If the proportion is suitably fixed the people will work hard, so that even if they are not urged on old and young will not neglect their work.[10]

The problem is that 'whether the passage represents a system which was already in existence, or whether it was merely the expression of a desirable reform is impossible to tell.'[11] What is of interest, certainly, is that whereas, if we are to accept the standard view, Solon viewed the *eradication* of sharecropping as a 'desirable reform', here its *introduction,* or its perpetuation, is seen in that light. One could find today economists prepared to argue either case (see Pearce's paper), but usually in thoroughly ahistorical terms. That is not how it should be conceptualised.

In order to pursue this point it is worth pausing to consider the *Guanzi* passage. It is of considerable interest to see sharecropping portrayed as part of a system which will ensure the maximum application of labour to the land (which will maximise labour input per acre so that 'the people's labour will be fully used'); and, indeed, as an incentive mechanism (cf. Martinez-Alier and Caballero below) i.e. 'if the proportion is suitably fixed the people will work hard'; and as certainly progressive by comparison with what had preceded it (cf. Adam Smith's views on sharecropping [*Smith,* 1976: 389-90]). Sharecropping, and what it implies, must be seen in genuinely historical terms: that is to say, in relation to the mode of production within which it is inserted, the 'laws of motion' of that mode of production, and the 'moment' of its development. In a particular, pre-capitalist/pre-socialist mode of production, it will represent, for the dominant class, the most appropriate means of surplus appropriation: the means that will maximise labour input, land productivity, and surplus. That would have appeared to have been the case in the Spring and Autumn Period – or so the *Guanzi* argued – and it is instructive to be alerted to that. At other conjunctures, however, it will become a 'fetter' upon the development of the productive forces (to use Marx's famous phrase [*Marx,* 1971: 5]), and must be swept away. Such may be the case with the full realising of capitalism's potential (as Pearce argues below), although that is not inconsistent with sharecropping continuing to exist, as a means of formal subsumption of labour, in the early phases of capitalist development (as Pearce argues, and as Patnaik, also, points out in her paper – although Patnaik does not, quite, limit her statement to the earlier phases). It is the task of political economy to distinguish these distinct sets of circumstances and explore their differing logic.

To return to ancient China, whatever was the case with respect to the Spring and Autumn Period the first definite record of sharecropping appears to be in a memorial to the Emperor by Dong Zhongshu (179-104 BC), a leading Confucian statesman of the Han Dynasty (206 BC – 220 AD). The memorial in question dates, probably, from about 100 BC, and in it Dong recommends a policy which would limit private landholding, because 'the wealthy possessed numerous connecting fields while the poor had not even enough to stick an awl into', one of the results of which was that 'some cultivate the land belonging to the rich and give them half of their yield in rent.'[12] This is manifestly sharecropping, with the archetypal 50:50 division (of which more anon). Sharecropping's ancient lineage is clearly established, if not in China of the Spring and Autumn Period then at least in China of the early Han Dynasty.

c. Ancient India

For ancient India, attention has been drawn [*Kosambi,* 1970: 150 and 152] to the mention of a form of sharecropping in the *Arthaśāstra* of Kautīlya, a celebrated manual of statecraft of the fourth century BC.[13] Kosambi describes the system as follows:

> Unfarmed lands could be . . . assigned on half-shares, generally to those who had nothing to supply but their bodily labour. The seed would be deducted and the womenfolk of the assignee subject to work at milling the grain taken by the state as its portion. Apparently the representative of the state arranged even for tools and oxen in such cases [*Kosambi,* 1970: 150].

So here we have landless labourers, separated totally from the means of production but given access to land (and to the instruments of production) by the state on a 50:50 cropsharing basis, but with the seed deducted before the share was made.

There were, also, more privileged tenants, soldiers and ex-soldiers, 'who would be granted easier terms if they could not take up land even on a fifth-share payment to the state.' [*loc. cit.*]. They, obviously, were not involved in a true sharecropping relationship, and were, indeed, to become ultimately 'a special class which supplied recruits to the army' [*loc. cit.*].

But the first category are clearly examples, to be set beside those of ancient Greece and ancient China, of sharecropping's ancient lineage.

3. THE SPREAD, PERSISTENCE AND OCCASIONAL DEMISE OF SHARECROPPING

Whatever its precise origins in time, sharecropping was certainly established in a variety of places long before the onset of the Christian era, and was sufficiently widespread, or appealing as a potential means of surplus appropriation, to attract some comment upon its rationale and implications. The Chinese and Indian evidence, at least, points to that. The fascination with sharecropping, exhibited thus early, has continued down through the centuries: as sharecropping has spread, flourished (if that is the right word), survived continuously, often for a millennium or more, in particular places (as Gill's paper testifies for Italy); diminished in others only to flourish again (as Keyder suggests for Turkey); had a relatively brief sojourn in still other places; and finally disappeared elsewhere, driven out in some areas by the relentless advance of capitalism and the spread of capitalist relations in the countryside or extirpated by land reform and collectivisation in socialist countries. Some day the full story may be written. For the moment it is instructive to consider some of the manifestations of the spread and persistence of sharecropping, in some of the places in which it has thrived.

a. The Roman Empire

In Rome of the second century AD we find Pliny the Younger (c. 61-112) contemplating a shift from a slave-run estate to sharecropping, to what was to be later called *colonia partiaria,* in response to 'declining returns from his north Italian farms' [*Ste. Croix,* 1981: 257]. What is fascinating about this is that it may have represented a stage in

> the intermediate period . . . between the general use of *slave labour* as the principal way in which the propertied class obtained its surplus, and large-scale *serfdom,* which . . . did not come into existence until the very end of the third century and in some areas was not complete until the late fourth century (as in Palestine) or even the early fifth (as perhaps in Egypt) [*Ste. Croix,* 1981: 256].

Indeed, part of such an instituting of sharecropping was the installing of Pliny's own slaves as overseers, to ensure that his interests were properly served *(operis exactores, custodes fructibus)* which in today's jargon (see Pearce's paper) would be subsumed under 'transaction costs' and 'supervision costs'. But the significant aspect is that here we have the possibility that sharecropping may have been a *transitional* form of surplus appropriation, transitional between a slave mode of production and classical feudalism; between the appropriation of surplus via slavery (in which the direct producer, the slave, is himself no different from an instrument of production, and is explicitly regarded as such), and via labour rent supplied by serfs (who, unlike slaves, possess the means of production).

We use the notion of a 'transitional form' not in a teleological sense (that is to say, there was nothing *inevitable* about the shift from one form to another: indeed, the task of the historian is precisely to delineate why and how the shift took place, without the intervention of any *telos*); and not in any sense that might imply any short-run movement (on the contrary, as indicated by de Ste. Croix the movements in question took place slowly over long stretches of time). The more usual way in which sharecropping is portrayed as a transitional form is as part of the transition from a slowly disintegrating feudalism to a fitfully and unevenly developing capitalism (as in Marx's brief treatment of *métayage* [*Marx,* 1962: 783 et passim], which is pursued by Pearce in his paper). But here we have it appearing in another long and drawn-out transition. Sharecropping would make its appearance again later, as feudalism broke down. But we see it, too, in the nineteenth century arise again in the wake of slavery, in, for example, the American post-bellum South and in Brazil (as explored in the papers by Mandle and by Stolcke and Hall): in the former case as part of a transition to capitalism, finally secured in the 1930s when sharecropping was swept away by a powerful upsurge in the forces of production; and in the latter as part of a *possible* transition to capitalism (as argued by Stolcke and Hall), and sometimes seen as a crucial element in enduring 'semi-feudal' relationships. However one views these latterday embodiments of sharecropping, it would appear that the dilemma faced by Pliny and Younger and the plantation owners discussed by Mandle and by Stolcke and Hall was similar. It is significant that each should reach towards sharecropping as a way of resolving the dilemma. It is a comparison that would repay closer examination. Equally significant, of course, is that sharecropping in Rome should yield not to wage labour, but to serfdom; while in the USA wage labour (capitalism) emerged triumphantly (or, if not that, highly mechanised 'family' farms), and in Brazil less triumphantly. Here, too, are comparisons that are potentially fruitful.

Of those sharecropping arrangements which must have emerged during and after the time of Pliny the Younger, but which certainly existed before his time, (along with a variety of other forms of tenure among peasants of the ancient Greek and Roman world[14]), it has been said that they 'might vary a good deal in the division of the crop between landlord and tenant: half-and-half was common, but the landlord's share (often depending on the nature of the crop) might be as much as two-thirds and was hardly ever less than one-third' [*Ste. Croix,* 1981: 216]. That is a statement that could, equally, be made of a whole number of contemporary poor countries. So, too, could the following account of the 'shares allocated to each party', which has a strikingly familiar ring for anyone conversant with the modern literature (apart, that is, from the possible provision of slaves, as part of the farm's equipment although slaves were, indeed, regarded simply as an instrument of production): 'the shares . . . would naturally differ according to how much the landlord provided of the slaves, animals, tools, corn and other elements in what the Roman lawyers called the *instrumentum* (the equipment) of the farm' [*Ste. Croix,* 1981: 216-7]. Obviously, the greater the proportion of the *instrumentum* supplied by the landlord the greater, *ceteris paribus,* would be the share taken by him. With the sharecropping agreements that one knows about for other places at other times indicate a tendency for the landlord's share to diminish with the quality of the land: since, one would suggest, an insistence upon a relatively high landlord share from poor land, if it could be imposed upon the sharecropper, would put his very existence in jeopardy).

In ancient Rome, as in more recent times, attention was drawn to sharecropping's 'partnership' character and to its peculiar ability to disperse risk for both landlord and tenant, with the second century AD jurist Gaius observing: 'The share-cropper [*colonius partiarius*] has a sort of partnership, and shares both profit and loss with the landlord' [*Ste. Croix,* 1981: 217]. But, for anyone tempted to see the sharecropper's condition in too rosy a light, it is well to recall that if this was a 'partnership' it was hardly a 'partnership of equals' (just as those modern theorists who portray the sharecropping agreement as the outcome of 'bargaining' between landlord and sharecropper should pause to consider that to conceive of 'bargaining' between 'partners' who are grossly unequally endowed is to evoke an ideological notion, which distorts more than it illuminates: as, in different ways, Bhaduri, Pearce and Cooper point out in their respective papers, and as is implicit in most of the other papers). Thus de Ste. Croix points out:

> In the event of a near-total crop failure even the share-cropper, who would then have to give his landlord virtually nothing, would himself soon be left with nothing to eat, and he would be just as much at the mercy of the landlord, or some usurious lender, as any tenant who defaulted in payment of a fixed rent. In a moderately bad year the sharecropper's position, and whether or not he was driven to borrow from his landlord or a moneylender, would depend as much on the size of his plot as on the share of the crop he was allowed to keep [*Ste. Croix,* 1981: 217].

These are important truths whose relevance rings out down through the centuries. We may consider them a little further.

Indebtedness in bad years has proved a powerful additional means of appropriating surplus from poor peasant sharecroppers and has been a condition with a strong tendency to reproduce itself, in good years and bad. And Bhaduri, in his paper, centres attention upon size of plot as an indirect means of control of the labour process by landlords: with land productivity maximised by parcelling out plots in such small sizes that the tenant is just able to meet subsistence requirements, and so, because of the threat to his survival caused by the tiny size of the leased-out land, applies the maximum amount of man hours per acre. Bhaduri demonstrates mathematically the lesson that de Ste. Croix draws from his historical material. The historical perspective is again illuminating. Not does it end at that point.

Just as we are aware, in what we know of sharecropping in the more recent past, of the possibly fictive nature of the sharecropper's share (see, in particular, Cooper's paper), so in the Roman Empire a similar reality is revealed. Our attention is drawn to an episode of c. 181 AD, in the first years of the reign of Commodus, which is recorded on an inscription found at Saik el-Khmis in North Africa (present-day Tunisia). The inscription refers to

> the *saltus Burunitanus,* an imperial estate let out to head lessees, *conductores,* who had sub-let to small peasants, *coloni.* The inscription records a petition by the *coloni* to the emperor, complaining of collusive action to their detriment between their head lessee and the imperial procurator, who were responsible to the emperor for managing the estate. (This situation is likely to have been very common throughout the Greek and Roman world.) The *coloni,* describing themselves as 'most unhappy men' and 'poor rustics', object that more than the proper share of their crops and the prescribed number of days of labour services (six per year) have been exacted and that the procurator has sent in troops and had some of them seized, and tortured, fettered and flogged, simply because they had dared to make a complaint to the emperor[15] [*Ste. Croix,* 1981: 215-6].

So much for 'partnership' (and so much for 'bargaining') and so much for any agreed share for the peasant. It is a situation which, in almost every respect (including that of the extraction of labour services), might have been replicated in Bengal in the inter-war years (as Cooper's paper shows most vividly).

b. Western Europe and France the locus par excellence: métayage and its features

The gradual emergence of large-scale serfdom in the Roman Empire from the late third century onwards, and the possibility that sharecropping acted as a 'transitional form' in this process, have already been noted. The 'transitional form' hypothesis is supported by Duby, for France, but for a later period – the beginning of the ninth century. Thus, parts of the estates of the Abbey of St Germain-des-Près in Paris, in the early ninth century, had been lately obtained as gifts, and:

> Almost all the *manses* recently acquired by gift remained economically dependent farms; forced labour had not been imposed on them, and they did not therefore participate in the general operations of the estate. The lord obtained their assistance in providing for his household in the simplest possible way, that is by demanding from them a portion of their own harvest each year. Here recourse to *métayage,* a farming process fundamentally different from the classic demesne system, was perhaps transitory and gradually prepared the way for the establishment of labour dues.[16] [*Duby,* 1968: 48].

Another source tells us that sharecropping is found elsewhere in France 'in the seventh century formularies of Angers, and probably goes back to Roman times' [*Postan, ed.,* 1966: 199].

But sharecropping, whether as a 'transitional form' or as a survival from Roman times, must, in the end, have given way, along with other forms of surplus appropriation, to labour rent: the essential mode of surplus appropriation in classical feudalism. That cannot be our concern here, fascinating as it is. Our interest, rather, is in the emergence (or re-emergence) of sharecropping in its different forms in western Europe, as classical feudalism began to break down (say from the thirteenth century on): forms which, once established, would show astonishing persistence in some social formations, and even now have not totally disappeared. Nowhere in western Europe has it been more prevalent or more deeply-rooted than in France.

In France the thirteenth century (or roughly 1180-1330), with its 'penetration of money and exchange into rural life' [*Duby,* 1968: 232] was the crucial period. *Métayage* (from *moitié* = half: *medietas* = half), had crept in before that era (and may have survived from an earlier period), being noted, for example, in the middle of the twelfth century on one of the manors of the monks of Cluny[17] [*Duby,* 1968: 275], and favoured by the mayors of Notre-Dame of Chartres.[18] Here was a possible seigneurial response to the 'fall in fixed land rents . . . [and] the general movement in prices which was gradually eroding the value of their money incomes' [*Duby,* 1968: 257]; which would afford some protection to the lords and would, indeed, permit them 'to profit from the hoped-for growth of demesne productivity, as well as from the rise in agricultural prices' [*Duby,* 1968: 276]. As a mode of surplus appropriation it appeared to have considerable advantages, and 'in the thirteenth century the contract of métayage or *facherie* as it was called in Provence, was everywhere in great favour' [*Duby,* 1968: 275]. *Métayage* would become, in English at least, the word which identified the phenomenon of sharecropping for those who wished to describe it.

Despite the proclaimed 50:50 division, embodied in the word *métayage,* it is significant to note that:

> Its stipulations were very diverse. It gave the master a widely differing share in the produce according to the quality of the soil and the condition of the 'farmer': sometimes it gave him two-thirds of the crop, more often a half or a third, but occasionally much less, especially when the soil was not very fertile. In 1338 the Provençal Hospitallers were unable to find a peasant willing to take and farm certain land on their demesne, even when he was offered nineteen twentieths of the produce to keep for himself [*Duby,* 1968: 275].

Again, the fertility of the soil and the capacity of the peasant to survive are crucial: a dead, or severely weakened sharecropper ceases to be a source of surplus, and would hardly be an encouragement for others to take on the lease. The trick was to push the level of surplus appropriated to the point that was just consistent with reproduction of the peasant household. That, one assumes, must have been the seigneurial aim. The share taken by the landlord would, therefore, depend upon whether the lord supplied the necessary instruments of production and inputs (plough, animals, seed). Where he did, his share would be greater. We are told:

> Even where the lord's participation was minimal, the contract assured him an important share in the net profits. For we must not forget that the *métayer* had to deduct seed and sometimes tithes from the portion that was left to him, and this was a heavy charge burdening the normally low yields of agriculture [*Duby,* 1968: 276]

In other words, the question of the agreed share was not necessarily of primary importance. There were other ways of appropriating whatever surplus he might have. The dice were heavily loaded against him.

Leases, moreover, were insecure and liable to termination by the lord [*Duby,* 1968: 257-8, 274, 324]. This might be taken (as contributors to the theoretical literature of the twentieth century have suggested – see Pearce for references) as a way of maintaining the diligence and good husbandry of the peasant, who might be fearful of ejection at the end of the lease if he failed to meet the seigneurial requirements. Whether that would work would depend, of course, upon the availability of alternative leases: it would work far more effectively if leases were scarce, which seems, indeed, to have been the case in the second half of the thirteenth century, with its 'growing overpopulation of the villages . . . [and] immense number of peasants prepared to take land on almost any condition' [*Duby,* 1968: 257]. But the lords were at pains to safeguard their interests more precisely:

> Contracts . . . carefully laid down the duties of the cultivator, obliging him not to break the rhythm of cultivation, to use on the demesne all the manure and straw, to work all the animals and maintain the tools. A system of obligations and safeguards guaranteed the masters against 'farmers' who broke their agreement [*Duby*, 1968: 276]

Such contracts might, for example, impose a minimum number of ploughings on the *métayer*, so ensuring adequate manuring, as in the Toulouse region in the fourteenth century [*Duby*, 1968: 343]. 'Bargaining' there may have been, but of a very one-sided nature.

The lords seem to have found in *métayage* a mode of surplus appropriation which, suitably circumscribed, gave them maximum returns in the given objective circumstances. The *métayers*, according to one account,

> seemed to be wretched individuals settling on the lord's lands empty-handed and expecting immediate help in the shape of an advance of seed, provision of tools and even sometimes of food for their families during the first year. The contracts . . . seem to have held the defenceless peasantry, working on land whose productivity had probably not been increased by improvements in technique . . . , tightly in a net of economic dependence [*Duby*, 1968: 327].

It is a description which fits well the condition of sharecroppers in many countries more than half a millennium later. One recalls James Agee's devastating record of sharecroppers in the American south (Alabama) in 1936, and those remarkable photographs by Walker Evans which accompanied it. Agee wrote of 'an undefended and appallingly damaged group of human beings . . . ignorant and helpless,' and of 'the nakedness, disadvantage and humiliation of their lives' [*Agee*, 1965: 7]. Neither description suggests a 'bargaining' relationship in a sense other than one which 'mocks the condition'[19] of the sharecroppers, whether of fourteenth century France or twentieth century America.

The description of *métayers* quoted in the previous paragraph refers to the *métayers* of the Midi (the south of France), and here we observe a phenomenon that recurs in any examination of sharecropping in a particular geographical area: its spread and rooting in some regions, and its relative absence or disappearance in others. It is a phenomenon which an adequate theorising of sharecropping should be able to encompass (and which is further discussed, for other parts of the world, below). In France: 'it became firmly established in Artois [in the extreme north] and over a great part of Western and Southern France, notably Anjou, Maine, Limousin, Poitou, Roussillon, Quercy and Provence; it was also popular in Burgundy' [*Postan, ed.*, 1966: 324], but 'soon fell into disuse' in parts of northern France. Sharecropping emerged, also, in parts of western Germany (the German terms are *halbpacht* and *halfmann* or *halfen* or *halfwinner*, and *teilbau*), the eastern Netherlands (the Dutch terms are *helftwinning* and *deelpacht*) and Belgium (in Hainault and the Namur and Liège districts: in Hainault the term being *tierce part*, where the landlord's share was supposedly one third) [*Postan, ed.*, 1966: 324; *Slicher van Bath*,

1963: 148], but seems to have lapsed into desuetude in all of these places. Of a later period (the eighteenth century) we are told that sharecropping:

> seems to have been employed more extensively in western, central and southern France. An estimated three quarters of all leased land in these regions was held by sharecroppers. It was rarely found north of the Loire valley except in Lorraine. It was also common in the Rhine province of Prussia, in Savoy and in a few localities of Switzerland [*Blum,* 1978: 102].[20]

Why there should have been these variations remains problematic.

In the passage on the condition of the *métayers* of the Midi, attention is drawn to the failure of the productivity of the land to grow via improvements in technique. The author, in fact, contrasts the south of France, where this was true and where sharecropping was widespread, with the north, where it was not and sharecropping was unusual. Duby does not hazard any analysis of the reasons for this divergence, being content to observe, cautiously, that 'it is obvious that only a vast extension of the present limited researches in social history will reveal to us the real reasons' [*loc. cit.*]. It is a pity that a little of that caution were not present among some of our modern theorists (and especially those writing in the neo-classical tradition) who are willing to make pronouncements about sharecropping on a basis of knowledge of sharecropping situations far more flimsy than that possessed by Duby.[21] But Duby's views on the subject, however tentative, would have been welcome. We are faced with the problem of 'what comes first?' Are the forces of production constrained by a particular set of backward social relations (centring on sharecropping and the 'web of dependency' of which it is an integral part), but in a manner that maximises the surplus for the landlord: so that their continuance seems desirable to him, and any technical progress that might change them undesirable (as argued by *Bhaduri* [1973])? Or, are the forces of production capable, independently of the relations of production, of a development that will sweep away a mode of surplus appropriation, like sharecropping, which interferes with their further expansion? This is one of the many puzzles that political economy must confront in this area. It is inherent in the historical material; and it remains an enigma of considerable significance today, in the analysis of the socio-economic trajectory of poor countries.

c. Southern Europe: Italy and the mezzadria

If France has been the locus, par excellence, of sharecropping in western Europe – *métayage* being the example of sharecropping to which Adam Smith, following the Physiocrats, drew attention in his celebrated attack on that form of tenure[22] – then in southern Europe, it was Italy which could claim distinctiveness in this respect. It was the Italian *mezzadria* system (*mezzadria,* sharecropping and *mezzadro,* sharecropper: from *meta* = half) which John Stuart Mill defended in his *Principles of Political Economy,* following the arguments of the sharecropping landlord, Sismondi.[23] The growth and the persistence of the *mezzadria* in Italy is of considerable interest.

The *mezzadria* made its first known appearance in the ninth century [*Postan ed.,* 1966: 414]. Thus, 'the first recorded contract of this type which has survived dates from 821, and was signed by the Abbots of Monte Amiata in Southern Tuscany' [*Pratt,* 1980: 5] (cf. Gill in his paper below, who mentions the existence of a contract in 819, these two – for 819 and 821 – being the only contracts discovered for the ninth century). For two centuries after 821 no such contracts have been uncovered, and then, in the eleventh and twelfth centuries, they begin to appear intermittently [*Pratt,* 1980: 5]: for example, in 1183 at the Venetian monastery of S. Maria della Carita [*Postan ed.,* 1966: 414]. As in France, and for similar reasons, the period of significant expansion started in the thirteenth century [*Pratt,* 1980: *loc. cit.; Duby,* 1968: 258, 324; *Postan ed.,* 1966: 414]. The *mezzadria* spread, particularly, in Lombardy and Tuscany [*Duby,* 1968: 258, 328]. We are told that by the fourteenth century: 'more than three-quarters of the leases drawn up in the country surrounding Siena in 1316 specified associations of *mezzadria*' [*Duby,* 1968: 324], and that in the second quarter of the fourteenth century, 'the references to *métayers, massarii,* and *partiarii* increased in the communal statutes of Lombardy' [*loc. cit.*]. What manner of relationship was this?

At this time, *mezzadria,* significantly, was 'linked to a short-term lease [of usually five years]' [*Duby,* 1968: 258], with implications I have noted already with respect to the French case. It has been said that the practice was that 'for a few years the peasant received the farm fully equipped on condition that he gave the owner half of the produce' [*loc. cit.*]. The short-term nature of the lease and the prevalence of the 50:50 division seem clear enough, but the notion that 'the peasant received the farm fully equipped' is misleading. Of the early Tuscan contracts it has been noted, indeed, that 'some capital is always provided by the *mezzadro'* [*Pratt,* 1980: 5]. This is inherent in the 'insistence on the residence of the family on the land cultivated, sometimes in a house to be built by the cultivator' [*loc. cit.*]. That would require some capital, as well as labour. The supplying of capital by the *mezzadro* is further implicit in 'the importance given to oxen . . . which may be provided by either side, or the expense may be shared' [*loc. cit.*]. It has been observed that 'some contracts seem built round the beasts . . . The improved farming techniques implied by the use of oxen, especially the substitution of the plough for the hoe, and the use of manure, is a conscious preoccupation of these contracts' [*loc. cit.*].

So, sharecropping, at its inception, was a means of introducing some development of the forces of production, whatever was true of its later implications. But whether this was of any advantage to the *mezzadro* must remain doubtful. This was a mode of surplus appropriation introduced by the lords to 'compensate . . . for the decline of the older type of due' [*Duby,* 1968: 257]; which represented 'changes . . . exceptionally unfavourable to the small cultivator' [*loc. cit.*], changes made possible by 'the depressed condition of a large part of the peasantry consequent on the growth of population and the circulation of money' [*loc. cit.*]. Duby has also noted the tendency of the lords to keep the size of the leased land as small as possible, limiting it to that amount that could be worked by the labour of a single family and which would supply its subsistence needs: this being true of both Italy and France [*Duby,* 1968: 325].[24] I have already had occasion to comment upon the significance of size of holding in the context of the sharecropping relationship, and pointed to its strategic importance in the model developed by Bhaduri in his paper below. Historical observation and theory are gratifyingly congruent, in this instance at least.

The *mezzadria* system extended to other parts of Italy, and was flourishing when John Stuart Mill wrote his *Principles,*[25] apparently in uninterrupted operation in the intervening centuries between its early expansion and then (cf. *Pratt,* 1980: 2 on Tuscany). It has, indeed, continued right through to the present [*Pratt,* 1980: 1-2] and has been described as 'a legal contract [which has] remained virtually unchanged for a thousand years' [*Pratt,* 1980: 2]. Its tenacity is not in doubt.

I have already drawn attention to certain of its features as exemplified by its early manifestations. Among them the 50:50 division (embodied in the name *mezzadria* itself) is frequently given prominence – as so often in writing on sharecropping.[26] But we do well to consider the situation more closely: here and in any case where the 50:50 division is given emphasis. The 50:50 split is sometimes taken to imply a true 'partnership' between landlord (who supplies 'his' land, and perhaps some of the capital and inputs) and the tenant (who supplies 'his' labour, and, again, perhaps some of the capital and inputs): equitable in its exactly equal division of the product, and symbolic of a 'justice' which is more 'feudal' than 'capitalist',[27] but which is superior to the relationship involved in either feudalism or capitalism. We have already noted the partnership notion as enunciated by the jurist Gaius in the second century AD. Now it is given further justification by the apparent presence of a 50:50 division. Some eighteen hundred years after Gaius, indeed, we find a 'theoretician of the Fascist state' asserting:

> The *mezzadria* contract contains within itself the essence of the corporative concept which seeks a harmonious co-ordination between capital and labour. Fascism has the great merit of having seen this essential content of the institution of the *mezzadria* and of having defended it [cited in *Pratt,* 1980: 12].[28]

The reality, however, is very different. In the case of the *mezzadria* – as practised in Tuscany, at least – the level of surplus appropriation was raised by a whole series of 'extras', which were, in fact, an integral part of the system. These were:

> (a) gifts made to the landlord by the *mezzadro* over and above the half of the produce. These included chickens, eggs, first fruits and so on, often imposing a considerable additional economic burden on the *mezzadri,* [and] (b) labour owed to the landlord apart from that directly involved in production, for example, carrying the landlord's part of the produce to market, providing transport for the landlord, and a whole range of other obligations [*Pratt,* 1980: 7-8].

As has been observed,

> the additional obligations destroyed the 50:50 balance of the so-called 'classical' *mezzadria*, i.e. that the produce was divided equally along with the provision of the working capital [*Pratt*. 1980: 8].

Moreover, these obligations were to have profound political significance in Italy, in the twentieth century, when they 'became one of the foci of the political struggle to end the *mezzadria* ... both because of their economic significance, and as symbols of personal dependence' [*loc. cit.*]. All of this – both the 'extra obligations' and the political implications – finds strong similarities in Bengal in the first half of this century (as documented in Cooper's paper herein). Before leaving this issue, it is worth stressing that while the *mezzadria* (with its postulated 50:50 division) dominated central Italy (the regions of Emilia, Tuscany, Umbria and Le Marche), quite 'other types of sharecropping' were found in the south (in, for example, Puglia and Sicily), and were associated with latifundia [*Pratt*, 1980: 1]. There, the 50:50 division need not hold. When, indeed, Mill composed his chapter on sharecropping, he wrote of the share paid by the sharecropper that 'in several districts of Italy it is two-thirds' [*Mill*, 1891: 183].

d. The survival of sharecropping into the recent past: France, Italy and aparcería *in Spain.*

The remarkable thing about sharecropping in both France and Italy (and we observe the same uneven regional spread in Italy that we did for France -- which cries out for explanation) is its survival into the very recent past: to a greater extent in Italy than in France, but still significantly in the latter. Table 1 reveals that in 1892 in France almost 11 per cent of the total farm area was still sharecropped land, a proportion which, given the unequal regional incidence, indicates far higher proportions in some regions; and it was a proportion which remained stubbornly constant in 1946. In Italy, in 1947, the figure was an astonishing 33.3 per cent (Table 1), with, again, the uneven regional incidence meaning that there were far higher proportions in some regions.[29] The figure for

TABLE 1

INCIDENCE OF SHARECROPPING IN SELECTED COUNTRIES: PER CENT DISTRIBUTION OF FARM AREA BY TENURE STRUCTURE

Country	Year	Owner-operated	Rented	Sharecropped	Other modes of tenure
France	1892	52.9	36.3	10.8	–
	1929	60	30	10	–
	1946	53.9	33.5	10.5	2.1
Spain	1952	63.6	22.3	14.1	–
Italy	1947	48.1	18.6	33.3	–

Source: [Dovring, 1960: 150].

Spain in 1952 for *aparcería* (sharecropping), 14.1 per cent, was also very high. The Spanish regional variety is shown in Table 2, where very high proportions of sharecropped land are revealed for certain regions (for example, more than 30 per cent for Levante and Cataluna y Baleares, and more than 20 per cent in Extramadura and Canarias). This particular historical perspective, showing a quite remarkable persistence for what is portrayed, often, as a 'pre-capitalist' or 'transitional' form, suggests that for contemporary poor countries – and especially those in which sharecropping is deeply rooted – those who predict the imminent demise of this mode of surplus appropriation, where capitalism appears to be developing in the countryside, may have longer to wait than their predictions would have one believe (that may be true of Halim and of Pearce, for example, in this collection). Even where capitalism *is* developing, sharecropping, it seems, has the capacity to survive. Patnaik may be correct when she

TABLE 2

INCIDENCE OF SHARECROPPING IN THE DIFFERENT REGIONS OF SPAIN IN THE EARLY 1950s: PER CENT DISTRIBUTION OF CULTIVATED LAND BY TENURE STRUCTURE

Region	Owner-operated	Rented	Sharecropped	Other modes of tenure
Andalucia Occidental	66.1	8.8	7.1	18.8
Andalucia Oriental	62.2	9.2	8.8	19.8
Castilla la Vieja	71.1	4.1	15.5	9.3
Castilla la Nueva	69.7	7.5	8.8	14.0
Aragon	75.2	12.7	4.6	7.5
Levante	45.8	38.8	4.6	10.8
Leonesa	71.3	1.0	10.2	17.5
Cataluña y Baleares	53.7	32.2	8.4	5.7
Extremadura	55.2	21.1	1.4	22.3
Rioja y Navarra	66.3	9.7	14.4	9.6
Galicia	67.3	11.0	21.4	0.3
Vascongadas	60.6	0.5	36.5	2.4
Canarias	66.2	24.0	3.4	6.4
Asturias-Santander	60.0	1.7	37.8	0.5
TOTAL	63.6	14.1	8.8	13.5

Source: [Dovring, 1960: 456].

argues, in her paper (below), that sharecropping is perfectly 'compatible with transitional and capitalist productions as well'. If we draw that conclusion, then the lesson that history teaches is that we need to explore rigorously an altogether more varied set of capitalist paths in the countryside than has hitherto been our practice.

A little further exploration of Spanish sharecropping, and a comparison with the sharecropper of the American South,[30] are revealing. Of Spanish sharecropping in the inter-war years it has been written:

> In many countries (the southern parts of the United States, for example) the sharecropper has traditionally been more akin to a permanent hired

> hand than to an independent entrepreneur in the sense that the owner provided all the instruments of cultivation, closely supervised the work that was done, extended credit on which the sharecropper lived until harvest, and then marketed the crop, from the proceeds of which he paid the sharecropper a fixed portion. The sharecropper contributed only his labour. In Spain, although sharecropping of this kind was not uncommon, owners usually contributed far less, and the sharecropper far more, of his time and equipment to the process of production [*Malefakis*, 1970: 117].

This small contribution of the owners, in time and to costs of production,

> did not spare the sharecropper entrepreneurial risks that confronted the small owner or tenant, nor justify in his eyes the large portion of the crop – usually half – that had to be paid to the owner at harvest [*Malefakis*, 1970: 118].

The arguments we have developed above, moreover, suggest that Spanish sharecroppers, like their counterparts elsewhere, would have had appropriated from them even more than the suggested 50 per cent of the crop. The foregoing description, however, does underline the *variety* of sharecropping situations of which we must take cognisance.[31] That is another lesson that history has to teach. We must carefully distinguish the recurring similarities, which are striking, from the substantive differences – such as between the different kinds of sharecropping in Italy, or between the Spanish sharecropper and the sharecropper of the American South – which are no less striking. Our analytical framework should be capable of encompassing both. If it cannot, then it is seriously defective.

e. *Russia: the* polovniki

So far, in our historical survey, we have concentrated on western, southern, and south-western Europe. We may move on, briefly, to eastern Europe. Space forbids anything other than the most cursory of treatment (and, anyway, the material available in English is sparse).

It emerges that in Russia of the Mongol era (roughly from the middle of the thirteenth century until the end of the fifteenth) there existed among the 'seignorial peasants' a category of sharecroppers, the *polovniki* (singular, *polovnik:* sharecropping = *polovnichestvo*). The *polovnik:*

> agreed to remain with the lord for an agreed upon number of years and to pay a fixed proportion of his crop to the lord as rental. The size of this share varied, ranging from one fifth to as much as one half. On his arrival the *polovnik* received a loan of cash or grain from the landlord in order to get started on his farming, and to carry him over until he brought in his first crop. As a rule the sharecropper was either excused from paying the obligations due to the prince, or else paid a lesser amount than did other peasant renters. The fact that the sharecroppers were given loans, that

> they were willing to work on shares, the most oppressive type of rental agreement, and that they paid little or no taxes to the government, are indications that they must have been in very reduced circumstances. Sharecropping was not limited to large holdings. The Novgorod land registers of the late fifteenth century show that sharecroppers sometimes worked land that belonged to *svoezemtsi,* the smallest proprietors, and even rented part of the holdings of seignorial peasants. [*Blum,* 1961: 100-1].

The *polovniki* were the largest category of 'landlord's peasants' [*Blum,* 1961: 94], existing alongside (and sometimes even working land from) on the one hand the *starozhil'tsy,* the largest category of seignorial peasants (literally 'old inhabitants' [*Blum,* 1961: 98]), who had fixed obligations in cash and kind (*obrok*) and in labour services (*barschina*) – largely, it seems, in terms of kind [*Blum,* 1961: 101-2]; and on the other the *novoprikhodtsy* (the 'newcomers'), who, during a grace period of initial settlement on an estate, were exempt from some of the obligations that the *starozhil'tsy* had to pay [*Blum,* 1961: 98-9]. Obviously, landlords must have competed sharply for new settlers, in circumstances of labour shortage [*Blum,* 1961: 100]. But, for whatever reasons, the lot of the *polovnik* was so miserable that he could not take advantage of this competition for 'newcomers', and he had to settle for the inferior and more oppressive sharecropping relationship.

The tale is a familiar one; except, that is, for the very low rental share of 20 per cent (one wonders how common so low a proportion was?). Here we have poor peasants, caught up in a web of indebtedness; occasionally working the land of richer peasants as well as that of the lord (the *kulak* tradition, it seems, is a very old one in Russia; and maintained 'in very reduced circumstances').

Sharecropping arrangements, we are told, continued into the first half of the sixteenth century, being particularly common then in Novgorod Land; and in the north, too, it persisted 'where it was much employed by monastic landlords in the seventeenth century' [*Blum,* 1961: 221]. *Polovniki* are found, too, in the eighteenth and nineteenth centuries, in the north and mainly in Vologda guberniia, renting 'from other peasants, or from nobles or merchants' [*Blum,* 1961: 484], paying from 40 per cent to 50 per cent of the crop, and sometimes more, as rent, and being subject, in addition, to 'a number of other payments in kind, labor, and sometimes in cash' [*loc. cit.*]. Again, a familiar story. On the verge of becoming serfs to their landlords in the early eighteenth century, they were saved by a decree of the tsar in 1725; they were supposedly protected from undue exploitation by regulations issued from the Ministry of the Interior, one hundred years later, in 1827; and, eventually, they were settled by the government on state land [*Blum,* 1961: 484].

They were, in Russia, however, – unlike, say, France or Italy – a remnant from the past, declining from 15,000 male *polovniki* in the 1740s to under 3,000 in the late 1850s [*Blum,* 1961: 484-5]. For, of course, in the eighteenth century the Russian peasantry were thoroughly enserfed [*Blum,* 1961: 414-441]. Sharecropping, as a mode of surplus appropriation, existed, anachronistically and only rarely, in the interstices of the new order. It was not until 19th February, 1861, with the Act abolishing serfdom, that this order came to an end [*Gerschenkron,* 1966: 706 et passim]. For the eventful years between then and 1917 there are few references in the English literature on Russia to *polovniki,* or to sharecroppers by any other name.

f. China: sharecropping's shifting fortunes

If, on the eve of her revolution, Russia's 'agrarian question' did not include sharecropping as a burning issue, the same could not be said of China. There, sharecropping loomed large.

I have discussed, already, sharecropping's appearance in ancient China. We may resume the story in mediaeval China. A recent writer directs our attention to the period of the Song Dynasty (960-1280 AD), and tells us that 'we know from official documents that sharecropping had existed in previous dynasties, whereas fixed rents are not mentioned' [*Bray,* forthcoming]; and she argues that in the early Song period the *predominant* system of tenure was likely to have been sharecropping. She tells us that during the Southern Sung (1127-1278) all three forms of rental contract known in later times – i.e. (a) sharecropping, (b) fixed rent in kind, and (c) fixed rent in cash – were to be found, but by then sharecropping seems to have been on the decline in southern China, with fixed rents paid in grain common. It seems, indeed, that 'by the 14th and 15th centuries tenants in Fujian and most other parts of central and southern China always paid their rents as a fixed quantity' [op. cit.]. This takes one through the Yuan or Mongol Dynasty (1280-1368) and the Ming Dynasty (1368-1644). By the time we come to the Ching Dynasty, which stretched from 1644 to 1911, sharecropping, however, had re-asserted itself, so that, of the three rental systems which have been identified, and which continued to exist side-by-side, sharecropping predominated [*Myers,* 1980: 104]. In China, sharecropping's shifting fortunes present a fascinating problem for the agrarian historian.

For the Republican Period (1911-1949), threading one's way through a complex labyrinth of practice, one may say the following for China as a whole (recalling that all-China generalisations, for this period as for any other, given the vast size of the country, must be regarded with caution):

(a) By the 1930s, about 30 per cent of total farm area was under tenancy [*Feuerwerker,* 1977: 58, table 16, part 1].[32]

(b) With respect to the peasants themselves, about 55 per cent of farm families [*Feuerwerker,* 1977: 57, table 16, part 1][33] were involved in tenancy relationships of one kind or another: between 30 and 33 per cent as 'pure' tenants and just under 25 per cent as part-owners/part tenants [*loc. cit.*].[34]

(c) Of these tenancy relationships, in terms of the three rental systems, fixed rent in kind was most common, with 51 per cent of rental agreements falling into this category; cash rent was least common, with 21 per cent; and sharecropping represented 28 per cent [*Feuerwerker,* 1977: 59, table 16, part 3].

It could not be said, on a basis of these figures, that share-crop rental still *predominated* in China as a whole among rental forms. Obviously, however, it continued to be of considerable significance for the Chinese tenantry.

What emerges from the wide sweep of Chinese agrarian history is a fascinating alternation of sharecropping and, it seems, rents fixed in kind as the dominant mode of surplus appropriation for landlords. The logic of that alternation presents an important puzzle for the analyst of sharecropping. In Keyder's paper on Turkey, we are presented, for a far briefer span, with a *cycle* of alternation between sharecropping and owner-cultivation: a cycle which is theorised in terms of the periodic cycles of the world economy and the corresponding movements of agricultural prices, and which postulates usury as an essential mechanism ordering the cycle. One guesses that the Chinese alternation has been mediated by largely *endogenous* influences (or contradictions). A treatment which attempted explicitly to theorise it would be of great interest.

The true significance of both tenancy in general and sharecropping in particular for specific regions of China – which are likely to be larger than many European countries – is, however, masked by treatment at the all-China level. Thus, tenancy was far more prevalent in the rice region of southern China, where 40 per cent of the land was rented, than in the wheat region of the north, where the figure was 13 per cent [*Hou,* 1979: 164];[35] and not only that but, as one might expect, it is clearly the case that there is a 'much greater incidence of *pure tenancy* in the rice-growing provinces of the Yangtze Valley and the southern coast than in the wheat-growing north' [*Feuerwerker,* 1977: 56, emphasis mine].[36] Yet, curiously, where tenancy did prevail in the wheat-growing north, the incidence of sharecropping was far higher than it was among rental contracts in the rice-growing south. Buck's massive survey, covering the years 1929-33, revealed that 'the Wheat Region has one-third of its lease contracts on the share-rent system, compared with only one seventh for the Rice Region' [*Buck,* 1937a: 198]. It is, of course, the case that, given the far higher rate of tenancy in the south, sharecropping must have been more common there than in the north.[37] That demands an explanation, as do the considerable regional variations in sharecropping we have already encountered for other countries. Why was sharecropping more common as a mode of surplus appropriation in south China than in the north?[38]

This, however, is not the question addressed by Cheung, in one attempted explanation of regional variation in China [*Cheung,* 1969: 70-72]. Instead, he focuses upon the greater incidence of sharecropping among rental contracts in the north than in the south, and asks why *that* should be so. His suggested explanation runs in terms of risk aversion: given the far higher proportional variance in wheat yields than in rice yields, sharecropping is the preferred rental contract, for both landlord and tenant, since it disperses risk. That I cannot pursue in detail here. I would simply note that it is not true (given the figures cited above) that 'in China share tenancy reportedly was more frequent in the wheat region than in the rice region' [*Cheung,* 1969: 70].[39] In fact, let us reiterate, it was more frequent, in terms of the proportion of land so worked and the proportion of peasants so covered, in the south; even if its incidence among actual rental contracts (far less common in the north than in the south) was greater in the north. Apart from that, we note an explanation for regional variations in the incidence of sharecropping in India which suggests the precise opposite to Cheung [*Rao,* 1971]. Rao, on a basis of Indian data, says that, indeed, it is 'situations of *high uncertainty* [which] may necessitate fixed cash rents' [*Rao,* 1971: 582, emphasis mine]. In Rao's case (in West Godavari district of Andhra Pradesh) sharecropping is found to be common in the rice zone, while cash leases predominate in the tobacco zone. The data on irrigation, rainfall and prices reveal a far higher degree of uncertainty in tobacco than in rice. Large size (and cash rentals), it is postulated, are the preferred form in tobacco, because this reduces risk for the landlord; while in rice, small size associated with sharecropping, in conditions of relative economic certainty, maximise the return for the landlord. Again, we cannot pursue the details of the argument here, but clearly the question remains an open one. It may, of course, be the case that the true explanation is not to be pursued in such terms (risk, uncertainty etc.) at all. This is argued explicitly in the Pearce paper, and is suggested in several of the other papers.

g. India

Just as in China, so in India any treatment of the 'agrarian question' in the inter-war years (as, indeed, today) would have to include sharecropping as an important element. In the present collection, the papers by Bhaduri, Cooper and Patnaik, in their different ways, cast valuable light upon this: in both theoretical terms (Bhaduri and Patnaik) and with specific, concrete analysis (Cooper). But in India's case, the reconstruction of the historical record, after the noting of the existence of sharecropping in the fourth century BC, appears to present greater difficulties. Whatever the reasons (be they the absence of the necessary evidence or the failure of historians to pursue the issue) we cannot make for India historical statements of the kind that are possible for China: at least not for before the nineteenth century. We have no reliable estimates, for the past, of sharecropping's precise incidence in India, of the kind we have noted for other countries. For the very recent past, we have an estimate that suggests that sharecropping covered 20 per cent of the cultivated land of Bengal in 1940 (see Cooper's paper), and another suggesting 20 per cent to 25 per cent [*Khusro,* 1973: 23]; while other evidence indicates:

> that in Punjab before partition the *batai* [the name for sharecropping in the Punjab] system prevailed over as much as a third of the cultivated area; and that in Madras too it accounted for substantial areas [*Khusro,* 1973: 24].

But such evidence is fragmentary. We are hardly better off with respect to contemporary India. There are no national estimates (for a heroic attempt to calculate the extent of tenancy in general, using the 1961 Census data, see Khusro [1973: 26]: Khusro estimates this at about a quarter of the cultivated area, and if we assume sharecropping to have constituted about half of this we may hazard a guess that sharecropped land may have been about $12\frac{1}{2}$ per cent of total cultivated land in India as a whole). An estimate for Bengal in the 1960s puts sharecropping's incidence at around 35 per cent of arable land [*Khusro,* 1973: 23-24]. There is a large body of impressionistic and localised evidence pointing to the prevalence (and perhaps even the growth in the 1950s and 1960s, at least [*Khusro,* 1973: 24]) of sharecropping in India. The pity is that the evidence does not exist on a systematic basis.

Kosambi boldly postulates a degree of continuity between India of Mauryan times and the present that is startling even by the standards of the Italian *mezzadria.* He says, à propos of Magadha, ancient Bihar:

> This half-share cropping, incidentally, persisted in Bihar through feudal times and was recognised by the British as the landlord's prerogative wherever sanctioned by custom [*Kosambi,* 1970: 150].[40]

It continues to exist pervasively in Bihar [*Prasad,* 1973, 1974, 1979], but whether we can realistically assume unbroken continuity between the time when the *Arthaśāstra* was composed and now, without supporting evidence, seems doubtful. Kosambi appears to wish us to accept that proposition, although he insists that its acceptance should not lead one 'to the view that there was no change in India'. He implies that, although this mode of surplus appropriation has survived, the class relationships with which it has been associated have changed fundamentally: from a direct relation with the state in Mauryan times and before, through a relationship with 'feudal intermediaries' [*loc. cit.*]; to (extrapolating his argument) what in recent times has been described as 'semi-feudal' production relations [*Bhaduri,* 1973: 120][41][42]. The implied position that sharecropping can, indeed, be embedded in *quite distinct modes of production* is an important one, upon which we have already commented. An historical perspective allows one to grasp this. Of crucial importance is the point upon which Kosambi insists (and which emerges from an historical approach, properly pursued): that although sharecropping persists, it may do so in circumstances of dramatic change (of transition from one mode of production to another right through to capitalism itself, as Patnaik argues and our French and Italian examples demonstrate). Kosambi's essential point, in this respect, does not, in fact, require the postulating of *unbroken* continuity. It is perfectly consistent with the early appearance, the disappearance, and re-appearance (perhaps several times over) of sharecropping as a major mode of surplus appropriation: whether in alternations over very long stretches of time (as in China), or with greater cyclical regularity (as, apparently, in Turkey). The Chinese (and the Turkish) example suggests that investigation of Indian agrarian history from ancient times until the present with such a hypothesis in mind might be fruitful.

Kosambi's argument relates to eastern India. But, as we have suggested, sharecropping has existed pervasively in areas other than Bihar and West Bengal. The *bargadars* (*barga* = share lease), *adhiars* (*adhi* = half) and *bhagdars* (*bhag* = share, part) of eastern India (see Cooper's paper for these terms and [*Thorner,* 1956: 32]) are matched by the *bataidars* of the Punjab and Uttar Pradesh (see Patnaik's paper and [*Thorner,* 1956: 34; *Thorner and Thorner,* 1962: 7]), the *sajehdars* of Uttar Pradesh (*sajeh* = partner)[43] [*Kotovsky,* 1964: 175; *Thorner and Thorner,* 1962: 7], the *vāramdars* of Tamil Nadu (*vāram* = share) [*Gough,* 1981: 47; *Mencher,* 1978: 84-94], and those sharecroppers of western India whose sharecropping relationship in the 1950s Daniel Thorner could describe as 'one of the most common arrangements under which land is still worked . . . today' [*Thorner,* 1956: 46]. They, and many others, deserve careful treatment: of the differing arrangements which their sharecropping relationship has entailed, of their historical evolution, of their contemporary manifestations. Occasionally they make an appearance in the historical record, identified by alert historians,[44] but we know astonishingly little about them. When, with the operation of the 'new technology' in the 1970s, we suddenly encounter a transition from 'traditional' forms of sharecropping (as in the deltaic area of Andhra Pradesh) to new arrangements – 'cost-share leasing' – which are still recognisably sharecropping arrangements, but which represent a qualitative change in the sharecropping relationship ([*Parthasarathy and Prasad,* 1974: 170], discussed in Byres [1981: 431]) it is startling to realise the extent of our ignorance of those 'traditional' sharecroppers. Our understanding of the new, and of the process of transition from old to new, is inhibited by our lack of comprehension of the old.

India's sharecroppers have escaped precise investigation by historians and economists. Historians have, as yet, not captured them in their investigative net (which is why Cooper's paper herein is welcome: demonstrating, as it does, what can be done). Economists have preferred, for the most part, to theorise *about* them, rather than examine them at first hand (with honourable exceptions: for references see Pearce's paper). When anthropologists go and live in Indian villages for protracted periods sharecroppers seem not to cross their path, or, if they do, to be unworthy of mention. On the rare occasions when they do attract attention, the results can be surprising. Thus, the distinguished sociologist (anthropologist?: the distinction is not always clear) André Beteille notes 'the widespread practice of sharecropping in states like West Bengal, Tamil Nadu and Kerala' [*Béteille,* 1972: 22]. And how does he explain this? It is widespread, it seems, firstly because landowners prefer to

> avoid manual work, particularly when it is arduous and onerous . . . [an] aversion . . . [which] is particularly strong in areas of wet paddy cultivation where the work is both more difficult and more unpleasant than in areas of wheat cultivation (*p. 22*).

This would explain a preference for using wage labour. But sharecropping is preferred, it is posited, since 'even the work of personally supervising paddy cultivation in the monsoon is something which many landowners prefer to avoid' [*loc. cit.*] While delightful in its simplicity, it is not an explanation which even the slightest acquaintance with the kind of historical perspectives I have been trying to delineate would allow. Apart from anything else, sharecropping is extensively practised *both* 'for the rice crops in eastern and southern regions in India and for wheat in the northern regions' [*Rao,* 1971: 583. see also *Lade-*

jinsky, 1965]). Moreover, the postulated aversion of landowners would be equally consistent with other forms of tenancy such as fixed cash rentals, or fixed kind rents. It may have been a stray hypothesis thrown out casually (although I think Béteille meant it more seriously than that), but it does not survive a moment's reflection in the light of any historical awareness.

h. Persia

If, indeed, one were seeking out deep historical roots, remarkable continuity of existence, and persistence as a mode of surplus appropriation, one would have to include Persia, where the crop-sharing contract, the *muzara'eh,* between the *muzari,* the landlord, and the *amil,* the tenant is 'of undoubted antiquity . . . It was recognised by the early Islamic jurists and would appear to have been in existence in Persia in Sasanian times' [*Lambton,* 1953: 206]. The Sasanian dynasty was founded in 226 AD and lasted into the first half of the seventh century. Those early Islamic jurists discussed various aspects of the sharecropping relationship, and when, some 1,300 years after the end of the Sasanians, in the early twentieth century, the Civil Code appeared, it included a 'body of law concerning land . . . [which was] largely a restatement of Islamic theory' [*Lambton,* 1953: 194], which assuredly contained provisions with respect to crop-sharing, including the nature of the share itself [*Lambton,* 1953: 206-9].

When Lambton completed her *magnum opus* in 1951, she could say: 'the relations between the peasant and the landowner are, for the most part, based on a crop-sharing agreement' [*Lambton,* 1953: 306], a statement based upon meticulous scholarship and quite unique fieldwork experience. Sharecropping, as in other places, has displayed a signal tenacity, and, Lambton wrote in 1951, 'from time to time there has been talk of fixing by law a minimum share for the *amil,* i.e. the crop-sharing peasant' (p. 209). But, she hastened to add, 'no government has had the courage to tackle the question' [*loc. cit.*]. That is not surprising, where the landlord interest has been so patently dominant.

The question of the share is of considerable interest, and, as in other places, its consideration reveals the divergence between ideology and practice. We have already discussed the ideology of the 50:50 share. In Persia, we find a basis for division yet more plausible in its apparent embodying of a 'just' division:

> Traditionally, five elements are taken into account in dividing the crop: land, water, draught animals, seed, labour; theoretically one share is allotted to each element and goes to whoever provides that element [*Lambton,* 1953: 306].

There is an almost neo-classical purity in the insistence that each input should attract its due reward, and, indeed, Alfred Marshall, in his celebrated treatment of sharecropping, that has been so influential, noted this very Persian system with apparent favour, citing the observation that since 'the landlord generally owns two [of the elements], so he gets two-fifths of the harvest' [*Marshall,*

1949: 536, note 1]. Lambton, however, wastes little time on what is so manifestly an ideological construct:

> In fact, however, this is little more than a theoretical abstraction, and the actual division, although it is materially affected by the ownership and provision of these different elements, is seldom made on the basis of the allocation of five equal shares for each element [*Lambton,* 1953: 306].

She observes that: 'In general the scales are weighted in favour of the [landlord], and little or no protection is afforded to the tenant' [*Lambton* 1953: 209]. She notes, in fact, a variety of division: including 50:50, but, also, two-thirds to the landlord where he provides only seed, four-fifths to the landlord where he provides only seed, and many other variations [*Lambton,* 1953: 308-29]. The variety is, indeed, remarkable, and sometimes bewildering. But it seems not to follow, in any regular pattern, division according to the provision of the five elements.

In a sense, however, the discussion of the division becomes almost irrelevant, when one is told:

> In addition to the share of the crop which goes to the landlord . . . there are also a number of dues deducted from the crop . . . Only after all the various dues have been taken into account can an accurate estimate be made of the share of the crop remaining to the peasant. In general, it would appear that in those areas where the peasant receives a larger share of the crop under the crop-sharing agreement he has to pay heavier dues under other heads, while in those areas where his share is smaller the extra levies made on him are fewer [*Lambton,* 1953: 309].

Here we have sharecropping's catch 22. The 'dues' or 'personal servitudes', as Lambton calls them (p .330) are various: they include liability to transport the landlord's share from threshing floor to granary; labour service or *bigari* ('the most onerous of the personal servitudes', p. 330) – so many days of free labour or provision of an ass, for building construction, irrigation works, road building, agricultural labour in the fields; dues of clarified butter, or hens, cheese, eggs, fuel, partridges, a lamb, a kid; a poll tax; the obligation to meet the expenses of entertaining government officials; and so on [*Lambton,* 1953: 330-36]. One might, indeed, hypothesise that the surplus appropriated *in toto* by the landlord class from the sharecropping peasant, via both rental share and 'dues', was such as to drive the peasant family to the barest minimum consistent with its reproduction as a cultivating unit. It is a phenomenon with which we are by now familiar.

4. CONCLUSION

The foregoing historical excursus has ranged far and wide. We have moved from ancient Greece and Rome, ancient India and ancient China, through western Europe (France, Germany, the Netherlands, Belgium), southern

Europe (Italy), south-western Europe (Spain), eastern Europe (Russia), south-eastern Europe and Asia Minor (Turkey), to sharecropping's persistence in India, China and Persia. Yet I have encompassed only the merest fraction of what would be desirable in a full treatment. We might, for example, have sought sharecropping's early origins in the Byzantine Empire from the seventh century on [*Postan, ed.,* 1966: 211]. And one should, perhaps, comment upon Britain before concluding. When Smith wrote his *Wealth of Nations,* he could say of *métayers* that 'they have been so long in disuse in England that at present I know no English name for them' [*Smith,* 1976: 389], although he did note of his native land that 'this species of tenants still subsists in some parts of Scotland. They are called steel-bow tenants' [*Smith,* 1976: 391]. The implication is that sharecropping had existed in both England and Scotland as a significant tenurial form, but the secondary historical sources are strangely silent. The quest for England's (and Scotland's) lost sharecroppers would be a fascinating one to pursue.

I have had little to say about one crucial area in which sharecropping has been of critical importance: the American post-bellum South. That is treated in Mandle's paper (and further references are given by Pearce in his contribution). I have said nothing about Africa, where the manifestations of sharecropping have been of considerable interest; or about Latin America and South-East Asia, where, too, its operation has been significant. Fortunately, South Africa (or part of it) is covered by Keegan's paper, Malaysia by Halim's, and Brazil by Stolcke and Hall. The papers by Keegan and Stolcke/Hall are of especial interest since, in their different ways, they are able to examine the actual instituting of sharecropping, its rise, operation and demise. There is, also, some brief discussion of West Africa (Ghana and Nigeria), Latin America (Brazil), and South-East Asia (Malaysia, the Philippines and Vietnam) in Pearce. Other parts of South-East Asia[45] and South Asia,[46] where sharecropping exists as an important institution, might have been covered; and so, too, the many countries of Latin America where it has been of considerable prominence – for example, Bolivia, Chile, Colombia, Guatemala, Peru, Venezuela, Mexico.[47] The Middle East, too, is notable by its absence: countries like Iraq [*Warriner,* 1969: 78, 85, 87, 103] and Egypt [*Warriner,* 1969: 51, 54, 419]. Far more might have been said about south-eastern Europe – for example, Romania [*Blum,* 1978: 436] [*Warriner,* 1969: 49]. One might have mentioned Japan [*Smith,* 1966: 152-4]. And so one might go on. The list is by no means exhaustive.

What emerges very clearly from our brief historical survey is that sharecropping has existed since remarkably early times (Slicher Van Bath's comment [1963:148] that 'the share-cropping lease [is] the oldest kind of leasehold agreement' is manifestly true); has been extremely widespread geographically; has shown an often astonishing historical continuity and tenacity; has, in some pre-capitalist/pre-socialist societies, such as China and Turkey, displayed a capacity to disappear and re-appear. It continues to exist pervasively in the so-called Third World. It is to be hoped that this collection illuminates both the historical record and the contemporary situation. But much remains to be done.

NOTES

1. See, for example, *Andrewes*, 1956: 84-6; *Forrest*, 1966: 147-54; *Ehrenberg*, 1973: 56-65. I am indebted to G.E.M. de Ste. Croix for drawing my attention to these sources. See also Hammond and Scullard [1970: 492].
2. The *hektēmeroi* are mentioned in Aristotle's *Athēnaion Politeia* (which dates from the 320s BC in its present form) and in Plutarch's *Life of Solon* (c. 100 AD), and in one or two other sources. G.E.M. de Ste. Croix was kind enough to provide this information with respect to Aristotle and Plutarch and to point out Aristotle as the apparent source of their identification as lease-hold tenants. The words cited are those of Dr de Ste. Croix.
3. Adi Cooper's phrase. See her paper below.
4. Ehrenberg goes on to say:

 > Even one sixth is anything but a light burden, especially in view of the generally poor land the small farmer held. The common charge of a tithe (= one-tenth) was often bad enough.

 It is the case that the quality of the soil has been singled out as a possible variable in the determination of the share (cf. Cooper in her paper), with the share paid as rent declining with quality. Certainly, however, by the standards of those sharecropping arrangements that are recorded historically, one sixth is a very small share (see below).
5. The interpretation that follows in this sentence is his, provided in a personal communication.
6. Ste. Croix, personal communication.
7. Ste. Croix comments immediately after the sentence quoted: 'and I can't remember any [evidence] for sharecropping until later still.'
8. Ste. Croix's phrase.
9. Charles Curwen has kindly provided the information on ancient China that follows.
10. Guanzi (730?-645 BC) gave his name to the compilation in question. He was made Prime Minister of the State of Qi (modern Shandong Province) in 685 BC and was responsible for several important economic and administrative reforms The book in question, however, does not represent his writings, but represents his 'school' (he was revered for several hundred years in the State of Qi and elsewhere). It seems to be argued among scholars of ancient China that the book, *Guanzi*, was compiled during the Warring States Period (403-221 BC), although some of the individual pieces may be from a much earlier time. The piece quoted, we are told, could well date from the Spring and Autumn Period, when a system of land tenure appeared and evolved, in which land was allocated to peasant families, who remained virtual serfs, in return for rent. The term *zheng*, translated tax in the passage, was a form of tax or rent, varying according to the quality of the land, which was first imposed in the State of Qi, of which Guanzi was Prime Minister. (I am indebted to Charles Curwen for the foregoing; the translation of the passage is his.)
11. Charles Curwen (personal communication).
12. The passage cited is from *Han Shu, 24 A,* and has been translated by Charles Curwen. Charles Curwen writes (personal communication):

 > Dong Zhongshu attributes such changes in the rural scene to the reforms of Shang Yang (about 350 BC). Shang Yang (died 338 BC) was a statesman of the state of Qi . . . In fact, the Shang Yang reform which, *inter alia,* officially recognised private land ownership, merely legalised the free buying and selling of land, which had been going on as far back as the Eastern Zhou (722-403 BC).
13. The *Arthaśāstra* is 'a book in Sanskrit rediscovered in 1905 after centuries of total eclipse' [*Kosambi*, 1970: 142], which has as its theme 'state policy and management', the title

meaning 'The science of material gain'. There is some dispute as to its full authenticity and its precise dating [*Basham,* 1959: 50, 79], although none as to its great value in giving 'very detailed instructions on the control of the state, the organisation of the national economy, and the conduct of war' [*Basham,* 1959: 80]. It has been described as 'a most precious source-book for many aspects of ancient Indian life' [*Basham,* 1959: 80]. The great Marxist historian of ancient India, Kosambi (whose authority is considerable) tells us that any doubts as to its authenticity, which were first entertained on its discovery, 'have since been removed by long discussion' [*Kosambi,* 1970: 142]. For an exhaustive treatment of the considerable debate on authorship and date see Kangle [1965, vol. 3: 59-115]. Kangle concludes that 'there is no convincing reason why this work should not be regarded as the work of Kautilya, who helped Candragupta to come to power in Magadha' (*p. 106*). Magadha is the present-day Indian state of Bihar. Candragupta Maurya acceded to power at some time between 324 and 313 BC [*Basham,* 1959: 50], and 'was the chief architect of the greatest of India's ancient empires'. He seems to have ruled for 24 years [*Basham,* 1959: 53].

14. Ste. Croix tells us:

> ... one can distinguish several categories ... according to the forms of tenure by which they hold their land, for example:
> 1. Freeholders who had absolute ownership of their plots.
> 2. During the Hellenistic period, men who in practice were virtually absolute owners for the duration of their lives, but who held their land on condition of performing military service, and who could not transmit it directly to their heirs without the endorsement of the king. (In practice, such lots often became eventually equivalent to freeholds.)
> 3. Tenants who either (a) held on lease for their lives or (much more commonly) for a term of years (which might in practice be renewable at the option of one party or the other or both) or (b) were what English lawyers call 'tenants at will', subject at any time to the possibility of being ejected or having their terms of occupation made more onerous (e.g. with a higher rent). These tenants, of either class, would fall into four broad groups, according to the nature of the landlord's return, which might be (i) a fixed rent in money, (ii) a fixed rent in kind, (iii) a share of the crop (the Roman *colonia partiaria,* modern *métayage,* or share-cropping), or (iv) labour services. Combinations of these alternatives were of course possible: in principle, a share of the crop could be combined with a fixed rent in money or kind or both; a rent could be made payable partly in money and partly in produce at a predetermined price ...; and labour services could be exacted in addition to rent in money or kind – although in point of fact there is surprisingly little evidence ... for labour services on anything more than a very small scale (about six days a *year*) until we reach the sixth century [*Ste. Croix,* 1981: 213-4].

One observes that his second category of tenure is similar to the one which appears in the *Arthaśāstra* and is identified by Kosambi (see above). The *variety* of tenurial forms is also worthy of note. When one considers that slavery, too, was in existence, then alone of the possible forms of surplus appropriation which is absent is that via wage labour. The critical issue, however, is which of these *predominated* at any particular time.

15. It is not perfectly clear from the passage that the *coloni* in question were, indeed, *coloni partiarii* (i.e. sharecroppers), but from the context it seems likely.

16. Duby's statement is based upon the evidence of a famous and detailed inventory, or *polyptyque,* 'made for the estates of the abbey of St. Germain-des-Près ... in the time of Abbot Irminon at the beginning of the ninth century' (*p. 12*) a document which 'enumerates 2,088 households and 10,026 individuals' (*p. 13*). This must be taken as quite remarkably detailed for the period, and evidence for a large estate.

17. Thus:

> Already in the mid-twelfth century the monks of Cluny had on one of their manors concluded an agreement of this kind with a villager, granting him temporarily a few

arpents of arable, providing him with the seed and the two oxen he needed in order to equip a suitable plough, and in exchange reserving half the crop [*Duby,* 1968: 275].

18. It is of interest that knowledge of this preference comes from an attempt to *prevent* the mayors of Notre-Dame of Chartres from using a sharecropping contract. They 'had to promise their lords not to use the contract of *medietates* in dealing with peasants' [*Duby,* 1968: 275]. We are not told how successful the attempted proscribing was, nor the reasons given for it by the lords in question. It may have had a fate similar to those attempts in recent times to legislate sharecropping out of existence (in, for example, India), which have simply resulted in driving sharecropping 'underground' and transferring it from a written to an oral lease, thereby rendering the sharecropping agreement more insidious and making its terms even more onerous for the sharecropper.

19. The phrase is Isaiah Berlin's [*Berlin,* 1958: 9], made with respect to peasants, but in a different context.

20. One recalls Adam Smith's observations in the *Wealth of Nations*: 'in France, where five parts out of six of the whole kingdom are said to be still occupied by this species of cultivators' [*Smith,* 1976: 391]. The editors of the bicentenary edition tell us [*loc. cit., note 35*] that according to Quesnay 6 parts out of 30 of the lands of France were cultivated by *fermiers,* and the rest (83%) by *métayers*; while Turgot added that

    > in Picardy, Normandy, the environs of Paris, and in the majority of the Provinces of the North of France, the land is cultivated by farmers. In the Provinces of the South they are cultivated by *métayers*; therefore the Provinces of the North of France are incomparably more wealthy and better cultivated than those of the South [*cited in loc. cit.*].

    It was the position of the Physiocrats that Smith expanded upon in the *Wealth of Nations,* in his famous passage on sharecropping [*Smith,* 1976: 389-91]. Smith located sharecropping in *historical* terms: identifying it as part of a system superior to slavery (which, he says, sharecropping succeeded); but clearly inferior to fixed money rent tenancy, inasmuch as unlike that system it destroyed the incentive to 'improve' (i.e. make productive investment on the land taken on share-crop).

21. Among the neo-classicals, Marshall, of course, is exempt from such a charge. He provided the first neo-classical analysis of sharecropping, in a few pages of pellucid exposition, considerable scholarship, and undogmatic awareness of a variety of possible outcomes [*Marshall,* 1949: 534-7]. One must also single out Clive Bell, among those in the modern neo-classical camp, for his detailed knowledge of sharecropping in the Indian state of Bihar [*Bell,* 1976 a, 1976 b, 1977].

22. See note 20 above.

23. See Mill [1891: 183-93]. Mill did note the French situation, but preferred to focus on Italy for his defence of sharecropping: arguing that in Italy (or at least in Tuscany) farms are of reasonable size, there is security of tenure, and the landlord is 'willing to provide capital for improvements' (*p. 185*). Mill describes Sismondi as a 'resident proprietor, intimately acquainted with rural life' (*p. 189*), and likely to be possessed, therefore, of more accurate information than mere travellers. That Sismondi's testimony might, on the contrary, be rendered suspect by his interested position is not contemplated by Mill. Sismondi's account, we note, related especially to Tuscany, where his own property lay. On Sismondi cf. Johnson [1950: 113]. Johnson describes Sismondi as 'a landowner and *métayer* landlord, who favored the system . . . Sismondi described the life of *métayers* in romantic terms' [*loc. cit.*]. Mill does, however, point out (*p. 184*) that Sismondi drew attention to the absence of secure tenure and consequent 'inferior condition of the *métayers* in some provinces of Naples, in Lucca, and in Riviera of Genoa' [*loc. cit*]. Mill further comments (*p. 184*) upon Châteauvieux's observations on sharecropping in Piedmont in his *Letters From Italy,* in Lombardy (*p. 187*) and in 'the valley of the Arno, in its whole extent, both above and below Florence' [*loc. cit.*].

24. It is of interest to note, in this context:

> Indirectly, following a rise in the productivity of land, the size of holding necessary to maintain a peasant family declined sharply from about a hundred acres a manse on an average in the ninth century to anywhere between twenty and thirty in the thirteenth ... The manse comprised the amount of land necessary to maintain one household at subsistence level; it therefore differed from region to region according to soil-fertility and customary subsistence needs of the peasant family unit [*Mukhia,* 1981: 278 and 297 note 33].

(On this Mukhia cites [*Hilton,* 1973: 28; *Latouche,* 1961: 82; *Gurevich,* 1977: 8].)

25. See note 23 (above).

26. Thus, for example, Bell: 'The division of the gross produce into two equal parts holds sway widely in India and elsewhere' [1977]; or, 'it is widely observed that the landlord and peasant split the produce in proportions which are close to, if not exactly, 50:50' [1976b].

27. Bell (who is not, one hastens to add, a proponent of the 'partnership' view, at least not in its 'romantic' form) avers: 'The division of the produce into two equal parts reflects a feudal rather than a market valuation of the inputs provided by the two parties to the contract' [*Bell,* 1972: 1]. And in another place he writes of the tendency to 'settle for that magic number, which has also the advantage of sounding equitable. Thus the particular rental share is elevated to the status of a social norm' [*Bell,* 1976b: 587]. As I point out in the text, the critical consideration, in practice, is the amount of surplus appropriated by means other than the 'rental share'. If this is high, then nonsense is made of the supposedly 'equitable' 50:50 division.

28. Pratt further points out that this 'principle' was embodied in the 1933 national *codice,* in which the terms of the contract were stated, in its second article, thus: 'the [*mezzadria*] is a special contract of an associative character, based on reciprocal trust, designed to guarantee solidarity and collaboration between the contracters' [*Pratt,* 1980: 12]. One notes, with interest, sentiments very similar to those contained in the text, uttered by Alfred de Musset of sharecropping in Lower Maine in the nineteenth century:

> Musset saw sharecropping in Lower Maine not as a source of backwardness but, on the contrary, as the starting point of all improvements: 'instrument of progress', 'alliance of capital and labour'. The resident owners (especially after 1852) initiated improvements in all fields, and their sharecroppers slowly followed their example [*Weber,* 1979: 12].

29. Pratt tells us that a survey done in 1951 (on a different basis to the one from which the figures in Table 1 are taken, let us stress) showed that in the four central regions of Italy (Emilia, Tuscany, Umbria, and Le Marche) the *mezzadria* constituted 62% of cultivable land. For the rest of Italy, the proportion under all forms of sharecropping was 17.5%: roughly half of that (8.7% of cultivable land) under the *mezzadria* and half under other forms (8.8% of total cultivable land). The 'other' forms occupied 18% of cultivable land in Puglia and 33.3% in Sicily [*Pratt,* 1980: 1].

30. On the American South see Mandle's paper below.

31. Martinez-Alier's theoretical approach to sharecropping (in his paper below) derives from his actual experience with sharecroppers in Spain: 'This way of looking at sharecropping I learnt from rural workers themselves, in Andalusia in the mid-1960s.' The references to his earlier published work may be seen in his paper.

32. Feuerwerker cites two estimates: that of John Lossing Buck [*Buck,* 1937b: 55-56], which comes out at 28.7%, and that of the National Land Commission's 1934-35 Survey, which is 30.7%.

33. According to Buck, 56% [*Buck,* 1937b: 57-59]; and according to the National Agricultural Research Bureau's figures 54%.

34. Buck's figures [*loc. cit.*] show 33% tenants and 23% owner-tenants; the NARB figures 30% and 24% respectively.

35. Hou cites Buck's figures [*Buck*, 1937a: Tables 19 and 21, ch. 6].

36. For the figures by provinces see Feuerwerker [1977: 57, Table 16, Part 1].

37. Thus, crudely, one seventh of 40% of the operated area (the situation in the south) works out at 5.7%, while one third of 13% (the northern case) is 4.3%.

38. There is far more valuable information on sharecropping in China during the Republican period than I have been able to draw upon here. I would direct the reader's attention to Ash's excellent study of Kiangsu province in the 1920s and 1930s [*Ash*, 1976].

39. See note 37.

40. Kosambi is one of those historians who see India as having passed through a feudal epoch, from about the end of the sixth century [*Kosambi*, 1970: 195] until it finally 'broke down ... [in] the middle of the 17th century under Aurangzeb' [*Kosambi*, 1975: 392], its lingering embers finally extinguished with the suppression of the 1857 Mutiny [*Kosambi*, 1946: 396]. Thereafter, it seems, came an epoch of 'semi-feudal agrarian production' [*Kosambi*, 1946: 396]. For a contesting of this – with respect to a whole range of Indian Marxist writers – see Mukhia [1981]. References to the relevant writing can be found there.

41. That Kosambi, too, thought in terms of 'semi-feudal' relations is indicated in the reference given in note 40.

42. Bhaduri's analysis is rooted in West Bengal: for an analysis specifically covering Bihar see Prasad [1973, 1974, 1979].

43. These *sajhedars* were, in fact, the product of legislation in U.P. (the U.P. Zamindari Abolition Act), which supposedly forbade sharecropping. The sharecroppers were known as *bataidars*, and the Act, as originally presented, stipulated that cultivators should work the land either with their own hands or with hired labourers. But the members of the legislature (themselves larger landlords, or their relatives) inserted a third category: so-called *sajhedars* ('partners' or 'helpers') [*Thorner and Thorner*, 1962: 7]. Daniel Thorner tells that the peasants of U.P. in the 1950s would say: '*Bataidari aur sajhedari eki ciz hai*, i.e. The system of *sajhedars* and that of *bataidars* are one and the same' [loc. cit.]. See also *op. cit., p. 62.*

44. For example in Mughal India [*Habib*, 1963: 77, 193, 195-8, 215, 218-28, 232-3, 236-7, 249, 314] or in the nineteenth century see Baden-Powell [1892: vol. 1, 269; vol. 3, 36, 119, 341, 371, 406] on *bataidars, varamdars* and *adhiars*.

45. For example, Thailand [*Turton*, 1976: 275], Indonesia [*Geertz*, 1963: 98-100] [*Schletema*, 1931].

46. For example, Bangladesh – see the papers by Pearce and by Cooper – Pakistan [*Stevens, Alavi and Bertocci eds.*, 1975: 323-5, 336-8, 341-2] and Sri Lanka [*Warriner*, 1969: 70] [*Gunasinghe*, 1980: 121-8].

47. See Pearse [1975: 7, 21, 34, 37, 49, 78, 85, 88-90, 92, 95, 107-8, 114,6, 180, 223-9], and Warriner [1969: 226-7, 230, 32].

REFERENCES

Agee, James (with Walker Evans), 1965, *Let Us Now Praise Famous Men*, London: Owen.

Andrewes, A. 1956, *The Greek Tyrants*, London.

Ash, Robert, 1976, *Land Tenure in Pre-Revolutionary China: Kiangsu Province in the 1920s and 1930s*, London.

Baden-Powell, B.H., 1892, *The Land Systems of British India*, three volumes, London.

Basham, A.L., 1959, *The Wonder That Was India*, New York.

Bell, Clive, 1972, 'Technological Change and the Demise of the Sharecropper', paper given to *Peasants Seminar* of the University of London, 3 November.

__________, 1976a, 'Production Conditions, Innovation, and the Choice of Lease in Agriculture', *Sankhya* C, 1976.

__________, (with Pinhas Zusman), 1976b, 'A Bargaining Theoretic Approach to Cropsharing Contracts', *American Economic Review*, September, vol. LXVI, no. 4.

__________, 1977, 'Alternative Theories of Sharecropping: Some Tests Using Evidence From North-East India', *Journal of Development Studies*, July, vol. 13, no. 4.

Berlin, Isaiah, 1958, *Two Concepts of Liberty*, Oxford.

Béteille, André, 1972, *Inequality and Social Change*, Delhi.

Bhaduri, A., 1973, 'A Study in Agricultural Backwardness Under Semi-Feudalism', *Economic Journal*, March, vol. 83, no. 329.

Blum, Jerome, 1961, *Land and Peasant in Russia From the Ninth to the Nineteenth Century*, New Jersey.

__________, 1978, *The End of the Old Order in Rural Europe*, Princeton, New Jersey.

Bray, Francesca, forthcoming, 'Patterns of Evolution in Rice-Growing Societies', *Journal of Peasant Studies*.

Buck, John Lossing, 1937a, *Land Utilisation in China*, Nanking.

__________, 1937b, *Land Utilisation in China. Statistics*, Nanking.

Byres, T.J., 1981, 'The New Technology, Class Formation and Class Action in the Indian Countryside', *Journal of Peasant Studies*, July, vol. 8, no. 4.

Cheung, Steven S., 1969, *The Theory of Share Tenancy*, Chicago.

Dovring, Folke, 1960, *Land and Labour in Europe, 1900-1950*, second edition, The Hague.

Duby, Georges, 1968, *Rural Economy and Country Life in the Mediaeval West* (translated by Cynthia Postan), London.

Ehrenberg, Victor, 1973, *From Solon to Socrates*, second edition, London.

Feuerwerker, Albert, 1977, *Economic Trends in the Republic of China, 1912-1914*, Ann Arbor, Michigan.

Forrest, W.G., 1966, *The Emergence of Greek Democracy*, London.

Geertz, Clifford, 1963, *Agricultural Involution*, Berkeley and Los Angeles.

Gerschenkron, A., 1966, 'Agrarian Policies and Industrialisation: Russia, 1861-1917', in M.M. Postan and H.J. Habakkuk (eds.), *The Cambridge Economic History of Europe. Volume VI. The Industrial Revolution and After*, London.

Gough, Kathleen, 1981, *Rural Society in South-East India*, Cambridge.

Gunasinghe, Newton, 1980, *Changing Socio-Economic Relations in the Kandyan Countryside*, Ph.d. thesis, University of Sussex.

Gurevich, A., 1977, 'Representations of Property During the High Middle Ages', *Economy and Society*, vol. VI, no. 1.

Habib, Irfan, 1963, *The Agrarian System of Mughal India*, London.

Hammond, N.G.L., and H.H. Scullard (eds.), 1970, *The Oxford Classical Dictionary,* second edition, Oxford.

Hilton, R.H., 1973, *Bond Men Made Free,* London.

Hou, Chia-chou, 1979, 'The Structure and Determinants of Tenure System in Modern China: 1900-1940', in Chi-ming Hou and Tzong-shiam You (eds.), *Modern Chinese Economic History,* Taipei, Taiwan.

Johnson, D. Gale, 1950, 'Resource Allocation Under Share Contracts', *Journal of Political Economy,* April, vol. 58.

Kangle, R.P., 1965, *The Kautiliya Arthaśāstra Volume III. A Study,* Bombay.

Khusro, A.M., 1973, *Economics of Land Reform and Farm Size in India,* Madras.

Kosambi, D.D., 1946, 'The Bourgeoisie Comes of Age in India', *Science and Society,* 10.

———, 1970, *The Culture and Civilisation of Ancient India in Historical Outline,* New Delhi.

———, 1975, *An Introduction to the Study of Indian History,* second edition, Bombay.

Kotovsky, Grigory, 1964, *Agrarian Reforms in India.* New Delhi.

Ladejinsky, W., 1965, *A Study of Tenurial Conditions in Package Districts,* New Delhi. Reprinted in *Mainstream,* March 12, 20, 27, 1965, vol. 3, nos. 28, 29, 30.

Lambton, Ann K.S., 1953, *Landlord and Peasant in Persia,* London.

Latouche, Robert, 1961, *The Birth of Western Economy,* translated E.M. Wilkinson, London.

Malefakis, Edward E., 1970, *Agrarian Reform and Peasant Revolution in Spain,* New Haven.

Marx, Karl, 1962, *Capital,* Volume 3, Moscow.

———, 1971, 'Preface to *A Contribution to the Critique of Political Economy',* in Karl Marx and Friedrich Engels, *Historical Materialism,* London.

Mencher, Joan P., 1978, *Agriculture and Social Structure in Tamil Nadu,* Bombay.

Mill, John Stuart, 1981, *Principles of Political Economy,* People's Edition, London.

Mukhia, Harbans, 1981, 'Was There Feudalism in Indian History?', *Journal of Peasant Studies,* April, vol. 8, no. 3.

Myers, Ramon H., 1980, *The Chinese Economy, Past and Present,* Belmont, California.

Parthasarathy, G. and D.S. Prasad, 1974, 'Responses to and Impact of H.Y.V. Rice According to Land Size and Tenure in a Delta Village, Andhra Pradesh', *The Developing Economies,* June, vol. XII, no. 2.

Pearse, Andrew, 1975, *The Latin American Peasant,* London.

Postan, M.M. (ed.), 1966, *The Cambridge Economic History of Europe. Volume I. The Agrarian Life of the Middle Ages,* second edition, London.

Prasad, Pradhan, 1973, 'Production Relations: Achilles Heel of Indian Agriculture', *Economic and Political Weekly,* 12 May, vol. 8, no. 9.

———, 1974, 'Reactionary Role of Usurers' Capital in India', *Economic and Political Weekly,* August, Special Number, vol. 9, nos. 32-34.

———, 1979, 'Semi-Feudalism: The Basic Constraints of Indian Agriculture', in Arvind N. Das and V. Nilakant (eds.), *Agrarian Relations in India,* New Delhi.

Pratt, J., 1980, *The Mezzadria,* Amsterdam (January, Euromed Working Paper No. 9).

Rao, C.H. Hanumantha, 1971, 'Uncertainty, Entrepreneurship, and Sharecropping in India', *Journal of Political Economy,* May/June, vol. 19, no. 3.

Rostovtzeff, M., 1926, *A History of the Ancient World, 1. The Orient and Greece,* Oxford.

Schletema, A.M.P.A., 1931, *Deelbouw in Nederlandsche-Indie,* Wageningen.

Slicher Van Bath, B.H., 1963, *The Agrarian History of Western Europe, 500-1850,* London.

Ste. Croix, G.E.M. de, 1981, *The Class Struggle in the Ancient Greek World,* London: Duckworth.

Smith, Adam, 1976, *An Inquiry into the Nature and Causes of the Wealth of Nations,* bicentenary edition, Oxford. First published 1776.

Smith, Thomas C., 1966, *The Agrarian Origins of Modern Japan,* New York.

Stevens, Robert D., Hamza Alavi and Peter J. Bertocci (eds.), *Rural Development in Bangladesh and Pakistan,* Honolulu, Hawaii.

Thorner, Daniel, 1956, *The Agrarian Prospect in India,* Delhi.

Thorner, Daniel and Alice, 1962, *Land and Labour in India,* London.

Tuma, Elias H., 1965, *Twenty-Six Centuries of Agrarian Reform,* Berkeley and Los Angeles.

Turton, Andrew, 1976, 'Northern Thai Peasant Society: Twentieth Century Transformations in Political and Jural Structures', *Journal of Peasant Studies,* April, vol. 3, no. 3.

Warriner, Doreen, 1969, *Land Reform in Principal and Practice,* London.

Weber, Eugen, 1979, *Peasants into Frenchmen. The Modernisation of Rural France, 1870-1914,* London.

Woodhouse, W.J., 1938, *Solon the Liberator: A Study of the Agrarian Problems in Attica in the Seventh Century.* London.

B. Theoretical Perspectives on Sharecropping

Sharecropping: Towards a Marxist View

R. Pearce*

I. INTRODUCTION

Sharecropping has proved the commonplace rental contract in many areas of the world during some period of the rural economy. It scarcely needs mentioning that this relationship is still an important facet of many agricultures, most notably where capitalism has not developed or has done so only tenuously. It is not surprising, therefore, that sharecropping has elicited much academic interest, discussion and investigation, and more significantly, a substantial volume of government legislation, often ineffective and of questionable intent.

There has in fact been a plethora of literature on sharecropping over the past fifteen years, sometimes commenting, with barely concealed surprise, on the widespread contemporary and historical pervasiveness of such arrangements, or, with thinly veiled quasi-disbelief on the ability of such contracts to realise levels of input intensity and output little different from those associated with allegedly more 'efficient' methods of organisation. A widely held view is that where it has not already been superceded, sharecropping is soon likely to be replaced by relationships more suited to the demands of a 'modern' agriculture.

Most of the literature on sharecropping approaches the issue from the perspective of neo-classical economics. This essay will attempt to chart a different course, navigated from outside the entrenched ideological perspective which characterises so much of the literature, a perspective preoccupied with technical relationships to the extent that social relations of production are mere parameters to be taken as found.

The concern here is with an attempt to outline and develop a framework through which it is possible to understand sharecropping as part of the changing fabric of history, as a relationship which is the product of certain social and economic contingencies and will disappear with such contingencies. The object is to approach such contracts from within the Marxist tradition, a tradition which until recently has applied itself all too little to the detail of rural change.

I am concerned with the incidence and persistence of sharecropping, with its relative technical performance and with the circumstances of its demise, but shall primarily focus on the production relationships which characterise the circumstances where sharecropping becomes the preferred form of contract, and how changes in these relationships precipitate changes in the manner in which surplus labour is appropriated. The overall purpose is not only to

**University of Reading.* *I wish to thank Terry Byres for extensive comments on an earlier draft of this essay.*

'explain' the occurrence of sharecropping, but also to understand the implications of such contracts in the context of contemporary rural change, more particularly in the context of emerging agrarian capitalism.

The opening section comprises a limited review of literature describing past and present manifestations of sharecropping contracts and some preliminary discussion of significant features. This is followed by a short critique of the existing theoretical literature, leading to the central feature of the essay: that is, an attempt to describe a methodology, from a Marxist position, which may allow such contracts to be understood as instruments of surplus labour appropriation rather than mere technical arrangements facilitating the allocation of productive resources.

II. SHARECROPPING, PAST AND PRESENT

That sharecropping has proved pervasive and often persistent over time and space has been noted. In order to provide an informed background to the theoretical discussion which follows, it is worthwhile reviewing, briefly, the circumstances, both historical and contemporary, which have led to the widespread occurrence of such contractual arrangements.

In many parts of Western Europe the challenge to seigneurial power and the break-up of classically feudal estates began before the thirteenth century, the process approaching completion by the year 1500. (On the notion of 'classical feudalism' cf. *Banaji,* 1977a). The release of the mass of the population from legal ties of bondage characteristic of serfdom facilitated the emergence of other manifestations of agrarian exploitation, paramount amongst which was the leasing of land on the basis of product-sharing rents reinforced by tenant indebtedness. Evidence suggests that these leases were frequently little more than labour contracts with the landowner controlling the productive operations with some precision [*Jones,* 1968; *Snowden,* 1979; *Bloch,* 1966: 146-149; *Blum,* 1978: 102; *Mitrany,* 1930: 588-592; *Jones,* 1831: 72-100].

The 'plight' of sharecroppers was a subject of concern to the early classical economists [*Smith,* 1976: 381-396; *Mill,* 1909: 302-317] with the condition of the French *métayer* unfavourably compared by Mill, with that of the Tuscan *mezzadrio,* the difference attributed to the latter's superior security of tenure. In fact, this mythical wellbeing of the Tuscan sharecroppers owes more to the propagandist accomplishment of the Italian landowning class (in particular through the writings of Sismondi), and their very thorough mechanisms of social control than to reality [*Snowden,* 1979], and the growth of Italian capitalism in the latter half of the nineteenth century saw a steady deterioration in their condition [*Gill,* 1981], until their virtual reduction to the status of day labourers during the period witnessing the emergence of Italian fascism.

A similar trajectory of events is revealed when considering the development of agrarian capitalism in the southern United States. The literature on the post-bellum South is prolific, but most commentators would agree with Holley [1980] that abolition of slavery and the consequent breakdown of the plantation system following the civil war severely disrupted both the labour supply and agricultural finance. In the event labour control in the form of closely specified and supervised sharecropping contracts filled the breach. Such arrangements were largely, although not entirely, associated with cotton production, and the demise of 'King Cotton' coupled with a changing pattern of productive forces all but drove the 'cropper' from the southern American scene by the end of the 1930s. The poverty and degradation of these people as they were swept aside by social relations more consistent with the advance of rural capitalism are poignantly chronicled by such novelists as Steinbeck [1951] and by Agee [1981].

Martinez-Alier [1971] describes the widespread use of share contracts in many areas of southern Spain as late as the 1960s, this reflecting a return to land relations prevalent before the Spanish civil war, the main motivation being a response to a shortage of cheap labour following organised action on the part of wage labourers leading to an increase in minimum wage levels. Such tenancies were particularly common with labour-intensive crops and are portrayed by Martinez-Alier as a form of incentive wage designed to increase the intensity of labour.

It is probable that sharecropping also became common in parts of Brazil for similar reasons: that is, to bring about an increase in the productivity of labour [*Martinez-Alier,* 1974]. The patriarchal dominance of the landowners of north-east Brazil is well documented by Johnson [1971: 114-131] although Goodman [1977] points to the relative decline of such forms of labour organisation in favour of fixed rental contracts combined with an increase in the employment of wage labour, in response to growing urban demand for agricultural commodities. Where sharecropping contracts do persist they are characterised increasingly by more rigorous supervision and surplus extraction.

Each of the above scenarios conforms to classical Marxist expectations as outlined in *Capital III* [*Marx,* 1981: 938-940]. Sharecropping emerges from more complete forms of labour control such as the slavery of the southern United States or villeinage of western Europe. It emerges as social relations underlying these forms of bondage prove unable either to accommodate demographic growth, or to secure the expansion of output necessary to meet a growing urban demand characteristic of incipient capitalism [*Jones,* 1968]. Production relations based on land leasing and small-scale peasant farming bring about an intensification of labour 'effort' leading to reduced costs of production and/or an increase in per capita output. It is in this latter context that sharecropping appears as a *transitory* phenomenon. It is transitory in the sense that the production relationships which it embodies themselves are swept aside by the 'engine of capitalist accumulation': that is, by the intrusion into agricultural activities of productive forces capable of revolutionising labour productivity, and requiring a scale of production and level of investment such that existing production relations become a constraint. The reasons why these relations act as a fetter on the development of productive forces and are displaced will be considered below.

Of course the dialectic of socio-economic change is always considerably more complex than this simplified outline suggests. Nevertheless, the point remains that sharecropping as a form of labour organisation can be viewed as intermediate between forms of agrestic servitude and the full commoditisation of rural labour itself. The gradual reduction of sharecroppers to the position of wage workers through intensification of the conditions of surplus extraction, coupled with the emergence of capitalist farming is, therefore, the expected pattern and one which has much historical precedent.

As one surveys the contemporary occurrence of sharecropping, in many cases the rationale fits, superficially at least, less tidily into the scheme outlined above. To what extent its apparent persistence in much of Asian and parts of African agriculture is due to a failure of capitalism to permeate agricultural production in these areas, or to other aspects of its rationale which persist after the transformation of farm organisation along what some might describe as capitalist lines [*Robertson,* 1980] will be discussed below. But, inevitably, it would appear that the problems of perceiving change and its direction are far fewer in retrospect than for current observation.

In the examples discussed above the most frequent pattern is for a single large-scale landholder to lease small parcels of land to numerous tenants, who probably possess little or no land or other means of production of their own. In India the situation is more complex. First, a substantial (though disputed) proportion of sharecroppers own some land of their own [*Narain and Joshi,* 1969; *Bardhan,* 1970]; and second, there exists a wide variation in landowner–tenant relationships, from the situation described above, where the position of tenants is little different from that of indentured labour, a situation more characteristic of eastern India [*Ladejinsky,* 1977; *Newaj and Rudra,* 1975], to the case of comparatively well-to-do landowning peasants leasing in land from small-scale landowners [*Bharadwaj and Das,* 1975; *Ray,* 1978].

In Bangladesh also, where up to 50 per cent of all farmers sharecrop some land [*Zaman,* 1973], many are part tenants part owner-occupiers, and landless sharecroppers are by no means the rule [*James,* 1981: 92]. A similar situation exists in Malaysia [*Bray and Robertson,* 1980], where sharecropping is most prevalent in the rice-growing Kelantan province. The association between paddy farming and sharecropping has also been observed in India [*Rao,* 1971], and in the Philippines [*Roumasset,* 1976], although in the latter case the large landowner/small-scale tenant scenario is more common [*Barber and Cardova,* 1978], as it was in pre-revolutionary China [*Buck,* 1937]. Ash [1976] in a study of Kiangsu province finds sharecropping to be most common in least productive areas, although here too part owners/part tenants are the most numerous. In the Mekong Delta of Vietnam as late as the 1960s nearly two thirds of the land area was held under tenancy, with large estates producing mainly export crops through small sharecropper units [*Paige,* 1975] – a situation more reminiscent of the West European model than much of the Asian context.

Sharecropping has become important in a very different context in parts of West Africa. With the growth of cash cropping, in particular of cocoa production, and a consequent expansion of peramanently cropped acreage, cocoa farmers faced problems in mobilising sufficient labour to care for the crops. Competition from other sectors, plus a reluctance by many young household members to undertake farm work, led to the increasing prominence of 'Abusa' or sharecropping contracts as a form of 'compromise' [*Robertson,* 1981] between landowners seeking a more reliable labour supply and migrant labourers searching for a permanent foothold in the rural economy. Such contracts are, nevertheless, essentially labour-hiring arrangements, a point made more apparent by the fact that often those leasing out on 'Abusa' terms are fixed rent tenants unable themselves to furnish the holding with the necessary labour supply [*Addo,* 1974; *Berry,* 1975]. This form of sharecropping is most common in Ghana and is found to a lesser extent in western Nigeria, although in the latter case there is evidence that as the use of chemical inputs becomes more widespread, share rents are being replaced by fixed cash payments [*Berry,* 1975].

From this brief survey of sharecroppers and their environments three important features may be singled out. First is the wide range of contexts, in terms of factor endowment, in which such contracts occur; from the relatively land extensive, labour scarce agricultures of West Africa and the Americas, to the relatively land scarce, labour surplus agricultures of South and East Asia. Secondly, the detailed circumstances of sharecroppers, and therefore their circumstances *vis à vis* those from whom they are leasing land, vary considerably from the landless, indentured quasi-labourer of North America, western Europe and parts of the Indian subcontinent, to the comparatively independent, landowning peasantry of many areas of Asia. Lastly, it is evident that sharecropping contracts are designed to facilitate the mobilisation of labour. The cases of Ghana and Spain are most obvious in this respect, but it will be argued below that this is the crucial feature of sharecropping contracts and the factor which, *more than any other,* explains their pervasive and persistent use.

III. THE THEORETICAL LITERATURE

Any discussion of the theoretical literature on sharecropping conveniently begins with Adam Smith. His interest focused on two central issues: the impact of product sharing on labour effort, and its implications for capital accumulation and on-farm investment [*Smith,* 1976: 389-391].

On the first count he viewed such systems as progressive, since, relative to the serfdom which they replaced, they brought incentive for greater or more careful application of the labourer's time, given that part of the proceeds from such increased effort now accrued to labour. Thus Smith saw the emergence of share contracts on the largely post-feudal estates of western Europe as a consequence of a desire to increase labour productivity. But, the ultimate disadvantage of such arrangements lay in the lack of stimulus towards investment on the part of these cultivators, since a proportion of any returns accrue to the landowner. These disincentives, he believed, would ultimately lead to the yielding of sharecropping to fixed rent contracts based on the 'English' model.

Later classical economists such as McCulloch [1907: 262-290] and Jones [1831: 101-108] were concerned with extolling the virtues of peasant proprietorship, and believed sharecropping practices to be the underlying cause of the poverty and stagnation characterising west European agriculture, although Mill [1909: 302-317] saw the widespread insecurity of tenure which existed, particularly in France, as the *root* cause of poverty and lack of investment. (In fact Mill was profoundly influenced in his view of sharecropping by the work of Sismondi and quotes him extensively on the subject. It was Sismondi's comparison of the condition of Tuscan sharecroppers to those in other Italian provinces which appears to have led Mill to stress the importance of fixity of tenure.) What distinguished Mill from many of his contemporaries, and from the majority of later commentators, was that he analysed sharecropping as an alternative to the employment of wage labour on large estates and in this respect, at least, found the practice not without merit in spite of its limitations.

It is useful to commence with a brief consideration of the views of these early commentators, inasmuch as many of the issues which were later to fill so many pages are here perceptively understood and outlined. While not presenting any formal analysis with regard to the implications of sharecropping, these writers enabled attention to be focused on two prime areas of concern: namely the dynamic rationale underlying alternative wage and tenure forms and the problems of resource allocation and productive efficiency. The two are seen by the classical economists as being fundamentally related, although it is the latter topic alone which claims the attention of most later writing.

The first formal statement concerning the allocative implications of sharecropping is presented by Marshall [1961: 644] and from this stems what has come to be known as the 'traditional' view.[1] If one assumes negotiation on the basis of mutual equality between landlord and tenant, and division of the gross product in equal proportions, the tenant will not invest resources beyond the level where the marginal cost of output is equal to half the value of the marginal product. Similarly the landlord will not invest unless the marginal product of such investment is equal to a minimum of twice the marginal cost. Thus equilibrium levels of resource allocation resulting from share contracts will be less than under systems of owner occupation or fixed rents. It follows, therefore, that total per hectare rent accruing to the landlord and total per hectare product will be less than optimal.

The model is internally consistent but of limited applicability owing to the nature of the assumptions underlying it. No one was more aware of these limitations than Marshall himself, who noted that if landowners possessed the opportunity to determine the level of inputs which the tenant supplies, then an 'efficient' solution would result. Thus when holdings are small, tenants poor, and landlords 'not averse to taking much trouble about small things' [*Marshall,* 1961: 645], the outcome will not be inferior to that generated by any alternative form of contract.

Marshall also registers the point that, though the landlord's share of produce may be set at some predetermined level, other dues and 'gifts' were often vague and subject to modification should circumstances merit:

> Whenever payments of all kinds made by the cultivator left him a margin beyond the necessaries of life for him and his family, together with those

> comforts and luxuries which were established by custom, the landlord was likely to use his superior strength to raise the payments in some form or other [*Marshall,* 1961: 640].

In spite of his own misgivings concerning the appropriateness of its implications, it is Marshall's formal model which became the focus of traditional thinking concerning sharecropping. A great many 'refinements' have appeared along the way, but the basic premise remains unchanged: that is that sharecropping is essentially 'inefficient'.

Schikele [1941] suggests that any tendency on the part of tenants to supply a 'sub-optimal' level of inputs will be partly offset to the extent that the tenant possesses an *insecure tenure,* since such uncertainty is likely to generate an intensification of inputs in order to mollify the landowner. Heady [1947] points towards full *cost-share leasing* as providing a formal solution to the 'efficiency' problem, whilst Johnson [1950] summarises the principal mechanisms through which landowners might ensure desired tenant behaviour: that is through input specification, cost sharing or short-term leasing. Further comments of note come from Drake [1952] formalising Johnson's stress on insecure tenure enabling the landlord to specify inputs (limit the tenant's degree of choice), reinforcing the point made earlier by Schikele [1941] that sharecropping will be least 'inefficient' where production is specialised and the level of technique low, thus reducing optimum levels of input intensity to a minimum. Adams and Rask [1968] anticipate later, more controversial, approaches in a paper emphasising both the risk-sharing properties of sharecropping contracts, and the requirement of a minimum level of tenant income. They expand on this latter premise to suggest that tenants may accept low levels of remuneration from sharecropping, where no alternative exists. Thus landlords may be in a position to bid up the rental share, which together with the division of land area into relatively small parcels may force tenants to increase inputs to the optimum.

If 'traditional' theory had been concerned with the essential 'inefficiency' of sharecropping and with methods whereby any sub-optimality might be ameliorated, Cheung [1968, 1969] was the first formally to outline not only how sharecropping might be *as productive* as other forms of contract, but also the contexts in which it might be the *preferred contract.* It had not escaped the notice of earlier commentators that not only was sharecropping very widespread in many parts of the world, in spite of the theoretical disadvantages, but also that the comparative agricultural performance of such contracts was by no means inferior. Given landowners' ability to limit the supply of land available to tenants, they are able to induce an optimum application of tenant inputs through manipulation of the rental share and size of the tenants' parcel of land: the constraint on landlords' bargaining strength being the alternative income tenants might earn in other activities. That is, variations in rental share or land area cannot be used to squeeze tenants' remuneration to the extent that the opportunity cost of tenants' labour is greater than tenant income. Thus the possibility of an 'efficient' equilibrium is established, the preference for share contracts being derived through discussion of transaction costs and risk.

Transaction costs, that is negotiation and enforcement costs, are borne by landowners and are heaviest with direct employment of wage labour and minimised under fixed rental contracts, with sharecropping being intermediate. (The nature of these costs will be specified and discussed below.) Risk, on the other hand, is entirely borne by landowners in the case of wage contracts and by tenants under fixed rents, with sharecropping distributing risk to the same extent as product. Cheung suggests, therefore, that should transaction costs alone be the determinant variable, sharecropping would be unlikely to be the preferred contract. On the other hand, it may be necessary to accept part of the risk burden in order to attract tenants, as occurs with share contracts. Thus:

> The choice of contractual arrangement is made so as to maximise the gain from risk dispersion subject to the constraint of transaction costs. [*Cheung,* 1969: 64]

Cheung's hypothesis has been subject to both criticism and extension, and has generated considerable debate amongst economists, much of it predictably esoteric in nature. The main protagonists can be usefully divided into those accepting and extending Cheung's analysis and those critical of his 'modern' variant, either from the perspective of asserting the validity of 'traditional' analysis or from that of alternative explanations of sharecropping 'efficiency'.

Currently, the chief architect of the 'modern' position is Newbery [1975] who notes that circumstances conducive to sharecropping arrangements according to Cheung could equally be met by a *combination* of fixed rents and wage employment. The additional assumption of uncertain alternative earnings (imperfect labour markets) is, therefore, attached to the conditional determination of sharecropping. Stiglitz [1974] stresses the importance of an incentive to 'effort' associated with share contracts *vis à vis* wage labour, and a formal integration of incentive/uncertainty is provided in Newbery and Stiglitz [1979]. Further attempts to integrate sharecropping 'efficiency' and competitive equilibria are provided by Braverman and Srinivisan [1981] and Mettrick [1981: 63-66].

The attack from the 'traditional' school is led by Bardhan and Srinivisan [1971], by Bell [1976, 1977] and by Bell and Zusman [1976]. The essence of their critiques revolves around the negotiating strength attributable to either party. Both criticise the 'modern' model in so far as maximisation is from the landlord's perspective only, thereby ignoring tenant interests. If the latter party is able to influence the bargaining process, then the theoretical outcome will adhere to Marshallian propositions.

Reid [1973, 1975, 1977, 1979a, 1979b], in a series of articles based on United States data, distinguishes between what he terms point and sequential uncertainty. He argues that the former, when 'the product of land and labour is randomly and immutably affected' [*Reid,* 1975: 430] cannot be reduced but only dispersed and that, *contra* Cheung, sharecropping is negatively associated with crops most subject to such uncertainty. Sequential uncertainty, on the other hand, implies the possibility of risk reduction. It implies an ability to take advantage of unforeseen circumstances. Thus where contracts are flexible, adjustments can be made *en passant* to ensure the best possible outcome for all concerned. Sharecropping, possessing such flexibility, is therefore preferred for its risk reducing rather than risk dispersing potential. In addition, Reid sees share contracts as enabling a distinction to be made between managerial and productive skills necessary to farm production. In the context, therefore, of a labour force with considerable experience in production but lacking managerial expertise, sharecropping will be the mutually preferred outcome, with the landowner providing detailed supervision of production. As such it provides a step on the farming ladder between wage-labour and fixed-rent contracts and eventual owner occupation, whereby the sharecropper can gain valuable managerial experience plus access to capital inputs.

It is not the purpose of this paper to demonstrate the adequacy of one of these approaches relative to another. In fact, it is usually possible to *select* sufficient 'evidence' to reinforce or denigrate most theoretical positions. Nevertheless, discussion of the empirical superstructure associated with the various points of view which make up this orthodox or neo-classical paradigm will aid illustration of what I take to be their overall inadequacy. The important issues in contention are the competitiveness of the bargaining procedure, the variability of the rental share, the comparative productivity of various contractual arrangements, and the implications for tenant and landlord investments.

Under Marshallian assumptions the supply of land to tenants is not restricted and tenants should theoretically be able to lease as much land as they wish. The Cheungian hypothesis, on the other hand, depends on the ability of the landowners to restrict land availability. Should the latter situation prevail, one would expect a single landlord to be leasing to many tenants rather than the reverse. In fact Bell [1977], in a study based on a research project in Bihar, finds the common case to be where a single tenant leases from more than one landowner. There is also ample evidence from other parts of south Asia [*Zaman,* 1973; *Bardhan,* 1970] and Malaysia [*Bray and Robertson,* 1980] that frequently sharecroppers own as well as lease land. Thus it is quite possible that tenants meet landowners from a position of comparable economic power. Even where they are landless and committed to one landowner the scarcity of labour may be such that the sharecropper is able to bargain on an equal footing [*Robertson,* 1981]. There are many other examples, of course, where this is not the case,[*Bhaduri,* 1973; *Bharadwaj and Das,* 1975; *Ladejinsky,* 1977] and where landlord dominance is paramount.

Crucial to the Cheungian analysis is the possibility that rental shares may vary rather than remaining fixed at some customary level. Apart from the possibility that other dues will vary with circumstances, as Marshall himself suggests, and as also found by other commentators [*Bharadwaj and Das,* 1975], a possibility which makes straightforward discussion of the rental share less than appropriate, there is some evidence, nevertheless, that rental shares are subject to adjustment [*Winters,* 1974; *Newaj and Rudra,* 1975; *Berry,* 1975] both in response to differences in land fertility and to technical change.

There is conflicting information concerning the relative productivity of sharecropped farms and others, with some commentators [*Bharadwaj*, 1974, *Bell*, 1977; *Hossain*, 1977 and *Jabbar*, 1977] finding sharecropped land to be less productive than either owned land or that rented through fixed contracts, while others produce contrary evidence [*Bray and Robertson*, 1980; *Ruttan*, 1966; *Zaman*, 1973 and *Huang*, 1975]. Much of the available evidence, however, is based on an analysis of sharecropped *land* rather than sharecropped *holdings*. Apart from the conceptual problems involved in such a substitution, the frequent exclusion of other factors influencing productivity, such as holding size and ownership of non-land assets makes any assertions concerning productivity dubious to say the least.

Productivity will, in part, be related to the degree of tenant or landlord investment and one implication of the Marshallian approach is that inherent disincentives will inhibit the growth in agricultural productivity considered necessary to industrialisation programmes.

A similar conclusion, but from a different perspective, is provided by Bhaduri [1973] who formally constructs a model to demonstrate the inhibiting effect of sharecropping on technical change, when this form of appropriation is reinforced by the indebtedness of tenants to the landlord, the argument being that landlords will oppose any innovation which will increase the earning capacity of their tenants, as this would remove the basis of their control (through debt bondage) and ability to maximise the extraction of surplus labour. The model is not without inconsistencies [*Griffin and Mabro*, 1978; *Ghose and Saith*, 1976; *Newbery*, 1975], and relates to a more general area of debate than is our concern here (the categorisation of agricultural modes of production), but it does have the implication that, in certain contexts, sharecropping will be incompatible with the development of productive forces. Certainly much of the discussion relating to the so-called 'green revolution' in India shows sharecropping disappearing as the 'new technology' is adopted [*Byres*, 1981; *Bardhan and Rudra*, 1980; *Bharadwaj and Das*, 1975] (although its persistence elsewhere merits further invesigation [*Bray and Robertson*, 1980]), such evidence appearing to contradict rather than confirm Bhaduri's main thesis that the relations of production associated with sharecropping will stifle such technological innovation.

To return to the neo-classical tradition: there would appear to exist evidence of a *partial* nature for any of the theoretical explanations of sharecropping contracts posited within it. In fact, the seeds of the major adaptions and extensions of Marshall's model can be found within his own writing. But a theoretical framework which starts from a given set of assumptions, yet which can only explain reality through constant adoption of assumptions contrary to the initial propositions upon which the theory was built, is surely suspect. Those within the Marshallian orthodoxy cannot adequately account for the rationale behind the widespread incidence and persistence of sharecropping. The Cheung-Newbery explanations on the other hand, are forced to make assumptions concerning the nature of the bargaining process, but fail to explain or come to terms with any theoretical understanding of circumstances which might produce relationships based on an unequal distribution of economic power.

Reid attempts to marry sharecropping 'efficiency' with 'competitive determination', but in basing his argument on the United States experience where sharecropping emerged from labour shortages following the abolition of slavery, and in a context characterised by extreme inequalities (social and economic), his case for a 'competitive equilibrium' seems flimsy in the extreme [*Wright*, 1973; *Holley*, 1980].

It is probable that in certain contexts sharecropping will inhibit investment and lower levels of input intensity and productivity. It is also probable that in others these contracts will facilitate the most efficient distribution of resources, given the objectives of one of the parties concerned. But the neo-classical tradition, with its emphasis on technical to the exclusion of social relationships, cannot provide a universally applicable understanding of sharecropping. To reach such an understanding, it is argued below, it is necessary to approach the subject from within the Marxist tradition of thought. But in one respect, at least, the neo-classical and Marxist approaches converge: and that is in their search for such universally applicable understanding. A suggested alternative is to seek only the specificity of anthropological explanation [*Robertson*, 1980]. This, however, I reject since it quite simply explains too little.

IV. AN ALTERNATIVE ANALYTICAL FRAMEWORK

a. Sharecropping as a Method of Surplus Appropriation

To an extent the diversity of view outlined in the previous section can be explained by a lack of clarity in definition. Sharecropping is a term used in the literature to describe a number of contractual circumstances which, while possessing some common attributes, also contain important and significant differences. The implications of arrangements outlined by Reid [1973], Paige [1975] and Bhaduri [1973], for example, where the landlord provides all inputs other than labour, can differ fundamentally from those where the landowner provides land only, or those where there is some pre-arranged sharing of input costs [*Roumasset*, 1976; *Rudra*, 1975; *Bray and Robertson*, 1980]. Bell [1977] is in fact more aware than many of this anomaly, and overcomes it by purposefully ignoring sharecroppers who do not fit his categorisation. Much of the theoretical literature focuses attention solely on one manifestation to the exclusion of others and thus coins explanations which are pertinent to a particular context only.

Sharecropping implies some form of contract, written or verbal. Share contracts refer to circumstances where access to some factor necessary to production is provided by one party, in return for a pre-arranged proportion of the resulting production. Since land is essential to crop production, sharecropping contracts entail the provision of, or access to, land. Apart from labour, other objects and instruments of production (means of production) will also be necessary to the activity, and the provision of all or some part of these may also be part of the contract. In general, then, sharecropping contracts constitute the provision of the means of the production, or some of them, by one party in return for a share of the crop yield. The extent of the means of production provided will vary as will the extent to which other exactions are added to the crop share.

It is necessary, though, to examine more closely the precise nature of such an exchange. The tenant receives access to means of production, whilst the landlord receives a part of the product. But Marx [1976: 668-672] maintains an equivalence between surplus product, surplus value and surplus labour, insofar as the first two are both reducible to the latter. The landlord, therefore, indirectly but effectively gains access to labour. The rationale behind such a transaction is obvious. The tenant possesses labour and requires access to means of production without which that labour cannot be utilised to the extent he desires. (The word 'he' is used as a shorthand expression for he/she throughout, and no sexist assumption is intended.) Similarly the landlord possesses means of production and requires labour over and above the quantity he is willing or able to supply, in order that these means of production are utilised to the extent deemed necessary.

Sharecropping, then, is one mechanism through which owners of means of production acquire access to others' labour. This acquisition being indirect, is less obvious in the case of cash or product rental contracts, whether proportionate or fixed, than with labour rents or wage labour, but occurs nevertheless. What it is necessary to theorise, however, is the circumstances in which sharecropping is likely to be the chosen method of such an acquisition; that is, what relationships between the two parties concerned generate the sharecropping contract.

Central to the Marxist position is the proposition that it is inadequate to view sharecropping merely as a form of rent or of labour remuneration. Rather, it should be seen as a particular *method of surplus appropriation*; a method through which surplus labour is transferred to the landlord in the form of surplus product. (The notion of surplus implies output, or the labour required to produce that output, in excess of that quantity necessary to provide some subsistence level of consumption plus the investments necessary to sustain it.)

Productive activity is characterised by class relationships when the agents involved in that process possess unequal access to the means of production [*Althusser and Balibar*, 1970: 212-216]. The class relations of *production* are a reflection of this unequal access and provide the possibility for the transference of surplus labour. The mechanism which facilitates this transfer, for example, rent or wages, constitutes the relation of *exploitation* [*Marx*, 1981: 1017-1024]. It is straightforward that the latter relations (exploitation) will be dependent on the former (production) since, for example, the payment of rent is conditional upon the transfer of means of production, which in turn implies some inequality of access to means of production.

The precise form which relations of exploitation take can vary: for example, rents can be fixed or variable, and paid in labour, produce or cash. Thus, given that rent is a general method of transfer, or means of appropriating surplus labour, it follows that sharecropping can be described as a *specific form of surplus appropriation*. The line of causality, therefore, runs from class relations of production and their determination to the particular method of facilitating the transfer of surplus labour. As these production relations change, therefore, it is possible, but not necessary, that methods of surplus appropriation will also change.

It is not pertinent in this paper to discuss the categorisation of modes of production and their attendant class relations, even if such an exercise would be likely to prove fruitful. The central concern is to identify those production relationships which will lead to sharecropping becoming the preferred mechanism or appropriation, and whether changes in these relationships, in particular the advent of agrarian capitalism, will lead to changes in such preferences. Of course, any such discussion must be predicated on an ability to identify capitalist production relations. A number of alternatives present themselves in the literature.[2] But all protagonists would agree that the existence of generalised commodity production, with its implicit notion that productive activity is mediated through the market place, provides at least a minimum necessary condition. The extent to which this is also a sufficient condition is a contentious issue, but some progress can be made along the lines suggested by Banaji [1977b] and Gupta [1980]. These authors, following the recently published 'new' chapter of *Capital* [*Marx*, 1975: 1019-1038], distinguish between the *formal* subsumption of labour by capital, where the wage relation and commodity production obtain, but the labour process remains unchanged (appropriation being in terms of absolute surplus value), and the *real* subsumption of labour by capital where the accumulative dynamic of capitalism has precipitated changes in the labour process facilitating the appropriation of relative surplus value. This distinction will be used below to argue that sharecropping can be consistent with agrarian capitalism in its early stages, but that there will be a tendency for such contracts to be superceded by others more appropriate to high rates of accumulation in agriculture.

b. Class Relations in Agriculture

Relations of production, defined by the separation from, or access to, means of production, describe control over those elements with which man interacts in the process of production. (They are not, though, necessarily consistent with property relations, cf. [*Banaji*, 1973].) They describe relations between people not only through raw materials, but also through the whole gamut of other accoutrements necessary to production. They describe control, therefore, over both the means and objects of labour.

Objects of labour in agriculture are described by the relevant natural resources. Broadly this category refers to those attributes of nature which are necessary to, or impinge upon, the production process, and which in the short term, at least, exist independently of man. (In the long term these elements will change through interaction with man.) Features such as terrain and climate will combine to determine the quantity of land available for agricultural use. More importantly, land fertility and water availability are particular natural phenomena with which man interacts through agriculture. Fertility, of course, describes a wide range of characteristics necessary to the production of crops and livestock, including especially the ability to retain nutrients and water, the latter including both surface and groundwater as well as rainfall.

Means of labour include as a major component the extent of capital utilised – tools, machines, fertilisers, herbicides, seeds, etc. It will also include the skills of the labour force. Control of the access to these elements can be as important a source of surplus labour as possession of the objects of labour themselves. In fact one can compound the exploitative impact of the other, as many commentators would concur, examples ranging from the landowner/merchants/processors in the cotton belt of the USA [*Holley*, 1980] to the landowner/moneylenders of the Gangetic plain [*Bhaduri*, 1973, 1977] and the Mekong Delta [*Paige*, 1975].

The owner, or person in effective control, of means of production, will face a range of alternatives when coming to a decision concerning the utilisation of these assets. Given that they will only become productive to the extent they are combined with labour, the choice relates to the manner in which the owner of means of production brings about such a combination: that is, determines the labour process. Apart from the owner's own labour this entails a number of possible optional methods of acquiring others' labour. These can be either direct, in the form of wage contracts, or indirect through some form of rental contract, the latter being either fixed rents, in the form of labour, produce or cash, or proportionate rents.

Given that any form of contract will be based upon an inequality in ownership or control of means of production and/or labour power, it is the extent of this inequality and the degree to which one party to the contract can exercise influence over the other through it that will determine the choice. Such influence, or class dominance, implies the degree to which one party is dependent on decisions made by the other for the reproduction of its own existence, and one can delineate, in the context of agricultural production, circumstances which will strengthen the position of one party relative to the other.

Central to the determination of such dependence will be the distribution of productive resources. Evidently the more unevenly distributed are means of production, the more dependent will one section of the community become upon another for the reproduction of its existence. Thus, for example, where the distribution of productive resources is very skewed, the opportunity arises for a greater exaction of surplus labour by those owning those resources. This is a fairly obvious point and one can find numerous examples. For example, Bharadwaj and Das [1975] find in Orissa that where tenants are leasing small parcels of land from large-scale landowners, the exaction of 'services' in the form of cultivating and domestic labour by the landowners, in addition to a rental share, is common, whereas where the two parties have a less unequal economic status such 'extra' payments do not occur. Similarly, Vyas [1970] found in Gujerat that the rental share decreased as the quantity of land held by tenants increased.

Equally interesting is the circumstance where the potential productivity of resources increases, for example through the advent of 'green revolution' technology. In some cases, the experience is that rental shares increase in such a context, as the landowner attempts to appropriate any increased surplus, especially where the landlord shares in the cost of the provision of the 'modern' inputs [*Bharadwaj and Das,* 1975]. Examples where a more equal landowner–tenant relationship obtains [*Bray and Robertson*, 1980] show the tenants more able to acquire the additional fruits of their own labour through downward pressure on rental shares. Of course, the issue is complicated by the extent of cost sharing, but the point still holds.

Class relationships, however, or the social relations of production, are subject to a wider variety of determinants than the technical distribution of resources. The extent to which one class is able to appropriate from another through effective possession of productive resources will depend on how complete and exclusive that possession is. Should the relationship involving access to productive resources be compounded by other relationships then the subjugation of the subordinate class will be extended.

A frequent example is where landlord–tenant relationships are reinforced by further mechanisms of appropriation. These can occur at either 'end' of the productive activity, namely through provision of inputs, physical or financial, and through control of the produce/marketing facilities. In either case the relationships may be cemented through indebtedness. There are many examples of such symbiotic methods of appropriation, as indicated in the previous section, such that the interlinkage of markets through these relationships makes it unrealistic to attribute priority as a method of appropriation to any particular source [*Bharadwaj and Das*, 1975]. It remains true, nevertheless, that any *combination* of exploitative mechanisms is likely to reinforce any one of them.

At the same time the dependence of the subordinate class will be mitigated by any possibilities for obtaining income/livelihood outside of the relationship in question. It should be stressed, however, that the existence of any potential opportunities does not necessarily assume the ability to take advantage of them, since the degree of subjugation may enable the appropriating class to prevent such a reduction in the exclusivity of the relationship. For example, any alternative employment opportunities outside of agriculture will improve the position of the tenant/labourer *vis à vis* the landowner. (Unfortunately, one of the major predicaments for many countries currently is the lack of employment possibilities both within and outside of agriculture, so obvious examples of such an attenuation of production relations are few.) One can cite the exodus of many former slaves to the northern industrial towns after the American Civil War, as producing circumstances where potential labour shortages arose; as did Spanish industrialisation more recently. Both of these cases saw the use of share contracts as a landowner response. Alternative employment opportunities in a general sense, though, imply some degree of choice facing those with inadequate control over productive resources, in the manner in which they dispose of their labour. It is the assumption of uncertainty attached to possibility of choice which is essential to the consistency of the Cheungian model. If the possibility of selling labour to, or renting land from, more than one landowner exists, then a degree of choice exists. In some contexts it is conceivable that such a possibility is excluded by location. More commonly it is prohibited contractually. Examples can be found of such contractual obligations [*Rudra*, 1975] and these would appear to be associated with a high level of landowner domination.

The arena of socio-political obligations and cultural constraints on economic activity frequently compounds the effectiveness of class control described above. One would expect any influences to be conservative in nature and designed to preserve the status of existing class relations. In this respect it is likely that such constraints will reflect the interests of the dominant class. At the same time such obligations are also likely to be orientated towards the preservation of social cohesion and this may involve suppression of class antagonisms. In some cases access to productive resources under preferential conditions may be offered to less privileged members of a social group in order to maintain such a cohesion. Thus in Sri Lanka [*Leach*, 1961:250], Malaysia [*Bray and Robertson*, 1980; *Horii*, 1975] and Bangladesh (personal communication from Peter James), there is evidence of an association between kinship and contracts relatively favourable to tenants. On the other hand caste division and custom in India can bring about a class of tenants/labourers customarily deprived of the opportunity to acquire productive resources [*Ladejinsky*, 1977 : 454-459].

Class relations in agriculture, therefore, are based upon the distribution of control of productive resources essential to the industry. Within this terrain, however, the nature and extent of class control or dominance may be reinforced or weakened by a wider range of attendant social relations.

c. The Nature of the Labour Process

The labour process describes the manner in which direct labour is combined with means of production in productive activity. This involves much more than the use of 'simple' or 'modern' techniques. Implicit also is the type and range of product and the organisation of labour itself. For example, given the particular qualities associated with the land, the way in which it is combined with direct labour (for example through wage labour or tenant labour) will influence cropping patterns and therefore the degree of certainty with which those involved may achieve a minimum level of remuneration. Thus there is an incentive for small-scale producers to concentrate on production of lower risk subsistence crops [*Bagi*, 1981] even though potential returns are lower than with other crops. The opportunity for choice, of course, may not always arise, as demonstrated by the work of Vyas [1970] who found small-scale tenants growing a lower proportion of staple crops relative to small-scale owner-occupiers, due to contractual obligation.

The question of uncertainty is a complex one which pervades much of the literature. Following Rao [1971] one might distinguish between *risk* and *uncertainty*. The former refers to change related to choice, where a decision is made which can influence the outcome of productive activity; whilst uncertainty implies a probability of failure brought about by circumstances beyond an individual's control. An obvious example of risk is the decision to grow one crop instead of another, in the hope of greater reward, based on climatic or market expectations. The latter might occur where no choice exists but where some probability of drought or flooding reduces the ability to predict the outcome of production. In effect, the distinction is spurious, since the advent of risk is in any case brought about by the existence of uncertainty, the only difference in the two situations lying in the *degree* of choice facing individuals involved in making production decisions. Since there is likely to be a continuum of choice varying from year to year and from decision to decision, a discrete division of circumstances seems inappropriate.

Equally inapposite is Reid's distinction between point and sequential uncertainty [*Reid*, 1975], the implication being that the former is 'immutable' whilst the latter can be reduced through responsive changes in methods of production. All this difference amounts to, in effect, is the *degree* of uncertainty; since a context where one is able to mitigate the effects of unforeseen or uncontrollable circumstances implies a context less characterised by uncertainty than one where mitigation is impossible. The ability to respond to such events may, in part, determine the form of contract, but Reid argues that it is the contract itself which makes possible the reduction of uncertainty. That is, that sharecropping, because of its greater flexibility, 'allows' such responses to occur. In fact, it is the *combination* of managerial expertise and the ability (power) to implement decisions which creates the possibility for risk reduction, and there is no *a priori* reason why such a combination is more likely to occur with one form of contract as opposed to another.

The uncertainty attached to the outcome of productive activity is, nevertheless, frequently an important influence in the organisation of agricultural production, as Cheung and others suggest. The treatment generally, though, views uncertainty as a technical relation, rather than one mediated by the socio-economic context. Farmers, whatever their tenurial status, will vary in their ability both to reduce and to cope with uncertainty. As a general rule, one can argue that the opportunity to ensure a predictable or acceptable outcome from any productive endeavour will increase with the quantity of productive resources at the disposal of the operator. For example, provision of irrigation facilities increases protection against a variable rainfall; a wider resource base increases the ability to produce a surplus to the requirements of self-consumption and offset the effect of poor harvest, through the market. In addition it increases the opportunity for growing low-value, low-risk crops as Vyas' [1970] comparison of tenants and operators demonstrates.

Not only is the level of uncertainty facing a producer greater the narrower the resource base upon which that producer operates, but so also is the 'felt response' to uncertainty. As Adams and Rask [1968] indicate, the disutility of a given income variance at levels of living approaching subsistence, where the major concern is one of survival, is likely to be higher than for those in more fortunate circumstances, i.e. the poor are more risk averse.

Uncertainty, then, is in part – and perhaps to a considerable degree – *socially* determined and therefore provides an opportunity for social control. This is a point of considerable importance, since when labour processes are characterised by uncertainty, the extent to which one class can extract surplus labour from another increases.

d. Factors Determining Methods of Appropriation

The primary purpose of this paper is to outline those circumstances wherein sharecropping contracts are most likely to obtain. The previous discussion has contributed to such an analysis to the extent that it has sought to identify the principal relations underlying agrarian production: namely 1) the level of class dominance; and 2) the labour process.

In the context of agricultural production the choice of contractual arrangement open to the participants has been noted. From a landowner's perspective one can assume that the contract preferred will be that which maximises the part of social product accruing to himself, subject to any operative constraints. Such constraints include the need to ensure the reproduction of the existing unit of production. For example, the source of direct labour, whether tenants or labourers, must be able to survive in numbers adequate to continue the productive activity, or with sufficient well-being to prevent defection elsewhere. In contrast the tenant/labourer will be concerned with assimilating the maximum part of the product of his own (and/or others') labour. The possibility that tenants employ wage labour is, therefore, incorporated. The choice of contract will be the outcome of these contradictory aspirations. The question arises, therefore, as to which party makes the decision, that is which party (if either) is in a position to enforce its interest. The importance of class dominance is therefore apparent, since whether or not sharecropping or some other method of appropriation is the outcome of any relationship, will depend upon the bargaining strength of the participants, as well as their objective preferences.

Contracts vary in the difficulty with which they can be implemented. There will be costs involving the time and effort necessary for landowners to ensure an acceptable outcome. I refer to these as supervision costs. These are broadly akin to the 'transaction costs' described by Cheung [1969a]. Cheung, however, distinguishes between 'negotiation' and 'enforcement' costs, which together constitute transaction costs. It is unclear from the literature precisely what one is to understand by the term 'negotiation' cost. That is: whether it implies solely the time spent in negotiation (and renegotiation) of contracts, or whether it also includes the cost of dividing, transporting and (where appropriate) selling the landlord's share. If it implies the former then it is likely that such costs are minimal and little different to those involved with wage or fixed rent contracts (even allowing for any greater complexity of contract); if the latter then they are, in part, indistinguishable from enforcement costs.

All such costs, therefore, are more usefully subsumed under a general title such as supervision costs. In effect these describe the time and effort necessary to ensure that the tenant/labourer maintains his side of the 'bargain'. For example, fixed cash rents paid in advance of production, or payment of wages, involve the landowner in less 'cost' in this respect than share contracts where measurement and division and possibly sale of the produce are necessary and the details of which have to be specified and agreed upon. They also refer to the time and effort necessary to ensure that the productive activity itself is appropriately executed: that agreed inputs are applied and output is distributed, according to the terms of the contract. At the same time, if one makes a distinction between labour 'effort' and labour 'time' and where the quality of human labour is crucial to the out-turn of production, the ability to ensure the necessary standard of work may be pertinent. Such costs are likely to be most pronounced for the direct employment of labour, whilst any form of rental contract will provide 'incentive' to the direct producer to carry out the requisite tasks satisfactorily.

In fact, these supervision costs are likely to vary inversely with the extent of class dominance. For example, where the labourer/tenant is extremely dependent on a particular landowner, and the competition for employment or tenancies is great, then such costs are minimised as tenants or labourers endeavour to maintain the landowner's favour. Hence the preference on the part of landlords, in the study by Bharadwaj and Das, for leasing out in small parcels due to the increased intensity of tenant effort forthcoming.

The magnitude of these potential costs will also vary with the labour process. In determining the type and range of product, together with how and when produced, the labour process will specify the level and range of skills requisite in production. Where it is such that a wide variety of skills is required – for example, where crop production is largely unmechanised and involves a large number of hand operations – then these costs are potentially high. On the other hand, if only a few simple operations are necessary they will be of less significance. It can be argued that, as a general rule, the greater the quantity of labour-replacing capital per unit of direct labour characterising any labour process, the narrower will be the range of skills required. The frequent replacement of a wide variety of hand and animal powered tasks by a small number of mechanical and/or chemical operations provides ample evidence.

It is premised above that class relations in conjunction with the labour process determine the method of appropriation most consistent with dominant interests. In the context of very high levels of class domination, the implication is that the landowner is in a position to appropriate the entire quantum of surplus labour; such that the tenant/labourer is left only with the means of survival. One would expect, in such circumstances, direct employment of labour, either through labour rents or some form of wage payment, to obtain. The latter may be in the form of cash or kind or, as is frequent in classical feudalism, the provision of meals, clothing and shelter. To the extent that labour rents potentially provide the tenant/labourer with greater opportunity to obtain the fruits of his own labour, these are superior from his perspective, providing the increased uncertainty attached to his income does not outweigh the advantage. Given the patriarchal obligation frequently associated with such heavily landlord dominated societies, for instance in Brazil as described by Johnson [1971: 114-131], the problem of uncertainty would be mitigated if the landowner had the final responsibility for the survival of his 'subjects'. Even so, assuming the landlord to be risk neutral wage payments would be the preferred outcome, since, assuming wages equal to bare subsistence this would ensure the maximum appropriation. If, however, such total domination was impossible through, for example, the possibility of migration, as in Latin America and feudal Europe, the landowner would be forced to offer incentives such as the provision of subsistence 'holdings', in order to maintain his labour supply.

Labour rents would be an acceptable solution to landowners provided the costs of supervision were not excessive. These, as argued above, will depend on the nature of the productive activity and the insecurity of the tenant. Where the skills involved are varied and their quality essential to the outcome, any threat of eviction may be an inadequate 'incentive'. The landowner will then have recourse to more indirect methods of labour appropriation, that is through share contracts. Share contracts would be preferred to fixed kind or cash leases, since they enable the landowner to obtain the fruits of the lessee's endeavour. The greater the ability to manipulate the rental share or to vary the incidence and extent of other exactions, the greater the opportunity for surplus appropriation.

Supporting evidence for the arguments can be found in the fact that labour rents have proved far less common relative to share rents in areas of tropical production as compared with temperate locations, even when levels of class domination have been similar. Tropical production (generally) involves a more intensive use of labour and many of the operations, such as weeding, thinning and transplanting, bear a crucial relation to output in terms of both timeliness and quality. The labour process is such that the incentive to labour effort provided by share contracts is necessary, to reduce what would become excessive costs of supervision. As Bharadwaj and Das [1975] and Rudra [1975] have demonstrated, even where labour dues are extracted, they frequently *accompany* and *reinforce* sharecropping. The use of share contracts as 'labour incentive' contracts is further documented by Martinez-Alier [1971, 1974] for Spain and Brazil and by Addo [1974] and Robertson [1981] for West Africa.

If it is the tenant class which is dominant, that is in a position to determine the choice of contract, then a different range of outcomes become probable. Given that such tenants will wish to retain the maximum part of their own or their employees' labour, the choice will lie between fixed and share tenancy. Supervision costs will not be of importance, and the onus of decision falls, therefore, on the extent to which the labour process is distinguished by uncertainty. Sharecropping, in shifting part of this burden on to the landowners, is likely to be the preferred outcome if the tenant is sufficiently risk averse. The partial acceptance of the uncertainties of production then becomes part of the 'price' the landowner must pay for gaining access to the others' labour. To the extent that the tenant is risk neutral, or the level of uncertainty is significant, the preference will be for fixed rent tenancies since these provide the tenant with the opportunity to reap more fully the product of his own labour.

That such relationships occur, in India at least, is demonstrated by Ray [1978] and Bharadwaj and Das [1975] amongst others. The most frequent example is where the landowner is a widow or is infirm and unable to cultivate. Increasingly, though, the practice has become more common as the owners of fragmented holdings find it more remunerative to lease to those better endowed with other productive resources, than to cultivate themselves [*Byres*, 1981].

When the economic position of landowners and tenant/labourers is more equal the implications stem primarily from the nature of the labour process. If the uncertainty attached to production is considerable and both parties are risk averse, the likely outcome is a share contract. This is particularly probable when both landowners and tenants are operating close to the margin of subsistence. If this is not the case, however, then one would expect a higher proportion of fixed rent contracts to ensue, since the incentive to share risks would no longer be present.

Certainly evidence from India [*Bharadwaj*, 1974] and Malaysia [*Huang*, 1975] corroborates this pattern. In the former fixed rents are most common in the north west (Punjab) where the incidence of marginal holdings is less, and occur least in the east (West Bengal) where mini-holdings abound. In Malaysia sharecropping is prevalent only in Kelantan province, agriculturally the poorest region. Further, the likelihood of such parity between tenant and landlord is not small, if one considers that many tenants in both of these areas are themselves landowners. One should not exaggerate this implication, however, since the ownership of small parcels of land is no guarantee of survival.

e. Changes in the Labour Process

Both in the United States and western Europe the advent of capitalist agriculture has seen the demise of sharecropping and its replacement by fixed rent tenancies and the employment of wage labour. The question arises, therefore, whether such contracts are incompatible with agrarian capitalism, or whether, in other contexts, they can persist without contradiction within capitalist agriculture. Discussion of this issue will make reference to the form of capitalist relations and the implications of these for the labour process.

Marx [1976: 1019-1038] distinguishes between the formal and real subsumption of labour by capital. It is argued below that this distinction is important for the subject matter of this essay, not only from the perspective of understanding sharecropping as it appears in the early stages of capitalism, but also in explaining its essentially transitional nature once the dynamic of agrarian capitalism is established.

Following Banaji [1977 b] one can argue that the *formal* subsumption of labour is contiguous with circumstances where the basic conditions of capitalism are met:

> The labour process becomes the instrument of the valorization process, the process of the self-valorization of capital – the manufacture of surplus value. [*Marx* 1975 : 1019]

but that appropriation is through *absolute* surplus value only. The existing labour process is subsumed by capitalist relations of production. This initial phase in the development of capitalism is one of primitive accumulation leading to eventual transformation of the labour process and the *real* subsumption of labour reflected in the appropriation of *relative* surplus value. In fact one can argue (although this issue will not be explored further here) that this transformation, the change from formal to real subsumption and from appropriation of absolute to relative surplus value, is the crucial stage in the transitional process, since prior to this transformation the accumulative dynamic of capitalism is not operative and the transition subject to blockage or reversal.

It follows that at the formal subsumption stage the process of valorization may impose itself upon labour processes in agriculture which are widely divergent and (accepting, as argued above, that forms of appropriation are determined by particular conjunctures of class power and labour process), that in certain cases such an imposition will be concomitant with sharecropping as a *relation of exploitation*. I would take issue with Gupta [1980] to the extent that he argues that the existence of sharecropping implies the formal subsumption of labour – a view dependent on a narrow perception of relations of production and confusion between relations and modes of production, a view roundly criticised by Banaji [1977 a]. In fact, sharecropping has most frequently been associated with circumstances where production has been effectively determined by a class intent on the maximisation of its own consumption rather than the valorization of surplus labour. Nevertheless, I would agree both with Gupta and Rudra [1978] in arguing the compatibility of sharecropping and agrarian capitalism, and with the former in identifying it, in this context, with early and *formal* manifestations of capitalist relations.

If capitalism is to obtain a secure foothold within agriculture it must transform the labour process. The appropriation of relative surplus value is dependent on the augmentation and replacement (in value terms) of labour by capital in production. This real subsumption implies the relative depreciation of labour and its skills (labour power) as the means of labour acquire greater significance. The method of labour organisation and the manner in which labour and means of production are combined will change as a consequence. It has been argued above that sharecropping is conjuncturally a consistent form of appropriation to the extent that it acts as either a labour incentive, that is increases the intensity of labour, where supervision costs are high; or as an incentive to tenants, that is increases the supply of tenants, where the labour process is characterised by uncertainty. With a substantial intervention of capital into the labour process, however, neither of these conditions obtain. The time and effort necessary to ensure a 'successful' or 'desired' outcome of production become less as the range and 'subjectivity' of skills necessary are reduced. Thus the greater is the integration of labour and means of production in performing any particular task, the more the performance of labour (intensity and quality of work) is determined by those forces with which labour is combined. Hence 'supervision' itself becomes part of the production process. Furthermore, the reduction of uncertainty is a *necessary* consequence of this intervention: given that the act of production is a relationship between man and nature, and that uncertainty, by definition, is a reflection of man's inability to control nature, it follows that any augmentation of man's labour which increases that control will reduce uncertainty. I would not argue that *all* changes in productive forces increase such control, but only that for changes concomitant with the development of capitalism this is the case.

Thus sharecropping can be consistent with capitalist relations, but only in a transitional sense insofar as it is associated with labour processes typical of non-capitalist modes of production, but subsumed under capitalist relations. The inevitability of its demise with the advent of real subsumpion is illustrated by reference to the so-called 'green revolution'.

The 'green revolution' technology consists of a 'package' of innovations primarily of a bio-chemical nature. That is: the use of high-yielding seeds reinforced by chemical fertilisers, insecticides and herbicides. These innovations may, or may not, be accompanied by mechanical ones such as pumpsets, tractors, harvesters, etc., but arguably the logic of the changes plus empirical observation suggests that they will [*Byres*, 1981]. The combined impact of these changes implies a transformation of the labour process.

One effect of the introduction of this technology is to render the outcome of production less dependent on the intervention of human skills and endeavour, and more a direct response to non-human inputs. To the extent that mechanical changes accompany the bio-chemical ones the dependence on the intensity of labour is further reduced. Thus the ability of the direct producer to influence the results of his own labour is diminished, and with this diminution the 'costs' of supervision fall. It follows that in contexts where class power primarily rests with landowners the advent of the 'green revolution' will cause a resumption of land and a substitution of wage labour for sharecropping [*Bhalla*, 1977]. Insofar as these innovations also reduce the uncertainty attached to agricultural production there will also be a tendency for share rents to be commuted to fixed rents where class power is more equally distributed and the landowner is no longer dependent on risk sharing to attract an adequate supply of tenants.

I have argued above that the overall effect of these developments in productive forces is to reduce uncertainty in production. In practice, however, there may be reasons why, *in the short run*, uncertainty, at least perceived uncertainty, is increased. In the first place the 'new' varieties are very dependent on the complete range of inputs if their use is to be successfully undertaken. An uncertain supply of any one of the necessary components of the 'package' can increase rather than decrease predicted output variance. Secondly, the adoption of the 'package' will initially be by farmers with little or no experience with such techniques. The risk of managerial error and consequent crop reduction or failure is therefore enhanced. (The high-yielding varieties are, in general, less disease resistant and more dependent on adequate water supply than traditional varieties.) The net result of these qualifications is that the process of proletarianisation may be more subtle and protracted than the above paragraph suggests [*Byres*, 1981].

Thus if landowners perceive the risks as well as the rewards of the 'new' technology to be substantial they may attempt to either share these risks with tenants through, for example, cost share leases [*Bardhan and Rudra*, 1980] or to off-load them entirely through fixed-rent tenancies. Bharadwaj and Das [1975] provide convincing evidence of the latter possibility with landlords substituting share contracts by fixed rents involving an increased level of payment in advance of production with no security of tenure – this in response to the increased profit potential associated with the 'package'.

Cost share leases can be regarded as an intermediate stage in the proletarianisation process, with the tenant taking an increasingly passive role in production [*Bardhan and Rudra*, 1980]. This form of lease enables the tenant to acquire the inputs necessary for the new production methods (and landlords to obtain a share of the greater profits resulting), when otherwise such acquisition may be inhibited either through an unwillingness of those existing close to the margin of subsistence to accept any extra burden of uncertainty, or through limited and distorted access to input markets. These same factors also explain the increasing phenomenon of small-scale landlords leasing to large-scale tenants in 'green revolution' areas. There the small owner-operator, having neither the access to resources nor the capacity to absorb risk which would make his own adoption of the 'package' feasible, finds that his return from leasing land to those able to take advantage of the greater profit opportunities afforded by the 'green revolution' is greater than that from self-cultivation along traditional lines [*Byres*, 1981].

Changes in the labour process associated with the 'green revolution', therefore, provide evidence that sharecropping will not prove consistent with production relations concomitant with the real subsumption of labour by capital. In this context sharecropping will appear as a transitory phenomenom of agrarian capitalism. Although its supercedence by other mechanisms of appropriation may be delayed for the reasons outlined above, the development of productive forces and attendant opportunities for increased profits will, in the long run, bring this about.

V. CONCLUSION

A very full conclusion is inappropriate for a paper of this kind. The approach has been influenced by the need to place sharecropping in an analytically workable perspective, one removed from the technicism of the neo-classical debate and its static notions of efficiency. In investigating such a perspective in such a short paper it has proved necessary to leave many issues unexplored which deserve fuller treatment. A further handicap is the inevitably eclectic attempts at empirical validation: inevitable because with few exceptions, the bulk of the information pertaining to sharecropping has been garnered by commentators working within a different perspective. (The principal exceptions being K. Bharadwaj and A. Rudra.)

It is suggested that share contracts must be understood as a particular method of surplus labour appropriation, and the preferred method in the following contexts: (1) where class relations are such that decisions concerning the nature of the contract lies with the landowning class, but where the costs of supervising production are potentially large; (2) where the tenant is in a position to have a decisive influence upon the contract and where the tenant is also insufficiently endowed with resources to allow him to discount income variance as a decision variable; (3) in contexts where class relations are less obviously determined and both landlords and tenants are existing close to the margin of survival.

I would suggest, finally, that Marx's original description of sharecropping as an intermediate method of surplus appropriation, is essentially correct, and that it will tend to disappear with the emergence of capitalist accumulation.

NOTES

1. cf. *Bell*, 1977 who distinguishes 'traditional' and 'modern' views.
2. Earlier contributions to this debate are brought together in *Hilton*, 1976; later views were expressed particularly in the pages of *Economic and Political Weekly*, for example *Patnaik*, 1972, *Chattopadyay*, 1972 and others.

REFERENCES

Adams, D. and N. Rask, 1968, 'Economics of cost share leases in less developed countries . . .', *American Journal of Agricultural Economics*, November.

Addo, N., 1974, 'Some employment and labour conditions on Ghana's cocoa farms', in R. Koley, C. Okali, B. Rowkes (Eds.), *The Economics of Cocoa production and marketing*, Legon: University of Ghana.

Adegboye, R., 1974, 'Land tenure problems and improved practices', in R. Koley, C. Okali, B. Rowkes (Eds.), *The Economics of Cocoa production and marketing*, Legon: University of Ghana.

Agee, J., 1981, *Let us now praise famous men*, London: Owen.

Althusser,L. and E. Balibar, 1970, *Reading Capital*, London: New Left Books.

Ash, R., 1976, 'Land tenure in pre-revolutionary China: Kiangsu province in the 1920s and 1930s', *Research Notes and Studies*, No. 1. Contemporary China Institute, S.O.A.S., London.

Bagchi, A., 1976, 'Cropsharing tenancy and neo-classical economics', *Economic and Political Weekly*, Jan. 17.

Bagi, F.S., 1981, 'Economics of Sharecropping in India' (Haryana), *Pakistan Development Review*, No. 1, Spring.

Banaji, J., 1973, 'Mode of production in Indian Agriculture', *Economic and Political Weekly*, April 17.

———, 1977a, 'Modes of production in a materialist conception of history', *Capital and Class*, 3.

———, 1977b, 'Capitalist domination and the small peasantry', *Economic and Political Weekly*, special number, August.

Barber, R., and J. Cordova, 1974, 'Labour utilisation in rice production', in *Economic Consequences of the New Rice Technology*, Manila: I.R.R.I.

Bardhan, P., 1970, 'Trends in land relations – a note', *Economic and Political Weekly*, Annual Number, January.

———, 1976, 'Variations in extent and form of tenancy', *Economic and Political Weekly*, Sept. 11 and 18.

Bardhan, P., and A. Rudra, 1980, 'Terms and conditions of sharecropping contracts. An analysis of village survey data in India', *Journal of Development Studies*, Vol. 17.

Bardhan, P., and T. Srinivisan, 1971, 'Cropsharing tenancy in agriculture: a theoretical and empirical analysis', *American Economic Review*, March.

Bell, C., 1976, 'Production conditions, innovations and choice of lease in agriculture', *Sankhya*, C.

———, 1977, 'Alternative theories of sharecropping. Some tests using evidence from Northeast India', *Journal of Development Studies*, Jan.

Bell, C., and P. Zusman, 1976, 'A bargaining theoretic approach to cropsharing contracts', *American Economic Review*, September.

Berry, S., 1975, *Cocoa, Custom and Socio-economic change in rural Western Nigeria*, Oxford: Clarendon Press.

Bhaduri, A., 1973, 'A study of agricultural backwardness under semi-feudalism', *Economic Journal*, March.

__________, 1977, 'On the formation of usurious interest rates in backward agriculture', *Cambridge Journal of Economics*, December.

Bhalla, S., 1977, 'Changes in acreage and tenure structure of land holdings in Haryana, 1962-72', *Economic and Political Weekly, Review of Agriculture*, March 26.

Bharadwaj, K., 1974, *Production conditions in Indian Agriculture – A study based on farm management surveys*, Cambridge University Press, University of Cambridge, Department of Applied Economics, Occasional paper 33.

Bharadwaj, K., and P.K. Das, 1975, 'Tenurial conditions and mode of exploitation. A study of some villages in Orissa', *Economic and Political Weekly*, Annual Number, Feb. and *Review of Agriculture*, June.

Bloch, M., 1966, *French Rural History: An essay on its basic characteristics*, London: Routledge & Kegan Paul.

Blum, J., 1978, *The end of the old order in rural Europe*, Princeton: Princeton University Press.

Braverman, A., and T. Srinivisan, 1981, 'Credit and sharecropping in agrarian societies', *Journal of Development Economics*, December.

Bray, F., and A.F. Robertson, 1960, 'Sharecropping in Kelantan, Malaysia', *Economic Anthropology*, Vol. 3.

Buck, J.L., 1937, *Land Utilisation in China*, London: Oxford University Press.

Byres, T.J., 1981, 'The new technology, class formation and class action in the Indian countryside', *Journal of Peasant Studies*, July, Vol. 8, No. 4.

Chandra, N.K., 1974, 'Farm efficiency under semi-feudalism . . .', *Economic and Political Weekly*, March.

Chattopadyay, P., 1972, 'Mode of production in Indian Agriculture – an anti-kritik', *Economic and Political Weekly*, March.

Cheung, S., 1968, 'Private property rights and sharecropping', *Journal of Political Economy*, November/December

__________, 1969a, 'Transaction costs, risk aversion and the choice of contractual arrangements . . .', *Journal of Law and Economics*, April.

__________, 1969b, *The Theory of Share Tenancy*, University of Chicago.

Drake, L., 1952, 'Comparative productivity of share and cash rent systems of tenure', *Journal of Farm Economics*, November.

Gosh, A., 1955, 'The Impact of commercial growth on agricultural tenure systems in India', *Manchester School*, May.

Ghose, A., and A. Saith, 1976, 'Indebtedness, Tenancy and the adoption of new technology in semi-feudal agriculture', *World Development* Vol. 4, No. 4.

Gill, D., 1981, 'Sharecropping in Tuscany, 1880-1920', Paper presented to Peasants Seminar, Institute of Commonwealth Studies, London.

Goodman, D., 1977, 'Rural structures, surplus mobilisation and modes of production in a peripheral region. The Brazilian North East', *Journal of Peasant Studies*, October, Vol. 5, No. 1.

Griffin, K., and R. Mabro, 1979 in K. Griffin (Ed.) *Political Economy of Agrarian Change*, London: Macmillan.

Gupta, D., 1980, 'Formal and Real subsumption of labour under Capital: the instance of sharecropping', *Economic and Political Weekly, Review of Agriculture,* September.

Heady, E., 1947, 'Economics of farm leasing systems . . .', *Journal of Farm Economics*, August.

Herring, R.J., 1978, 'Share tenancy and economic efficiency: The South Asian case', *Peasant Studies*, Fall.

Higgs, R., 1973, 'Race, tenure and resource allocation in Southern agriculture, 1910', *Journal of Economic History*, Vol. 33, March.

Hilton, R., (Ed.), 1976, *The Transition from Feudalism to Capitalism*, London: New Left Books.

Holley, D., 1980, 'The Sharecropper' in T. Peterson (Ed.), *Farmers, Bureaucrats and Middlemen*, Washington: Howard University Press.

Hossain, M., 1977, 'Farm size, tenancy and land productivity: An analysis of farm level data in Bangladesh agriculture', *Bangladesh Development Studies*, Vol. V, No. 3, July.

Horii, K., 1972, 'The land tenure systems of Malay paddy farmers', *Developing Economics*, Vol. 10, March.

Hsiao, J., 1975, 'The theory of share tenancy revisited', *Journal of Political Economy*, Vol. 83, No. 5.

Huang, Y., 1975, 'Tenancy patterns, productivity and rentals in Malaysia', *Economic Development and Cultural Change*, Vol. 23.

Ip, P., and C. Stahl, 1978, 'Systems of Land Tenure. Allocative efficiency and economic development', *American Journal of Agricultural Economics*, February.

Jabbar, M.A., 1975, 'Sharecropping and economic efficiency in Bangladesh', *Bangladesh Development Studies*, Vol. III, No. 2, April.

__________, 1977, 'Relative productive efficiency of different tenure classes in selected areas of Bangladesh', *Bangladesh Development Studies*, Vol. V, No. 1, January.

James, D.P., 1981, *Mechanisation and Institutions in Noakhali,* in H. Mettrick and D.P. James, *Farm Power in Bangladesh Vol. 2,* Department of Agricultural Economics and Management, University of Reading, September.

Johnson, A., 1971, *Sharecroppers of the Sertao – Economics and dependence on a Brazilian plantation*, Stanford University Press.

Johnson, D.G., 1950, 'Resource allocation under share contracts . . .', *Journal of Political Economy*, April.

Jones, P.J., 1968, 'From Manor to Mezzadria. A Tuscan case study in the medieval origins of modern agrarian society', in N. Rubenstein (Ed.), *Florentine Studies – Politics and Society in Renaissance Florence*, London: Faber and Faber.

Jones, R., 1831, *Essay on the distribution of wealth, and on the source of taxation,* London: John Murray.

Ladejinsky, W., 1977, *Agrarian reform as unfinished business* (Selected papers of Wolf Ladejinsky), Oxford University Press for World Bank.

Leach, E.R., 1961, *Pul Eliya: A Village in Ceylon – A Study of Land Tenure and Kinship*, Cambridge: Cambridge University Press.

Marshall, A., 1961, *Principles of Economics*, London: Macmillan. First published 1890.

Martinez-Alier, J., 1971, *Labourers and Landowners in Southern Spain*, London: George Allen & Unwin.

__________, 1974, 'Peasants and Labourers in Southern Spain, Cuba and Highland Peru', *Journal of Peasant Studies*, Vol. 1, No. 2.

Marx, K., 1976, *Capital, Volume I*, Harmondsworth: Penguin. First published 1867.

__________, 1981, *Capital, Volume III*, Harmondsworth: Penguin. First published 1894.

Mazumdar, D., 1975, 'The theory of sharecropping with labour market dualism', *Economica*, August.

McCulloch, J.R., 1907, *Principles of Political Economy*, Ward, Lock & Co.

Mettrick, H., 1981, 'Some aspects of the economics of animal power' in H. Mettrick and D.P. James, *Farm Power in Bangladesh, Vol. II*, Department of Agricultural Economics and Management, University of Reading, September.

Mill, J.S., 1909, *Principles of Political Economy*, London: Longmans, Green & Co. First published 1848.

Mitrany, D., 1930, *The Land and the Peasant in Roumania: The War and Agrarian Reform (1917-21)*, London: Oxford University Press.

Narain, D., amd P.C. Joshi, 1969, 'The magnitude of Agricultural Tenancy', *Economic and Political Weekly Review of Agriculture*, September.

Newaj, K., and A. Rudra, 1975, 'Agrarian Transformation in a district of West Bengal', *Economic and Political Weekly*, March 29.

Newbery, D., 1975, 'Tenurial obstacles to Innovation', *Journal of Development Studies*, July, Vol. II, No. 4.

__________, 1976a, 'Choice of Rental Contract in Peasant Agriculture' in L.G. Reynolds (Ed.), *Agriculture in Development Theory*, Yale University Press.

__________, 1976b, 'Risk Sharing, Sharecropping and Uncertain Labour Markets', Technical Report 202, I.M.S.S., Stanford.

Newbery, D., and J. Stiglitz, 1979, 'Sharecropping, Risk Sharing and the Importance of Imperfect Information', in J. Roumasset, J.M. Boussard, I. Singh (Editors), *Risk Uncertainty and Agricultural Development*, S.E. Asia Regional Center for Graduate Study and Research in Agriculture, Phillipines.

Paige, J., 1975, *Agrarian Revolution*, New York: The Free Press, Macmillan.

Patnaik, U., 1972, 'On the Mode of Production in Indian Agriculture' *Economic and Political Weekly*, 30 September.

Raj, K.N., 1970, 'Ownership and Distribution of Land', *Indian Economic Review*, April.

Rao, C.H., 1971, 'Uncertainty, Sharecropping and Entrepreneurship', *Journal of Political Economy*, May/June.

Ray, D., 1971, 'Agrarian Relations – the Small Lessor and the Big Lessee', *Economic and Political Weekly*, 23-30 December.

Reid, J.D., 1973, 'Sharecropping as an Understandable Market Response: The Post-Bellum South', *Journal of Economic History*, Vol. 33, March.

__________, 1975, 'Sharecropping in History and Theory', *Agricultural History*, Vol. 49, No. 2, April.

__________, 1977, 'The Theory of Share Tenancy Revisited – Again', *Journal of Political Economy*, Vol. 85, No. 2.

__________, 1979a, 'The Evaluation and Implications of Southern Tenancy', *Agricultural History*, Vol. 53, No. 1, January.

__________, 1979b, 'Sharecropping and Tenancy in American History' in J.A. Roumasset, J.M. Boussard, I. Singh (Editors), *Risk, Uncertainty and Agricultural Development*, South-East Asian Regional Center for Graduate Study and Research in Agriculture, Philippines.

Robertson, A., 1980, 'On Sharecropping', *MAN* (N.5) Vol. 15.

Robertson, A.F., 1981, *ABUSA – The Structural History of an Economic Contract*, Mimeo.

Roumasset, J.A., 1976, *Rice and Risk Decision Making among Low Income Farmers*, North Holland.

Rudra, A., 1975, 'Sharecropping Arrangements in West Bengal', *Economic and Political Weekly*, Review of Agriculture, September.

__________, 1978, 'Class Relations in Indian Agriculture', *Economic and Political Weekly*, June 3, 10, 17.

Schickele, R., 1941, 'Effect of Tenure Systems on Agricultural Efficiency . . .' *Journal of Farm Economics*, February.

Shlomowitz, A., 1979, 'The Origins of Southern Sharecropping', *Agricultural History*, Vol. 53, No. 3, July.

Smith, A., 1976, *The Wealth of Nations*, London: Oxford University Press. First published 1776.

Snowdon, F., 1979, 'From Sharecropper to Proletarian: The Background to Fascism in Rural Tuscany, 1880-1920' in J. Davis (Ed.): *Gramsci and Italy's Passive Revolution*, Croom Helm.

Sokoloff, S., 1980, 'Land tenure and political tendency in Rural France: The case of sharecropping', *European Studies Review*, Vol. 10, No. 3, July.

Steinbeck, J., 1951, *The Grapes of Wrath*, Harmondsworth: Penguin.

Stiglitz, J., 1974, 'Incentives and Risk Sharing in Agriculture', *Review of Economic Studies*, April.

Taylor, R.H., 1943, 'Post-Bellum Southern Rental Contracts', *Agricultural History*, Vol. 17.

Vyas, V.S., 1970, 'Tenancy in a Dynamic Setting', *Economic and Political Weekly*, September.

Winters, D.L., 1974, 'Tenant Farming in Iowa, 1860-1900: A Study of the terms of Rental Leases', *Agricultural History*, Vol. 48, January.

Wright, J., 1973, 'Comments on papers by Reid, Ransom and Sutch, and Higgs', *Journal of Economic History*, Vol. 33, March.

Zaman, M.R., 1973, 'Sharecropping and Economic Efficiency in Bangladesh', *Bangladesh Economic Review*, Vol. 1, No. 2, April.

Classical Theory of Rent and its Application to India: Some Preliminary Propositions, with Some Thoughts on Sharecropping

Utsa Patnaik*

1. CLASSICAL THEORIES OF RENT

Of importance in all countries with landlordism is the theory of absolute ground rent. Differential rent poses no particular problems: it merely relates to variations in the level of rent depending on varying soil fertilities and degrees of application of capital. But what determines the level of rent in the first place? In the case of pre-capitalist relations of production, in particular all cases where petty producers pay a rent/tribute/tax to the ruling classes commanding property in land, the entire surplus labour over necessary costs for petty producers is absorbed in rent. The normal form of surplus labour, of what becomes later surplus value, is ground rent in such cases: and the limit to rent extraction is set by the struggle over necessary labour, given the historically existing level of productivity. 'Necessary labour' itself is not biologically determined but carries a 'moral and historical element'. What constituted the customary concept of 'necessary labour' for the seventeenth century rent-paying peasant in Mughal India no doubt differed from 'necessary labour' for the thirteenth century serf in Western Europe, and both differed from biological subsistence. But attempts to raise rent by curtailing necessary labour led to revolts in both cases.

In the case of the capitalist organisation of production, rent is no longer the normal form of surplus value: it is only one of its components, profit being the typical form. It follows that a consistent theory of value must explain not only capitalist profit but also capitalist rent. In particular, it must explain the determination of the exchange values of commodities such that manufactured commodities prices give a profit rate (assuming the rent element to be negligible or non-existent) while agricultural commodities prices give the the same profit rate and rent. The value theories of Smith, Ricardo and Marx can be looked at in the light of the degree to which they provide a consistent explanation of capitalist rent along with capitalist profit.

Adam Smith quite explicitly distinguishes between absolute rent and differential rent: 'As soon as the land of any country has all become private property, the landlords, like all other men, love to reap where they never sowed, and demand a rent even for its natural produce.' [*Smith*, 1961: 56] The fact that rent does not arise *only* from better land (better owing to more application

**Jawaharlal Nehru University, New Delhi, India.*

of capital or higher fertility) is established by pointing out that rent exists even for that land which is not capable of any improvement, therefore exists owing to the fact of monopolised property ownership: 'The rent of land therefore considered as the price paid for the use of the land, is naturally a monopoly price.' [*Smith*, 1961: 162] Thus rent is explicitly seen as the consequence of the existence of landed property. This is the concept of absolute ground rent, distinct from that rent, which Adam Smith terms 'interest', which arises from the greater application of capital.

However, the relation of rent to the other components of value in Smith, remains obscure. Firstly, Smith holds an almost Physiocratic view that food production 'always and necessarily affords some rent to the landlord' [*Smith*, 1961: 180], while there are other branches of production within agriculture which do not necessarily yield rent though they may do so. Secondly, he introduces several concepts of price, inclusive of and exclusive of rent, which are mutually contradictory. These are:

(1) 'Natural price' which comprises value of stock (constant capital and wages), profit and rent.[1]

(2) 'Sufficient price', or the price sufficient to replace stock and give an average profit, i.e. the price of production [*Smith*, 1961: 180-197].

(3) 'Ordinary price'. When this is more than the sufficient price rent exists; when it equals sufficient price, rent is zero; when it falls short of sufficient price, profit is curtailed [*Smith*, 1961: 180-197].

The concepts of 'sufficient price' and 'ordinary price', taken together, contradict the concept of 'natural price'. First, the discussion is in terms of the 'natural price' at which commodities exchange, which is supposed to include rent in the same way that it includes wages and profits. Then the 'sufficient price' is postulated, which is sufficient to bring certain commodities to market by giving the average profit on stock alone: i.e. the 'sufficient price' is the price of production. For other commodities – for example, foodgrains – however, the sufficient price is not in fact sufficient and an 'ordinary price', higher than the sufficient price by the amount of rent, is required. If commodities exchange at varying prices, inclusive of and exclusive of rent respectively, clearly the original theory of exchange at 'natural prices' is contradicted.

The above contradiction is, however, a 'fruitful contradiction' in Adam Smith because it arises from a real problem. This problem was later posed and solved by Marx: the existence of profit plus surplus profit (rent) in the field of agricultural production coincided consistently with the existence of profit alone in other fields under competitive capitalist production.

Ricardo considers differential rent alone and has no concept of absolute rent: he ignores the economic effects of landed property *per se*. This is because in Ricardo's theory of value, the prices of production ('sufficient price' of Adam Smith) and values are identical. Given the (assumed) identity of values and prices of production, logically there is no room for a theory of absolute ground rent. The admission of the existence of absolute ground rent would imply that agricultural commodities sell at prices higher than their prices of production by the amount of rent, while other commodities sell at their prices of production. Since prices of production are identified by Ricardo with values, this implies that a larger quantity of embodied labour exchanges against a smaller quantity of embodied labour, and the entire theory of exchange according to labour values collapses. To maintain the consistency of his value theory, Ricardo implicitly denies that landed property gives rise to absolute rent. Adam Smith's 'fruitful contradiction' is thus not resolved, but only assumed away.

Ricardo's definition of rent and his criticism of Smith reveal very clearly the absence of any concept of absolute ground rent. 'Rent is that portion of the produce of the earth, which is paid to the landlord for the use of the original and indestructible powers of the soil' [*Ricardo*, 1970: 67] according to Ricardo. This leaves unanswered the obvious question: given two homogeneous stretches of territory with identical 'original and indestructible powers of the soil', why does one with private monopoly of land ownership give rise to the category of rent, while the other, without monopoly of property, does not. On the basis of his particular definition of rent which stresses 'the original and indestructible powers of the soil', Ricardo puts forward criticisms of Smith which are theoretically misconceived, in ignoring completely landed property which forms the basis for Smith's concept of rent.[2] In this, Ricardo is consistent in adopting the viewpoint of pure capitalist production, for which landed property is irrelevant. Differential rent (and only this form of rent) can be accommodated, however, within Ricardo's value theory. Differing soil fertilities and differing degrees of application of capital mean that individual capitals can produce at lower prices of production than the socially necessary price of production, and thus obtain surplus profit payable as rent. This is no different in principle from a similar deviation of individual prices of production in industry from the average, socially necessary price of production, yielding surplus profit. The only difference is that in industry surplus profits tend to be more or less quickly competed away; while in agriculture, landed property intervenes to pocket the surplus profit as differential rent, and thus tends to stabilise existing surplus profits over a longer period.

The ignoring of landed property *per se* and therefore of absolute ground rent by Ricardo leads to certain logical anomalies in his argument regarding differential rent (which is treated as the only form of rent). Firstly, the proposition that 'the worst soil under cultivation yields no rent' is clearly incorrect as a general proposition and can hold only in a situation where landed property does not exist. Ricardo's fondness for quoting as a general case the situation in the North American colonies (where there then existed virtually unlimited areas of cultivable land) arose precisely from the fact that absolute rent by definition could not exist in such a case since monopoly of landownership did not exist. However, as Marx points out this case is highly specific and is quite unrealistic in the context of Europe where landed property confronts capital as a historically given condition. [*Marx*, 1969: ch.xi. 240-250] Similarly it is a case not applicable to India or indeed any country with a history of landed property. The correct proposition here would be, that the worst soil under capitalist cultivation must not only replace stock with average profit but also yield rent wherever landed property exists. Secondly, it is not clear why the price of production in agriculture is determined by the price of production on the worst soil, i.e. under the most unfavourable conditions of production, as Ricardo states, and not determined by the price of production under the average, socially necessary conditions of production. If the latter is the case, then individual capitals invested in agriculture can yield less, the same, or more than the average profit after paying absolute rent, just as individual capitals invested in industry can yield less, the same, or more than the average profit. Ricardo's formulation, which makes the price of production under the worst conditions the determining price, implies that individual capitals – other than those earning the general average profit – can only give more than this profit, and never less. Ricardo's formulation, therefore, introduces a logically unnecessary asymmetry between agriculture and industry in this respect. Thirdly, Ricardo's assumption that, with expansion of demand for agricultural products, the transition is always from the 'most fertile' to the 'least fertile' land, is not valid. The factors of advantageous location and 'intrinsic' soil fertility may well work in opposite directions. Typically, in the history of settlement of colonies, the advantageously located lands were cultivated first even when intrinsically more fertile lands existed elsewhere. Hence it is possible for less fertile land to pay a rent when more fertile land does not: or for the former to pay a higher rent than the latter. Furthermore, depending on conditions of demand, barren, non-rent-yielding land may become rent-yielding because it is particularly suited to the production of a particular commodity: for example, wool production replacing crop production.

Marx's theory of absolute and differential rent is integrated consistently into his theory of value. His theory of value permits the integration of Adam Smith's concept of rent as payment for the use of land in monopolised ownership (absolute ground rent) and Ricardo's concept of differential rent (shorn of its logical anomalies) into a consistent system.

Ricardo's identification of values with prices of production is stated by Marx to be incorrect, and arising ultimately from the failure to ask the question: what is the origin of surplus value? Which leads to a failure in answering the question: what determines the general rate of profit? Prices of production must necessarily differ systematically from values precisely because values form their basis. This is because the capitalist mode of production (like all previous class societies) is based on the exploitation of one class by another. The struggle of the working class as a whole against the capitalist class as a whole determines the division of the net social product into the value of necessary labour and surplus value, thus establishing the general rate of profit. The competition of capitals distributes the total surplus value so appropriated by the capitalist class, over the different branches of production in accordance with the tendency to equalisation of individual rates of profit to this general rate of profit. Given the fact that the organic composition of capitals differs from branch to branch (so that the same total capital generates a higher surplus value where the organic composition is lower and generates a smaller surplus value where it is higher), this process of equalisation of individual rates of profit to the general rate necessarily implies that prices of production must deviate systematically from values. Branches of production with a lower than average organic composition (such as crop production) generate more surplus value on a given total capital compared to branches of production with a higher than average organic composition, so price of production must be below value in the first case while price of production must be above value in the second case, in order that the tendency for equalisation of the rate of profit be operative.

As soon as prices of production are seen necessarily to deviate from values in this theoretically specified manner, the explanation of absolute ground rent is no longer logically contradictory to the theory of value. Absolute ground rent is a tribute exacted by the class of landlords by virtue of their monopoly of landed property, from the capitalist class out of the total surplus value appropriated by the capitalist class from the working class. A part of this total surplus value is simply taken away by the landlords, and thus drops out of the pool of surplus value available for distribution over the individual capitals of the capitalist class. Since landed property confronts capital as a historically pre-existing phenomenon, the amount of surplus value appropriated by the landlord class is historically determined (everything else, notably conditions of production, remaining the same.)

There are two consequences of the fact that the landlord class exacts capitalist rent out of the total pool of surplus value, as compared to the situation where landed property is non-existent. Firstly, the general rate of capitalist profit, i.e. profit accruing to the capitalist class, is itself lower than would have been the case if rent had not been deducted. This is evident. Secondly, the rate of profit in agriculture is not equalised to this general rate of profit, but to a level higher than it by the rate of rent (or, the rate of surplus-profit), while the rate of profit in non-rent-paying branches of production is equalised to the general rate. This means that while the price of production in agriculture is lower than the value, the price at which agricultural products exchange is higher than the price of production, and becomes equal to value in the limiting case. (We abstract here from the question of specific monopoly of particular types of land suited for particular commodities, in whose case the price at which they exchange can exceed value.) price of prodn + rent = value

It also follows that the existence of landed property and hence of absolute rent operates as a barrier or hindrance to investment under capitalism. From the viewpoint of capital it is only necessary that profit be obtainable for a piece of land to be cultivated. If that piece of land is owned by a big landowner, however, it will not be leased out for cultivation at all unless it pays a rent in addition. Thus as long as landownership and the investment of capital are divorced (as was substantially the case in nineteenth-century England), landed property constitutes a barrier to the expansion of the field of investment.

Differential rent poses no deep problems for Marx's theory of value. As mentioned earlier, divested of its logical shortcomings, the Ricardian discussion of differential rent can be integrated into Marx's theory of value. Differential rent merely constitutes the case where individual capitals produce at prices of production which are lower than the average, socially necessary price of production. (Symmetrically there can be individual capitals which produce at prices of production higher than the socially necessary price, so that they obtain less than the average profit.) An individual capital can produce below the average price of production either because of the inherently greater fertility of the land on which it is invested or because of productivity-raising effects of previous investment. Thus, because they operate under better-than-average conditions, individual capitals can obtain a surplus profit over the average rate, which may be termed relative surplus profit to distinguish it from absolute surplus profit (paid as absolute rent) discussed earlier (which must be earned by *all* capitals invested in agriculture where there is monopoly of property in land).

While relative surplus profits in agriculture are more stable than in industry owing to their 'natural basis' (for example, more-than-average land fertility), at the same time the stability should not be overestimated. While such surplus profits may not be immediately competed away as in industry, in a capitalist agriculture there exists a constant tendency to create the conditions for relative surplus profit through productivity-raising investment. Today's 'best soil' yielding the highest relative surplus profit, can become, if not tomorrow, by the day after tomorrow the 'worst soil' yielding no surplus profit after paying rent, because yesterday's worst soil has undergone heavy productivity-raising investment, while with the expansion of output and fall in cost the average price of production has fallen.

So far we have discussed briefly the theories of absolute and differential rent under capitalist conditions of production. That is, we have considered to what extent the value theories of Smith, Ricardo and Marx explain capitalist absolute ground rent consistently with capitalist profit. However, absolute ground rent exists also under pre-capitalist conditions of production, and indeed constitutes the normal form of surplus labour under such conditions. The questions arise: what are the forms of pre-capitalist ground rent; and, what is the historical process of transition from pre-capitalist to capitalist ground rent?

On these questions no consistent theoretical analysis is put forward by the classical economists other than by Marx.[3] Hence on these questions, which are of especial relevance to countries like India, we may with justice speak of the applicability of the specifically Marxist theory of rent.

II. APPLICATION OF MARXIST THEORY OF RENT TO INDIA

Economies such as India's are characterised by a variety of production relations since they are economies in transition in some form or another, from pre-capitalist to capitalist production. Light can be thrown on this process of transition by a study of Marx's analysis of the forms of rent characteristic of feudalism, and other pre-capitalist formations, and the transition from pre-capitalist to capitalist ground rent.

The basic characteristic of pre-capitalist rent is that it constitutes the entire surplus labour of the petty producer working with his own and family labour and owning means of production other than land; and it is extracted for the use of land by a superior class monopolising property in land. This may be a territorial overlord as in European feudalism, to whom rent is paid directly in labour, in kind or in money. It may be a centralised state as in some periods of Asian history, in which case pre-capitalist rent and tax payment coincide. Or it may be a combination of rent to a territorial overlord and tax to the state as seems to have been the case in Mughal India. In every case, rent is the typical, normal form of the petty producers' surplus labour and (along with loan interest) accounts for the entire surplus labour. In their classical form, pre-capitalist relations are direct relations of domination and subordination between the ruling classes and the direct producers, unmediated through the market: and hence they are characterised by extra-economic coercion.

Under petty commodity production, the prices at which commodities exchange must be such as to cover necessary production costs under average conditions of production, including the customary consumption of the producer after paying pre-capitalist rent. For petty commodity producers, 'prices of production' are not relevant since they are not capitalists but operate with the objective of satisfying consumption needs. The positions of the rent-paying petty producer and the rent-paying capitalist producer may be set out schematically as follows:

1. *Petty Commodity Producer*	2. *Capitalist Producer*
Price must cover:	Price must cover:
1. Outlay on materials	1. Outlay on materials
2. Consumption of family	2. Wages
3. (Pre-capitalist) rent	3. Average profit
	4. (Capitalist) rent

Thus, pre-capitalist rent represents the surplus of output value over production costs including customary consumption of the petty producer, while capitalist rent represents the surplus of output value over the price of production (which includes average profit).

Under the commercial and land-revenue exploitation of British rule in India, new forms of landed property were created and production for the market expanded considerably. Direct relations between landlords and rent-paying producers gave way increasingly to relations mediated through the market. The petty producer also became increasingly the petty commodity producer. However, whether rent is extracted by custom backed by extra-economic compulsion as earlier, or it is extracted via the market mechanism given monopoly of landed property in a situation of land-hunger (as at present), rent remains pre-capitalist rent to the extent that it is the petty producer who pays it, and it constitutes the normal form of his surplus labour.

There are, historically, two paths of transition from petty production to capitalist production in agriculture, and hence from pre-capitalist to capitalist ground rent. Firstly, a section of the petty producers themselves develop into labour-hiring rich peasants and ultimately into capitalist producers. Secondly, landlords and at a later stage urban moneyed people turn into agricultural capitalists. In the first case, because the capitalist tendency originates from within the peasantry, embryonic forms of profit can exist. Owing to some specific reason, the costs of production including the family's consumption and rent payment taken together may not exhaust the price of the product but leave a small surplus with the peasant which can be further augmented over time. The specific reason may be that (1) the individual cost of production for particular peasants is less than the necessary cost: the resulting surplus in peasant hands constitutes embryonic profit which can be further augmented by hiring labour, trade etc. (2) the process of formation of a rich peasant stratum becomes especially rapid when there is a secular tendency of inflation in output prices combined with relative stability in the cash rent or in costs of production. This was the case during the sixteenth- and seventeenth-century price inflation in Europe. The fixity of cash rent combined with rising prices meant that profit emerged for the commodity-producing middle peasant.

In India, however, the process of formation of a rich peasant stratum, though present, was considerably 'damped' during the colonial period owing to heavy rental/revenue burdens. The main factors conducive to the growth of a rich peasant class were (i) the statutory limitation on rent in the case of favoured categories of tenants particularly in *zamindari* (landlord) areas, allowing an increasing element of surplus to be retained in their hands; (ii) production for export, of commercial crops and high-value food crops, where that handful of better-off cultivators able to undertake the necessary higher outlays without indebtedness, reaped profits.

In the post-Independence period the process of formation of a rich peasant class has received a marked stimulus from the effects of land reform legislation, and from the rapid inflation in agricultural, particularly foodgrains, price from the early 1960s. Under land reforms the rich peasant tenants succeeded in buying ownership of landlords' land which had been vested in the State Governments after intermediary abolition (while middle and poor peasant tenants faced eviction). The rapid inflation in agricultural prices since the early 1960s has also led to increasing profits in the hands of this class.

In economic importance, the rich peasant element in the developing capitalist tendency is, however, secondary compared to the element of landlord capitalism. The extremely limited character of redistribution of land as a result of intermediary abolition and tenancy reform means that the effective monopoly of landownership remains high though slightly moderated. In the case of the landlord class, the transition to capitalist production faces the barrier of pre-capitalist rent, which is paid by the ordinary mass of petty tenants.

Levels of pre-capitalist absolute rent in India continue to be very high. Given the existence of a large class of destitute agricultural labourers and the prevalence of unemployment and underemployment, there is a premium on land possession, for this guarantees a certain minimum level of income in a way that the labour market cannot. In their struggle to ensure a subsistence by renting land, petty producers pay not only their entire surplus labour as rent, but in many cases are obliged to depress necessary labour itself, by underfeeding the family, underfeeding livestock, etc. Typically, rent amounts to 50 per cent of gross output in the case of the major staple foodgrains. (In areas of higher fertility or more advantageous conditions of production differential rent is incorporated so that the rental share may go up to 70 per cent.) The remainder, according to existing studies on cost of production, just covers the necessary paid-out costs with very little return to family labour.

The landlord and the urban capitalist with investible funds already employ those funds in various profit-yielding ways: in industry, trade, money dealing. As distinct from the well-to-do peasant, who may be satisfied with embryonic profit (less than the general rate of profit obtainable on capital in the economy) because his production is still partly for satisfying direct consumption requirements, the landlord will invest money capital only if it yields at least the generally prevailing average return. The same consideration should apply to his investing money capital in agricultural production. In fact, however, the existence of monopoly of landownership and high pre-capitalist ground rent implies that investment of money capital in agricultural production must yield much more than the average rate of return on capital if such investment is to be undertaken in preference to alternative uses of the funds. The reason for this is amplified below.

The essence of monopoly of landownership in the Indian situation is that, merely by virtue of the claim to landed property and without making any outlays whatever, the landlord obtains high pre-capitalist ground rent, Rs R, from leasing out a given standard acre of land. If he cultivates directly instead, he must invest a given sum of money Rs M in the fixed and working capital requirements. Assuming this sum Rs M can yield him Rs P average profit in alternative uses, clearly it is not enough if by withdrawing an acre of land from tenant cultivation and investing Rs M on it for direct cultivation, only Rs P, the average profit, is obtained. For in the alternative situation Rs P profit was being earned, on the money capital, *plus* Rs R rent on leasing out the acre of land. Therefore, the return on investing Rs M money-capital in direct cultivation must give at least Rs P+R, the total return in the alternative situation, for direct cultivation to be economically attractive. In short, direct cultivation requires that a given capital earns surplus profit equal to pre-capitalist rent, over and above average profit.

In the case of a capitalist who does not own land but leases it (or buys it) the investment of the given sum of money Rs M in direct cultivation must also give at least Rs P + R; for in this case rent must be paid for the leased acre (or a return on the sum invested in land purchase obtained) and the average profit obtained on the money outlay in production. In both cases, therefore, investing a given sum of money in direct capitalist cultivation in agriculture is conditional upon that sum of money yielding a surplus profit at least equal to the existing level of pre-capitalist rent, over and above the average profit.

The above argument is a logical extension of the idea of capitalist absolute rent constituting a barrier to capitalist investment. In a situation where capitalist production is confronted by the existence of pre-capitalist absolute rent as a historically given condition (as in India), this must also constitute a barrier to investment in the manner analysed above. Furthermore, certain inferences regarding the character of technical change in agriculture in such a situation, can be drawn.

If direct capitalist production and hence capitalist investment by landlords in agriculture is to emerge given a milieu of petty production and payment of pre-capitalist rent, for given market conditions a necessary requirement is that capitalist investment embodies new methods of production which lead to a discrete rise in surplus per unit area. Only with such a sufficiently large discrete rise in surplus per unit area can the investment of a given sum of money give not only the rent which petty producers are already producing but a profit in addition. It may be borne in mind that technical change is all the more essential because there is no possibility at all of raising surplus by cutting costs further within the existing set of techniques operated by petty producers: they are already cutting costs to a minimum, indeed underpaying family labour, in order to pay pre-capitalist rent. In fact, a hypothetical labour-hiring holding employing the same methods of production as the petty producer is almost certain to end up with a *lower* surplus, to the extent that the market wage rate is higher than the return per labour day of the family workers. No labour-hiring holding in reality will produce with the same set of techniques as petty producers.

In the light of these considerations, it becomes understandable why such private investment in technical change as had taken place in Indian agriculture has been necessarily of a character such that a substantial rise in surplus per unit area is the result. In short, there has been land-augmenting investment. Irrigation, fertilisers, high-yielding varieties of seed, all combined permit the more intensive cropping of a given physical area, raising output and surplus per unit of area substantially. This is the only way in which private capitalist investment can overcome the barrier of pre-capitalist rent. On this understanding, there is a logical necessity underlying the specific form of technical change which has been taking place.

Land-augmenting investment is not incompatible with mechanisation and relative labour displacement in certain agricultural operations. If the requirements of raising surplus per unit of area demand it, mechanisation will be profitable for landlords. Thus multiple cropping, the main element of increased output per unit area, demands the speeding-up of the crucial agricultural operations of preparatory tillage and if the land area is large enough leads to tractorisation. Irrigation under private auspices leads to investment in pumpsets. If the multiple-cropping index is raised sufficiently it may become profitable to mechanise threshing and harvesting.

Forms of Rent Payments in Relation to Production

The form of rent appropriation by itself tells us very little about the underlying production relation on the basis of which that rent is produced. It is only the form of pure labour-rent which is specifically found to be associated with feudal production relations, representing as it does the most primitive form of rent payment. The forms of product rent and cash rent are, however, both found to be compatible with feudal, transitional and capitalist relations of production. Whether rent is feudal or capitalist, does not depend on its form, but depends rather on whether it is produced by the independent petty producer on the basis of family labour; or by the capitalist producer employing wage labour. Schematically:

	Form of Labour	*Form of Rent*
1. Feudal Production	Petty producer using family labour	Labour; Product; Cash
2. Transitional Production	Dependent petty commodity producer using family labour	Product; Cash
3. Captalist Production	Wage labour	Product; Cash

Sharecropping is a specific sub-form of product rent such that rent is not fixed but varies in proportion to the harvest. It follows from the above schematic presentation that the common tendency of identifying sharecropping with 'feudal' or 'semi-feudal' relations, is not really tenable. To establish whether a relation is 'semi-feudal' or not, we have to look not at the form of rent but at who is producing the rent: the petty producers, on the basis of family labour; or the capitalist entrepreneur, on the basis of wage labour. A share-of-crop rent in the first case is pre-capitalist rent; the same form of rent as share of crop represents, however, capitalist rent in the second case.

It is empirically true (in underdeveloped countries like India with a history of feudal landed property) that the sub-form of crop-share rent, which is widely prevalent, is generally associated with petty producers operating on the basis of family labour. But, empirical association is a different matter altogether from an analytical definition. (It is quite true, for example, that the development of capitalist manufacturing is associated with urban development, but it would be incorrect to identify capitalist production with urban production.) While sharecropping as a sub-form of kind-rent payment is empirically associated with petty production, it is not restricted to petty production, but can be compatible with transitional and capitalist production as well.

In an agriculture like India's, undergoing a process of highly uneven capitalist development under continuing land monopoly, we find a co-existence of pre-capitalist and capitalist absolute ground rent. Let us restrict ourselves to the sharecropping sub-form of ground rent for the time being. Typically share-of-crop rents are paid by a mass of petty producers to land owners, under names ranging from *barga* in Bengal to *batai* in Punjab. As we have noted, the levels of absolute rent are extremely high and generally absorb the entire surplus product of a given production cycle in the case of petty producers. The tenant is left with barely enough to cover the next cycle's production costs plus family consumption. However, share-of-crop rents are sometimes also found to be paid by labour-hiring rich peasants, or by capitalist entrepreneurs.

Let us consider some examples of the modal type, the petty tenant striving for subsistence. Our data relate to a sample survey of three hundred and sixty households, cultivating and landless in Haryana (formerly part of Punjab), in 1972-73. This is a region well known for the extensive adoption of 'green revolution' technology. Twenty-two households were found to fall in the 'poor peasant' class, defined as those working for others as labourers or tenants, to a greater extent than they were self-employed. All paid rent on leased land as share of crop, at 50 per cent: however, only six of these were pure tenants, paying rent on all plots. The class as a whole obtained from cultivation an average family labour income of Rs.814 per season, which was already below the lowest possible 'poverty line'. Rent payment, amounting to just over half of this income, slashed their disposable income to less than Rs.400, necessitating hiring out of family labour on wages and employment in crafts to make good the deficit, although even after adding on the wages earned and other non-agricultural incomes, however the average poor peasant failed to reach the 'poverty line' (Table 1).

By contrast, consider the case of the following capitalist tenant from Ahmedabad district, Gujarat. (Data from primary survey of big holdings carried out in 1969.) The capitalist tenant managed a one hundred and fifty acre holding initially entirely belonging to a landlord and financier (this area incidentally is nearly twice as large as the entire area operated by the twenty-two poor peasant tenants discussed above). The estate was cultivated on an

TABLE 1

LABOUR INCOME AND DISPOSABLE INCOME OF TWENTY-TWO POOR PEASANT HOUSEHOLDS, HARYANA 1972-73 (SIX-MONTH SEASON)

Item	Per household value in Rs.		Per household value in Rs.
1. Family labour income from cultivation	813.75	5. Income from wages	130.00
2. Rent payment on land leased in	411.94	6. Other non-agricultural incomes	215.45
3. Payment of loan interest, land revenue and marketing commission	8.68	7. Total family disposable income	738.58
4. Family disposable income from cultivation (1-2-3)	393.13	8. Deviation from lowest 'poverty line'	–259.33
		9. Total Acreage operated by twenty-two households	83.16

Note: 1. Family labour income = Gross output of crop and livestock products less all paid-out costs of production.

2. Lowest poverty line is the Dandekar-Rath estimate of Rs. 15 per capita per month at 1960-61 prices, which gives Rs. 196.2 for a six month season at 1972-73 prices.

extremely intensive basis, employing permanent hands as well as casual labour, with a wage-bill of over Rs. 50,000 annually; it grew high-value commercial and orchard crops (high-yielding cereals, cotton, tobacco, table grapes and other fruit) using pumpsets, tractors and threshers; the annual surplus was in the order of Rs. 2.45 lakhs. Of the one hundred and fifty acres, fifty had been purchased by the capitalist tenant from the landlord out of his share of the profits; he continued to pay 50 per cent of gross output as rent on the remaining one hundred acres. The capitalist tenant's share was equivalent to just over one-third of the total profits, nearly two-thirds going to the landlord. However, the landlord advanced a substantial part of the money capital required for cultivation on the land leased out: in this case the 'rent' therefore represents a combined category of profit and rent.

There are certain interesting implications of such a co-existence of pre-capitalist, and capitalist ground rent. Firstly, there are evidently qualitatively quite distinct types of lessees or tenants. At one extreme is the poor peasant or small peasant tenant, cultivating with family labour, wholly or mainly leased land on which he is dependent for a livelihood; at the other extreme is the capitalist tenant, who runs leased land as a business enterprise on the basis of hired labour for profit. The first type continues to make up the overwhelming majority of tenants, and a majority of them in turn work on sharecropping terms. The second type of capitalist tenant – and here we may include the rich peasant tenant – is to be found only in the agriculturally most advanced areas; and of these, the tenants leasing on sharecropping terms form an even smaller subset. Despite constituting a numerical minority, however, the second type of tenant may command a substantial share of area leased, particularly in the advanced regions.

Secondly, if we assume that for a given type of land growing a given type of crop, there is competition between the mass of land-hungry poor peasant and small peasant tenants, on the one hand, and the capitalist entrants into agriculture, and the rich peasants, on the other, it follows that the latter must pay at least the same level of absolute ground rent as the former are paying already. If an acre of paddy-growing land on average gives ten bags of crop-share rent in paddy under the customary methods of cultivation by petty producers, then the capitalist entrant into agriculture must also pay at least the same rent per unit area. The difference between the petty producer and the rich peasant or capitalist entrepreneur, as we have seen, is however this: the former produce for subsistence, and, therefore, the entire surplus produce can be paid as rent; the latter produce for profit, and, therefore, only a part of the surplus product can be paid as rent, the remainder being retained as the profit (which constitutes the economic rationale of leasing in, for this type of producer). It follows that the rich peasant and capitalist tenants can only lease in land at the same absolute level of rent as the petty peasant tenants if they succeed in producing a surplus per unit area, which is higher than the latter's surplus by the amount of profit (at the going rate of profit).

Rich peasants and capitalist tenants do produce a higher ouput and surplus per unit area than do petty producers, because they are able to employ productivity-raising, improved techniques of cultivation. Owing to the production of a higher output and surplus per unit of area, the same level of absolute rent which constitutes say 50 per cent of gross output for petty producers, amounts to a correspondingly lower proportion of gross output for the emerging capitalists. We would, therefore, expect to find that the large-scale, hired-labour-based producers in agriculture, pay a lower proportion of their output value as rent per unit of land leased in, than do the petty producers. Table 2 gives the result of applying a direct index for identifying agricultural classes, to farm household data for Haryana, 1972-73. The rich peasants and middle peasants – defined as primarily dependent on hired labour, and substantially dependent on hired labour, respectively – were found to produce a higher output value per unit of area, than the family labour based holdings of the small peasants and poor peasants. It may be noted, however, that the conventional classification of the same data by acreage operated shows the familiar 'inverse relation'. Table 3 gives the rents paid as proportion of output value for some large-scale capitalist farms, contrasted with rents received as percentage of output-value, from family labour based petty producers. In general, the latter percentages are higher than the former.

We may on the basis of the above discussion, put forward the following hypothesis.

Assume that land is homogeneous with respect to its intrinsic fertility, locational worth, etc., and differences in output per acre arise only on account of the different applications of capital by different types of producers, of whom

TABLE 2

OUTPUT PER ACRE BY ECONOMIC CLASSES AND ACREAGE GROUPS, HARYANA 1972-73

Economic Class	Combined Output (Rs.) Per holding	Combined Output (Rs.) Per acre	Acreage group	Combined Output (Rs.) Per holding	Combined Output (Rs.) Per acre
1. Rich Peasant	5533.3	746.1	1. 0-2.5	1494.8	928.1
2. Middle Peasant	4233.9	671.7	2. 2.5-5.0	2189.6	656.6
3. Small Peasant	2745.3	594.9	3. 5.0-10.0	4411.4	635.6
4. Poor Peasant	1764.5	431.6	4. 10.0-15.0	5675.9	497.0
All	3228.1	633.6	All	3228.1	633.6

Note: Combined output is the output of crop plus livestock products. The data relate to 242 cultivating households operating 15 acres or less. For the definition of economic class the reader may refer to my 'Class Differentiation within the Peasantry.' *Economic and Political Weekly* (Bombay), Vol. XI, No. 39, 1976. The results of analysing the sample are forthcoming in the author's 'Economic Characteristics of Agricultural Classes'.

capitalists constitute one extreme type and the poor peasant the other. We may then expect to find an inverse relation between output per unit of area and the rental share of output for a cross-sectional sample of holdings containing all types of tenants. Further, since the absolute amount of rent at a point of time is given a constant determined by the total surplus per unit area which is producible by the petty tenant, this inverse relationship will approximate to a rectangular hyperbola (Fig. 1):

FIGURE 1

HYPOTHETICAL RELATION OF OUTPUT PER UNIT AREA AND RENTAL SHARE

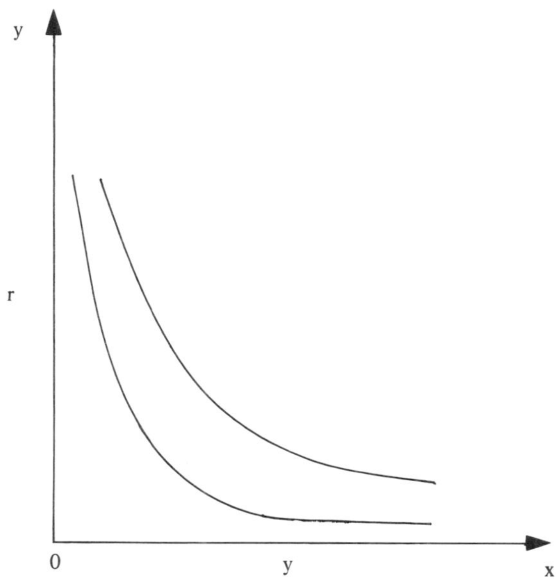

$r.y = R$ where $r =$ rental share of output
$y =$ output per unit of area
$R =$ level of rent per unit of area

For example, if the total surplus of petty tenants in a rice-growing area is say 10 bags paddy per season per acre, making up 50 per cent of gross output per acre, then this same level of rent will constitute less than the entire surplus and hence less than 50 per cent of gross output for the capitalist producer. If the latter's outlays in production embody improved techniques such that output and surplus are raised by 50 per cent, rent will constitute one-third of gross output.

Differential rent can be conceptually integrated into the above formulation. If we now allow intrinsic soil fertilities to vary and locational advantages to exist, this implies that more fertile farms or better-located ones are able to

TABLE 3

RENTS PAID AND RECEIVED BY LARGE-SCALE PRODUCERS PER UNIT OF AREA

Holding No.	Rents paid on land leased in as per cent of gross output	Holding No.	Rents paid on land leased out as per cent of gross output
5.	20.0	11.	49.8
8.	28.3	19.	72.0
16.	49.7	20.	40.1
21.	28.1	21.	39.1
39.	44.8	23.	33.3
47.	30.0	34.	64.0
		35.	41.7
		38.	54.0
	33.1		43.4

Note: The data relate to sixty-six large-scale holdings studied in 1969. Leasing out was done to small-scale peasant producers, except in one case (No. 23) where the owner leased out to a rich peasant. If this case is excluded, the overall percentage of rent received to output on leased out land would rise.

produce a higher surplus for the same outlay as before. Depending on fertility, location, etc. y will vary and therefore R and r will vary. The curve depicting the relation of r and y will shift outward from the origin while its shape will remain unchanged. If the acre of paddy-growing land leased by a petty tenant now produces 30 bags instead of 20 owing to its being of higher fertility, for the same outlay as before, rent will now be 20 bags or 66.6 per cent of output. This new level of rent will be paid and land leased in by rich peasants and capitalists only if their investment embodies sufficiently advanced techniques to raise output and surplus per unit area by the required margin of profit, implying a lower than 66.6 per cent share of rent in output value.

It remains to be established what will be the effect on petty producers of capitalist investment which yields a sufficiently higher surplus per unit area to make direct cultivation paying, compared to leasing out. On the one hand, it would lead to an increased level of eviction of tenants and a switch to direct cultivation, thus aggravating the problem of livelihood of petty producers. This problem would become the more acute if capitalist investment on balance is labour-displacing. On the other hand, there would also be a tendency for greater leasing in of land from petty producers by the capitalist producers wishing to expand the scale of production. Both factors together would indicate a trend towards increasing economic concentration in agriculture.

NOTES

1. 'When the price of any commodity is neither more nor less than what is sufficient to pay the rent of the land, the wages of the labour, and the profits of the stock employed in raising, preparing and bringing it to the market according to their natural rates, the commodity is then sold for what may be called its natural price' [*Smith*, 1961: 62]. As is well known,

Adam Smith often identified 'price' of a commodity with what today we would call 'value added' in its production, ignoring constant capital. Provided this error is corrected, the other propositions retain a logical unity. The corrected version of the different concepts is given here.

2. Thus, Ricardo criticises Smith for terming as 'rent' what is paid by a capitalist to a landlord for felling timber, because 'the compensation was paid for the liberty of removing the timber and not for the liberty of growing it' [*Ricardo*, 1970: 68], whereas the Smithian concept of absolute rent relates precisely to such payment extracted by virtue of property rights. If the latter did not exist, the capitalist would not have to make payment 'for the liberty of removing' the timber, even though the 'original and indestructible powers of the soil' in producing forests remains unchanged in both situations.

3. Richard Jones [*Jones*, 1831] presented a very important pioneering analysis of the different historical forms of landed property and of rent, including labour rent, *métayer* rent, *ryot* rent (paid in India by the peasant) and capitalist rent: but the theory of rent was not considered in relation to the theory of value.

REFERENCES

Jones, Richard, 1831, *An Essay on the Distribution of Wealth and on the Sources of Taxation*, London: John Murray.

Marx, Karl, 1969, *Theories of Surplus Value, volume 2*, London: Lawrence and Wishart.

Ricardo, David, 1970, *Principles of Political Economy and Taxation* (volume 1 of the *Works and Correspondence of D. Ricardo*, edited by Piero Sraffa), Cambridge: Cambridge University Press.

Smith, Adam, 1961, *The Wealth of Nations*, edited by E. Cannan, London: Methuen University Paperbacks. First published 1776.

Cropsharing as a Labour Process, Size of Farm and Supervision Cost

Amit Bhaduri*

The context of our present analysis is an agrarian economy which is commonly described as 'labour surplus'. Broadly speaking, it entails an agrarian situation where land is relatively scarce (primarily due to an extremely unequal distribution in the ownership of land) in relation to available labour. Consequently, a landlord wishing to lease out his land can find any number of potential tenants, provided an agreement can be reached between the two parties regarding the terms and conditions of contract.

Under these conditions, the essence of the problem is easy to see: a landlord would try to maximise the productivity of land, i.e., value of output per acre (denoted by x), because it ensures that his fixed share of total output from given land is also at a maximum. In contrast, a pure tenant owning nothing but his family labour would like to maximize labour productivity (denoted by z), i.e., the value of output per (equivalent) man-hour, as that procedure ensures maximum income for any given man-hour equivalent of the family labour that he may command.[1] Logically speaking, the two parties would then bargain with different objectives – the landlord trying to maximise land productivity (x) and the pure tenant trying to maximise labour productivity (z). However, bargaining is a meaningful concept only if both parties enjoy more or less symmetrical economic power. This is hardly the case when a landlord confronts a pure, landless tenant with meagre and uncertain alternative employment opportunities. Neither the neo-classical notion of alternative 'opportunity cost' for household labour of the tenant family nor bargaining based upon rough symmetry in economic power, makes a plausible starting point in the analysis.[2]

A more natural starting point is to postulate that the landlord dominates the situation and the labour contract gets shaped more or less entirely according to his interest. It is useful to characterise how labour contracts are imposed by a landlord under these circumstances. As far as the landlord is concerned, the main purpose of such a labour contract is to extract the maximum amount of man-hours of labour per acre leased out at minimum supervision cost. In this, cropsharing as a labour process begins both to resemble and to differ from the capitalistic labour process discussed extensively by Marx.[3]

Under any typical wage–labour contract, the employer would be interested in maximising productivity per labourer (y), because the wage contract typically relates to the labourer and not to man-hours of labour. Hence, there is the attempt on the part of the employer to extend working hours per day (under

**Jawaharlal Nehru University, New Delhi, India.*

daily wage) through direct supervision and the discipline of the factory system and modifications of technology.

In the absence of possibilities of direct supervision, the question naturally arises as to how a landlord enforces such an 'optimum' labour contract from his point of view. The existing literature on sharecropping has largely bypassed this question.[4] A central proposition of this paper is to suggest that the landlord's decision regarding the size of the farm to be leased out to each individual tenant is largely determined by such considerations of 'optimum' labour contract in the absence of direct supervision. In short, the 'optimal' size of the leased-out farm is guided by considerations of indirectly enforcing a labour contract that is most advantageous to the landlord.

In order to see how this may happen, consider first a wage-contract system. Let S = size of operational holding in acres leased out by the landlord to each tenant.

If x = land productivity, i.e. value of output per acre and, y = productivity per labourer, i.e. value of output per agricultural worker then, the landowner's profit (R), namely the value of output from the size of leased-out holding minus wage cost is defined as,

$$R = xS - w\left(\frac{xS}{y}\right) \qquad \ldots \quad (1)$$

where, w = given wage rate per labourer.

It is reasonable to assume in the light of the so-called 'size–productivity' debate,[5] that land productivity decreases with the size of operational holding, i.e.,

$$x = f(S),\ f' < 0 \qquad \ldots \quad (2)$$

while, productivity per worker tends to increase with the size of the operational holding,[6] i.e.,

$$y = F(S),\ F' > 0 \qquad \ldots \quad (3)$$

Consequently, in expression (1), wage cost is reduced through higher productivity per labourer (y) with increased size which, at the same time, depresses land productivity. With these two counteracting tendencies operating on the size of land leased out, a balance is struck and profit is maximised where,

$$\frac{dR}{dS} = 0 \text{ and, } \frac{d^2R}{dS^2} < 0 \qquad \ldots \quad (4)$$

In view of (1), (2) and (3), the first-order condition reduces to,

$$\left(x + S.\frac{dx}{dS}\right)\left(1 - \frac{w}{y}\right) + xS\frac{w}{y}.\left(\frac{1}{y}.\frac{dy}{dS}\right) = 0$$

which yields on simplification,

$$\frac{(y - w)}{w} = \frac{\eta}{(\mu - 1)} \qquad \ldots \quad (5)$$

or alternatively,

$$y = [1 + \frac{\eta}{(\mu - 1)}] w \qquad \ldots \quad (5a)$$

where, $-\mu = \frac{S}{x} \cdot \frac{dx}{dS}$, i.e., elasticity of land productivity with respect to size, defined by relation (2) [and hence negative, $\mu > 0$].

and, $\eta = \frac{S}{y} \cdot \frac{dy}{dS}$, i.e., elasticity of productivity per worker with respect to size defined by relation (3) above.

As is evident from (5), an economically meaningful notion of 'optimal' size from the point of view of the landowner requires the size-elasticity of land productivity to exceed unity in absolute value, i.e.,

$$|\mu| > 1 \qquad \ldots \quad (6)$$

And, so long as (6) holds, the 'optimal' productivity per worker (y) ensured by the landlord through choosing 'optimal' size is higher for higher wage rate, as exhibited by (5a). Thus, under our present theory, regions with a higher (real) wage rate will be expected to exhibit both higher mean size of operational holding and, hence, higher mean labour productivity under wage contract.

In contrast to a wage contract, under fixed cropsharing tenancy the landlord may be assumed to choose the size of leased out land in such a manner as to maximise the level of output on each tenant farm, i.e.,

$$\max_{S} (xS) \text{ implying, } x(1 - \mu) = 0$$

$$\text{or, } \mu = 1 \qquad \ldots \quad (7)$$

With a simple linear relationship postulated between farm size (S) and land productivity (x) as a special case of relation (2), we may now visualise the solution in the two cases of wage–labour contract and fixed-share tenancy as in the diagram opposite.

Since, $\mu = \frac{AP}{BP}$, at the mid-point P, the relevant elasticity $\mu = 1$ and fixed-share farms leased out to poor tenants would tend to have (mean) size around $\bar{S}$ and land productivity $\bar{x}$. But under wage contract, farms would operate with $\mu > 1$ [see (6) above] in the zone to the south-east of point P, so that (mean) farm size would be higher and land productivity lower in wage labour farms compared to cropshared ones. Unlike more conventional analysis of the size–productivity debate (which leaves farm size indeterminate), the present scheme of analysis therefore integrates farm size as one of the decision variables to be determined endogenously under alternative modes of organisations of agricultural production.[7]

The significance of treating the size of the leased-out farm as a decision variable lies in the control over the labour process which this entails. In an extreme case, land productivity (x) would tend to be maximised by parcelling out land in such small units that the tenant can just about meet his subsistence requirement only if he uses it most intensively. In other words, a threat of survival created by the tiny size of the leased out land ensures maximum labour use (in man-hours) per acre, instead of enforcing any labour contract through direct supervision. In this regard, cropsharing tenancy as a labour process in backward agriculture tends to be distinct from capitalistic labour processes. If this view is correct, then one would expect more fertile (e.g., irrigated) land to be leased out typically in smaller sizes.[8] One would also expect not only larger tenant families with more household labour to be favoured in tenancy contracts but perhaps also that a positive correlation would exist between tenant family-size and the size of leased-out farm,[9] so that only a more or less uniform per capita subsistence requirement is met across such leased-out farm sizes and the survival threat is maintained in controlling the labour process.

There has been a considerable amount of interest in recent years in trying to characterise the nature of interlocking between major agrarian 'markets'; in particular, the link between rural credit and land lease has been widely discussed.[10] The view of sharecropping as a labour process elaborated above, not only explains how the threat of survival can be used as an alternative

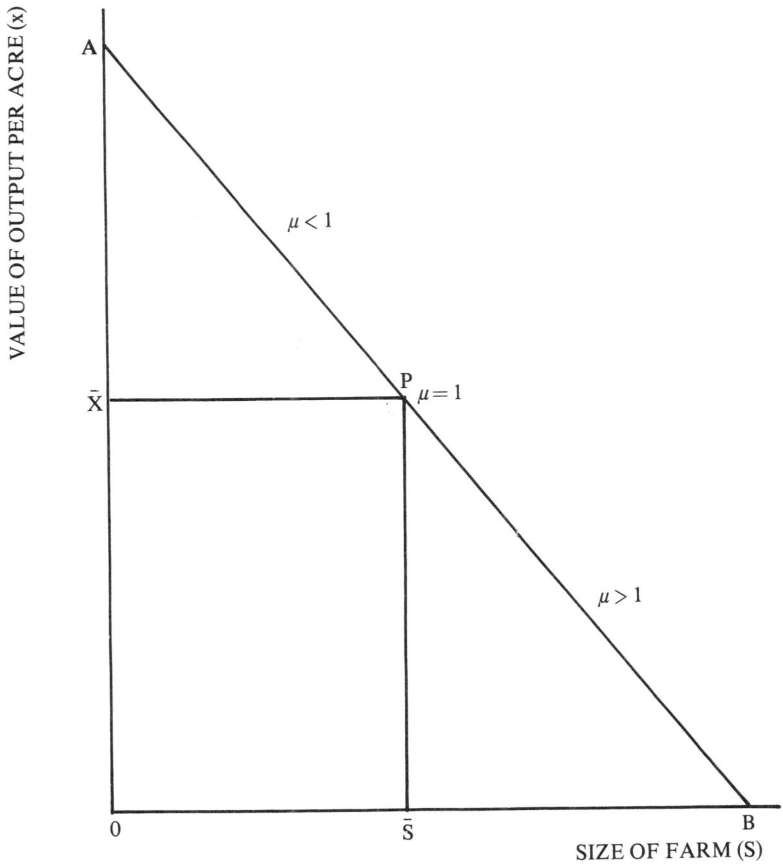

mechanism to direct supervision of labour but it also indicates how transactions in land and labour get interlocked by the same mechanism.

NOTES

1. The distinction between labour-service and labourer is important in this context. Definitionally, daily productivity of a labourer (y) = productivity per labour-hour (z) times hours of work performed during the day (h). A given man-hour equivalent of family labour implies a constant length of working day, which may not usually be the case under alternative modes of organization of agricultural production.
2. For example, the formulation of a cropsharing contract as a Nash-bargain requires one to imagine that the tenant has alternative employment opportunities as wage labour, i.e., the 'opportunity cost' of family labour is a useful concept. See, C. Bell and P. Zusman, 'A bargaining theoretic approach to cropsharing contracts', *American Economic Review*, Vol. 66, No. 4, September, 1976.
3. See, K. Marx, *Capital*, Vol. 1, especially, Part III, Chapters VII, VIII, IX and X for discussion of the capitalistic labour process in relation to the generation of 'surplus value'.
4. Compare S.N. Cheung, *The Theory of Share Tenancy*, Chicago: University of Chicago Press, 1969.
5. See K. Bharadwaj, *Production Conditions in Indian Agriculture*, (A study based on farm management surveys), Occasional Paper 33, Department of Applied Economics, Cambridge: Cambridge University Press, 1974, especially chapters 2 and 3 for a careful discussion and survey of the Indian literature. See also, P.L. Roy, 'Transition in agriculture: empirical indicators and results (evidence from Punjab, India)', *The Journal of Peasant Studies*, Vol. 8, No. 2, January, 1981 for a later discussion, particularly showing how the inverse relation may have broken down in 'green revolution' areas of the Punjab.
6. Although direct evidence on this relationship (3) is meagre, it seems reasonable to postulate that 'extensive' use of land overcompensates the productivity (per worker) loss due to less 'intensive' use of land on larger sized farms in general. Like the size of farm and land productivity discussion, there is scope for empirical examination of the proposed obverse relation between size and labour productivity, to extend our understanding further on these issues.
7. The explanation usually runs in terms of the difference in 'marginal supply price' or 'opportunity cost' of hired and family labour, e.g., A.K. Sen, 'Size of holding and Productivity', *The Economic Weekly*, Annual Number, February, 1964 and D. Mazumdar, 'Size of farm, and productivity: a problem of peasant agriculture', *Economica*, September, 1965. The conventionally simplified picture that, with zero opportunity cost of family labour, the 'marginal productivity of labour' is pushed to zero on cropshared or owner-operated farms, while 'marginal product' is equated to the wage rate under wage contract is inadequate insofar as the *position* of the marginal/average product curves depends on the size of the farm. Unless 'constant returns to scale' is arbitrarily assumed and one is permitted to normalise by working with only intensive magnitudes, size of farm must enter as an essential element in the discussion to define the relevant 'marginal product of labour'. We drop the assumption of 'constant returns', not merely as a technological characteristic, but because of considerations of control of the labour process through size discussed later in the text.
8. For example, it has been observed by Bharadwaj that, 'smaller farms generally have a higher percentage of their area irrigated', op.cit. p. 41. Whether this general feature holds with particular force in the land-lease market remains an empirically open question. It should also be noted that the causation underlying may go both ways: smaller farms use their 'surplus'

family labour to improve irrigation and/or better-irrigated areas are leased out in smaller sizes to ensure high labour utilisation per acre without direct supervision (as postulated in the text).

9. See, for instance, C. Pant, 'Contractual Arrangements in Agriculture: Some theory and empirical evidence' (unpublished Ph.D thesis, Indian Statistical Institute, 1980), Chapter V.

10. For example, A. Bhaduri, 'A study in agricultural backwardness under semi-feudalism', *Economic Journal*, March 1973; K. Bharadwaj, op.cit.. Chapter 8 and P.K. Bardhan, 'Interlocking factor markets and agrarian development: a review of issues', *Oxford Economic Papers*, March 1980.

Sharecropping: Some Illustrations[1]

J. Martínez-Alier*

Sharecropping, following Marshall's analysis [*Marshall*, 1920: 534-537] was usually seen as an inefficient way of using land and labour, and it is still sometimes seen in this way. It can now be argued, however, that sharecropping is more efficient than wage labour, in the sense that production will be higher, and both landowners' and sharecroppers' remunerations may also, therefore, be higher than under wage labour. This has always been my view [*Martinez-Alier,* 1965, 1968, 1971, 1977], and in this paper I wish merely to add a graphical analysis to two propositions which I take from my previous work. First: 'By using wage labour [the landowner] has to pay wages, by definition greater than zero, while at least in some seasons of the year marginal productivity under full employment conditions might fall below this wage ... Thus, by turning labourers into tenants, landowners are able to get extra returns from the otherwise unemployed labour' [*Martinez-Alier,* 1977:36]. Secondly: 'The supply of labour [depends] on the method of remuneration. Different land tenure and use of labour arrangements can be discussed in terms of their impacts on the labour supply ... sharecropping (or small cash-renting) are a means to make labourers (turned into peasants) work harder and better for only a small increase in total remuneration over that of wage labourers, thus increasing landowners' profit-rents' [*Martinez-Alier,* 1977: 10]. This way of looking at sharecropping I learnt from rural workers themselves, in Andalusia in the mid-1960s, who often said that sharecropping was akin to piece-work.

In the last few years a number of other authors have also reached the conclusion that sharecropping is a system of labour use and land tenancy which is consistent with profit-making motives: either because of risk diversification and lower transaction costs [*Cheung,* 1969]; or because the subjective opportunity cost of the sharecroppers' labour-time is less than the market cost [*Mazumdar,* 1975]; or because sharecropping provides an incentive for the workers to work better [*Reid,* 1976].

THE INEFFICIENCY OF SHARECROPPING

In Marshall's analysis, sharecropping was condemned, with some qualifications, as inefficient, on the grounds that sharecroppers would stop furnishing labour sooner than wage workers since the marginal return they would get from their labour would be less (say, one half) than the marginal return the landowners would get from wage labour. This is at least how Marshall's comment may be interpreted.

**Universitat Autonoma de Barcelona.*

FIG. 1

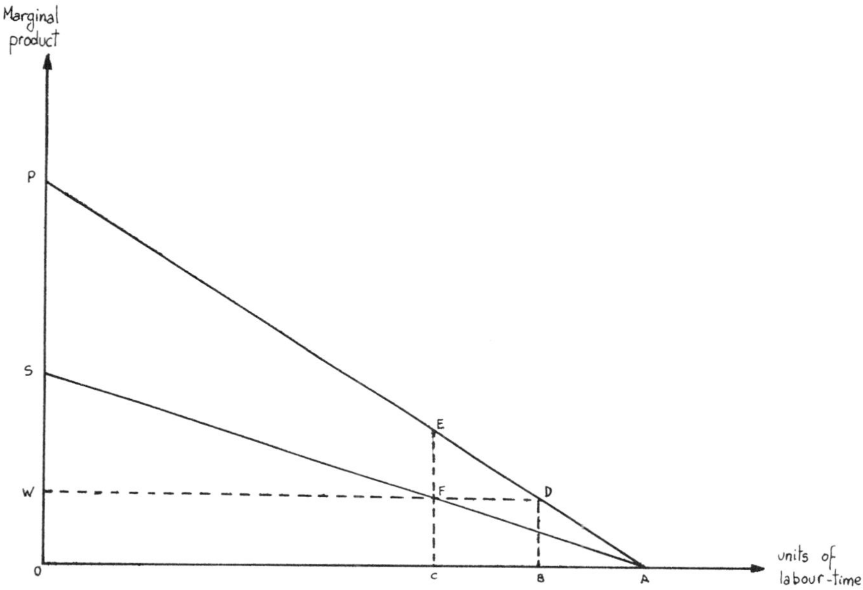

In fig. 1 (taken from Reid, 1976), if PA is the marginal productivity line (and OAP is total production under full employment conditions), and if OW is the conventional wage rate per unit of labour-time (either the legal minimum wage, or any wage rate set conventionally that even unemployed workers are not willing to undercut), then the profit-maximising landowners, under a wage-labour system, will employ OB units, will pay out OBDW as wages, and will make a profit of WDP (excluding other costs but labour). The ratio OS/OP is the terms of the sharecropping contract. Under full-employment conditions the sharecroppers would get OAS, and SAP would go to the landowners as profits/ rents. But the sharecroppers will not supply labour until point A or even until point B. If they impute to their own labour an implicit cost equal to the ruling wage, they will supply only OC units of labour-time (since their own marginal return is indicated by the line SA). Therefore, sharecropping is an inefficient way of employing labour because production is lower than under the wage system. Depending on the terms of the contract, the sharecroppers might earn possibly less or possibly more than they would earn as wage workers, but the landowners certainly earn less (SFEP is certainly smaller than WDP, whatever the terms of the contract). Landowners who use sharecroppers could not be proper capitalist, profit-minded entrepreneurs.

SHARECROPPING AND LABOUR MARKET DUALITY

However, if sharecroppers impute to their own labour not an implicit cost equal to the ruling wage, but a lower cost, since we can assume that they lack alternative employment opportunities, then the situation might be as in fig.2.

FIG. 2

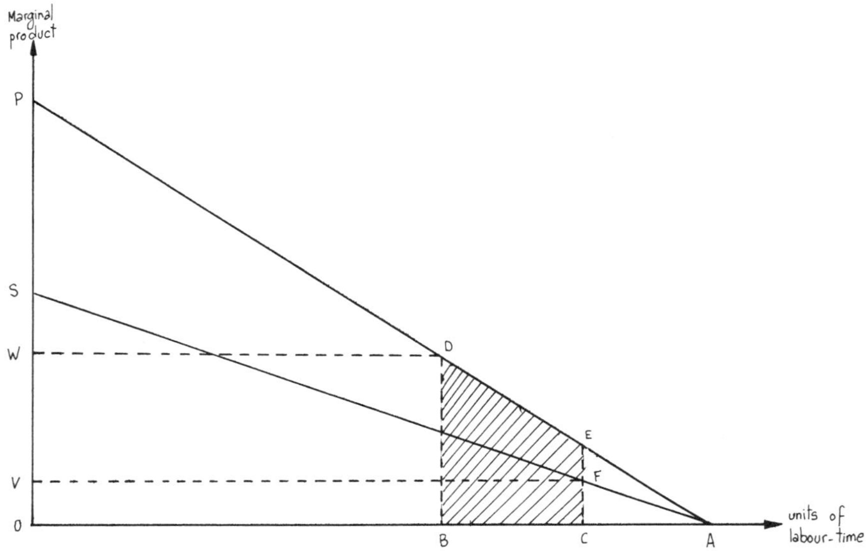

Here, sharecroppers impute a 'wage' cost of OV (instead of OW) to their own units of labour-time, and there is a gain in production. Production is OCEP which is larger (by the shaded area) than OBDP. Whether both sharecroppers and landowners profit from this gain (and are thus willing to change from wage labour to sharecropping) will depend on the terms of the sharecropping contract. It is in the landowners' interest to set the terms so that the sharecroppers' share of the harvest is only slightly greater than they would have earned as wage workers (enough to induce them to accept the new system). Perhaps not very much will be needed since we are assuming a situation of abundance of labour, with OA units of labour-time (or at least more than OC units) being in fact available. In fig. 2, OCFS is larger than OBDW, and SFEP (the landowners' share) is larger than WDP. The landowners also save in managerial costs of supervision, since the sharecroppers work more on their own.

SHARECROPPING AS AN INCENTIVE SYSTEM: THE SAME AMOUNT OF WORK IN LESS LABOUR-TIME

Up to now I have assumed that both wage workers and sharecroppers would do the same amount of work in the same amount of time. But, in fact, if sharecropping is an incentive system of payment by results, this means that the sharecroppers, in the same amount of time, will perform more work and will produce more. They work harder (they spend for instance 6 kcal. per minute instead of 4 kcal. per minute) and they work perhaps more carefully. This quality of 'carefulness' would not apply to pieceworkers who are employed for a single task, and who are uninterested in the final result of the harvest.

Sharecroppers might also put in an extra hour of work at the end of the day; they might work on holidays; they might use their own families' labour – but this possible addition of units of labour-time we shall leave aside for the moment.

If they work harder this can be shown as in fig. 3.

FIG. 3

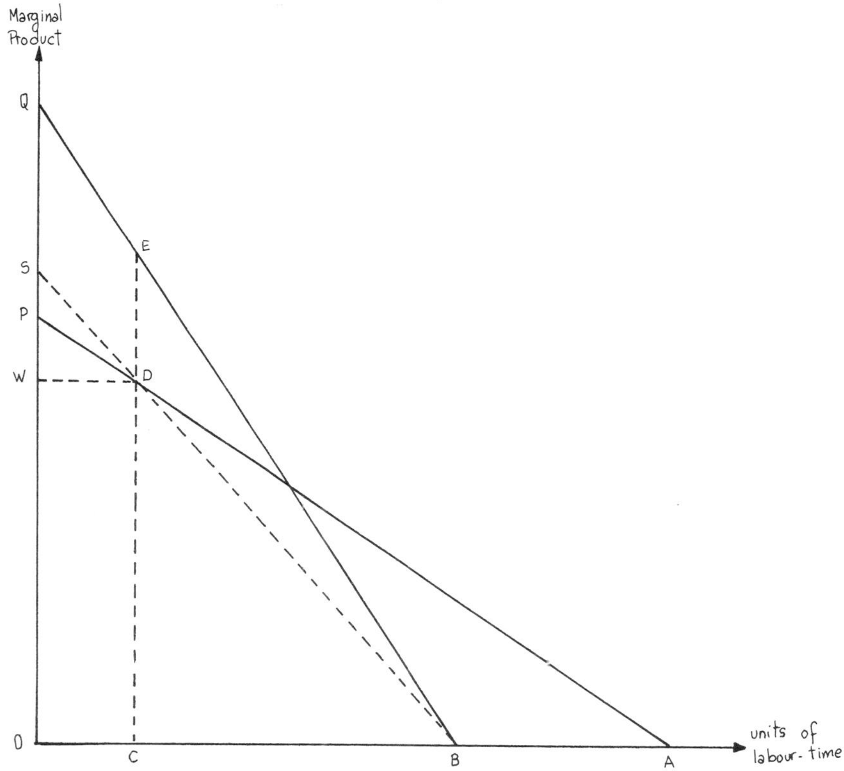

On the horizontal axis we have again units of labour-time, and on the vertical axis marginal productivity. But the labour-time of sharecroppers is more productive than wage-work paid by time, so that the same amount of production is done in less time by sharecroppers (which is a different way of saying they work harder). So, the sharecroppers would produce OBQ by working OB units of time and the wage workers would produce OAP by working OA units of time. OBQ is equal to OAP, since we assume this is the maximum production that can be achieved.

Now, it has often been observed that when wages increase, landowners introduce sharecropping. Fig. 3 illustrates what might happen under conditions of relatively high wages. This might come because of an increase in the legal minimum wage, or because of union pressure, or, if one wishes, because of scarcity of labour-time available (we could assume that OA units of labour-time are no longer available, but only OB, for instance, or even only OC, so that the last part of the line DA is hypothetical: it indicates how marginal productivity would be if such labour was available). Then, under wage labour, assuming OW to be the wage rate, OCDW are wages, WDP are profits, and production is OCDP. If the terms of the sharecropping contract are set in such a way that the sharecroppers get the ratio OS/OQ, then, even if they impute to their own units of labour-time the same 'wage' as they would earn as wage-workers, they will nevertheless supply the same units of labour-time as landowners would employ of wage workers' labour-time. Production will be greater, and both landowners and sharecroppers will profit (since OCDS is greater than OCDW and SDEQ is greater than WDP). If sharecroppers were to impute to their own labour-time a 'wage' cost lower than OW and if there were labour available beyond OC, then, of course, the benefits could be still greater. However, if landowners are too greedy and lower the share OS/OQ then they will perhaps induce a retraction of labour-time supply, but production might still be greater than with wage labour, and both landowners' profits and sharecroppers' remuneration too.

Let us now go back to conditions of relative abundance of labour and low wages, which may seem more relevant for poor countries with concentrated landownership (except for the fact that seasonal scarcities of labour might appear in most types of agriculture). In fig. 4 we assume that there are at least OA units of labour-time available, and that the conventional wage rate is OW. When using wage labour, OD units of labour-time are employed, production is ODFP, wages are ODFW and landowners' profits are WFP. Now, if wage workers become sharecroppers, and if they are willing to supply, for instance, OC units of labour-time, then sharecropping would be more productive than wage labour because the triangle CBE is smaller than the triangle DAF (same height and shorter base). The triangles CBE and DAF measure what is lacking up to maximum production, which is obviously achieved when marginal productivity drops to zero. If sharecroppers impute to their own units of labour-time the 'wage' rate OW, they will never supply OC units of labour-time. They would only do this if they got a 100 per cent share of the crop (along the line QB). However, if we assume that they impute a lower 'wage', say, OV, they will supply OC units of labour-time provided that the terms of the contract are set right, i.e. if their share is in the ratio OS/OQ. Since OS/OQ is equal to CG/CE (for instance, one half) and since CG/CE is equal to OV/OW, then it follows that if sharecroppers impute to their own labour-time an implicit 'wage' rate which is half the ruling wage, then the terms of the sharecropping contract must be such that they get half the crop. If the sharecroppers impute their own labour-time a cost which is one third of the ruling wage, then the landowners could impose a contract of only one third of the crop for the sharecroppers, and still get OC units of labour-time supplied.

In fig.4, sharecroppers would get OCGS and landowners would get SGEQ, and total production would be OCEQ (which is greater than ODFP). Since total production is greater, it is intuitively clear that there must be a way of arriving at a contract by which both parties will gain something, i.e. a contract

FIG. 4

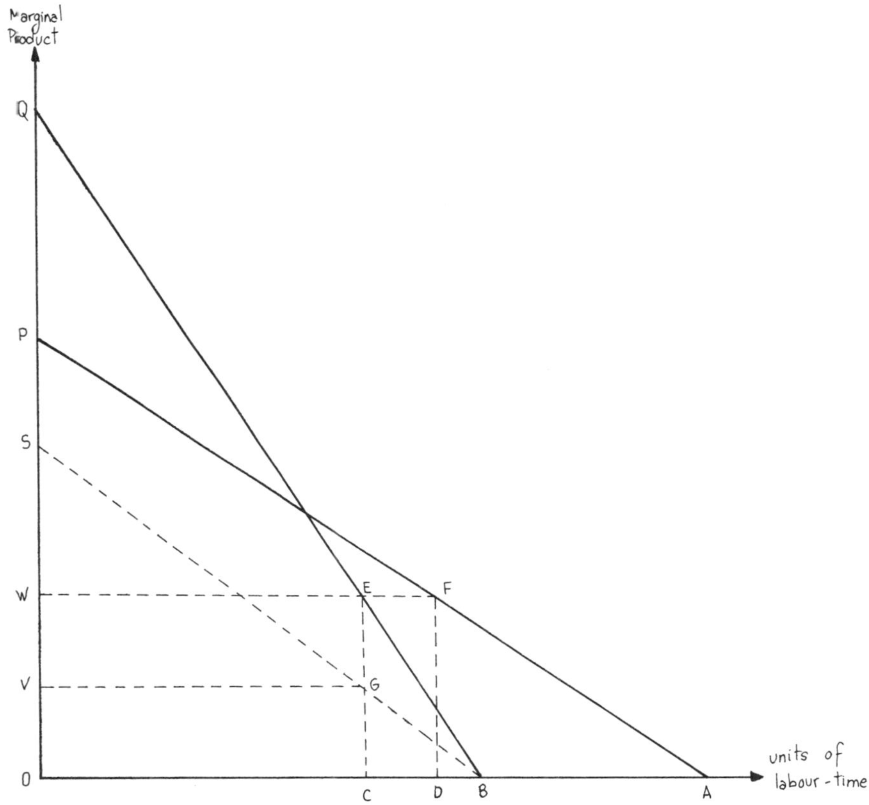

such that sharecropping has the necessary incentive effect, and that sharecroppers gain something more than as wage workers, and that landowners' rents, as 'rentier' landlords, are greater than their profits as capitalist entrepreneurs. Since the triangle CBE is smaller than the triangle DAF, a contract may be established so that OCGS be larger than ODFW and that SGEQ be larger than WFP. This will be easier to achieve the greater the incentive effect of sharecropping and/or the lower the implicit cost imputed by sharecroppers to their own labour-time.

In fig. 4, as it is drawn, the sharecroppers would get a share of the crop (OCGS) which is smaller than the part of wages in production when labour is paid by time (ODFW), and therefore they might be reluctant to accept the contract. However, per unit of labour-time (OC in sharecropping, OD in wage labour), their remuneration would be higher in sharecropping. Whether they accept the contract will depend on how the now unemployed 'units of labour-time', CD, will react, and social sanctions might appear against sharecropping, either formal (through union action) or informal. In any case, as the figure is drawn, the landowners are certainly interested in introducing sharecropping, since SGEQ is greater than WFP. We could also set the question as in fig. 5 which is perhaps easier to read.

FIG. 5

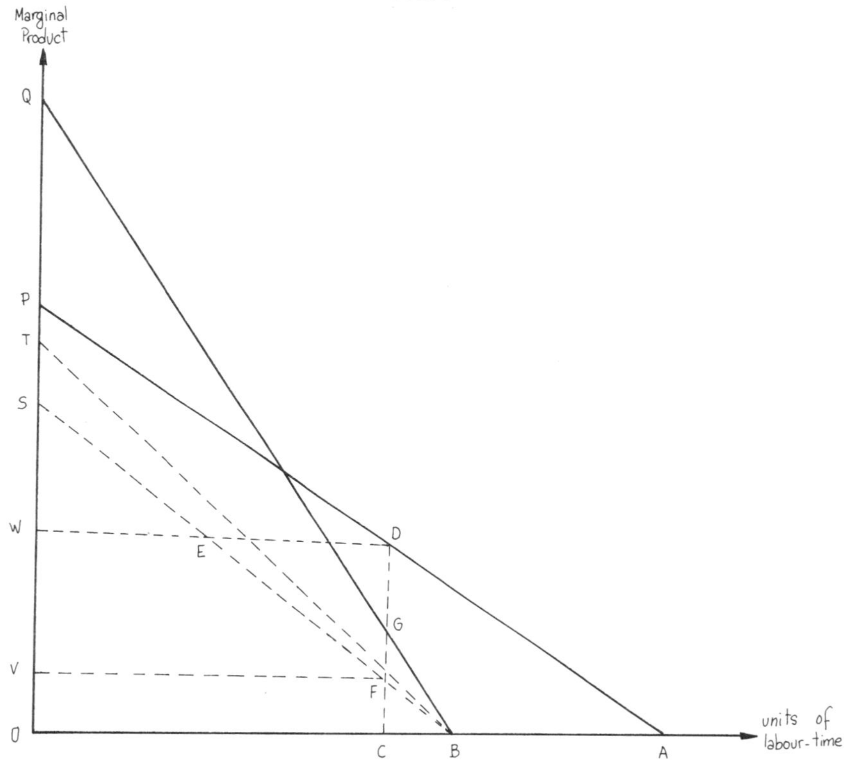

In this case, under wage labour, OC units of labour are employed when the wage rate is OW. If the landowners succeed in making sharecroppers supply this same amount of units of labour-time, without an increase in their remuneration, then the landowners make larger rents under sharecropping than profits under wage labour. This can be done by setting the contract so that the line SB intersects WD at point E, being WE = ED, so that SWE = EFD, and therefore OCDW = OCFS. For a supply of labour-time OC to be forthcoming from the sharecroppers, we have to assume that they impute a cost of OV to their own units of labour-time. Landowners would certainly gain in such a case, since the sharecroppers' remuneration has not increased, and production has indeed increased (the triangle CBG is smaller than CAD). Perhaps sharecroppers will insist on a larger share, say, OT/OQ, and then they will supply a little more labour-time (if they still impute OV to their own labour).

We know, however, that in conditions of great labour abundance and very low wages, the stimulus to introduce sharecropping diminishes. The loss in potential production is so small that the changeover from wage labour to sharecropping might not seem worth the trouble (leaving aside for the moment changes in crop-mix, or in rotation). The incentive effect of sharecropping becomes less important in such a situation. Sharecropping might nevertheless be adopted if sharecroppers, pushed perhaps by such bad circumstances in the labour market, are ready to work until near the point of marginal productivity equals zero, i.e. until point B in fig. 5. The lack of alternative employment opportunities makes them impute a negligible 'wage' cost to their own labour-time; they have in fact a certain amount of unused labour-time, CA or BA in fig 5. If they behave in this manner, then sharecropping is more productive and more profitable to landowners, who can easily get the sharecroppers' share of the (increased) harvest to be only slightly higher than they would get as wage workers.

MORE WORK FROM THE SAME AMOUNT OF LABOUR-TIME

It might be asked, however, what we mean by saying that marginal productivity would drop to zero. Some would argue that in a given area with no improvements and no new techniques (no change in the production function) the amount of production that can be obtained will always reach a maximum if enough units of labour-time are put on to the land. On the other hand, some would argue that there is never enough labour available to drive marginal productivity to zero. There is always in any farm some land not used that could be used, weeding could always be done more carefully, the harvest could leave no grain uncollected, all waste products could be fed to animals or used for composting, irrigation could take place just in time and could be distributed to the best effect, double cropping could be attempted in some areas by a speedier harvesting or at least (in non-irrigated farming) a more intense rotation could be tried. In fact, the notion that agricultural production cannot be increased by the application of some extra work (at least at some times of the year) is more a theoretical construction than a technical reality. Or so, at least, it could be argued.

What this means is that more production can be achieved from the same quantity of labour-time units. It is not a question, here, of achieving a given result in less time (i.e., of performing quicker a given amount of work, as when payment is by task), but of producing more in the same amount of time, by increased effort and increased quality of work. We assume in fig. 6 that there are only OB units of labour-time (so many hours per year, the same for both systems of remuneration), and the ruling wage is OW (so that we introduce the condition, unfavourable to sharecropping, that with wage labour the ruling wage is equal to, and not greater than, marginal productivity under full employment). The marginal productivity line for wage labour is PA and for sharecropping is QA. Does it make sense to say that two different production maxima can be achieved, by changing the system of remuneration (i.e. by changing effort and quality of work)? I think it does. Fig. 6 means that if there were OA units of labour-time, and if they would be paid by time and not by results, they would produce less, because of less effort and quality in work, than the same amount of units of labour-time paid under an incentive system. The practical manifestations of this increased production may vary.[2]

The fact that under both systems marginal productivity would drop to zero at the same point A is very unlikely – fig. 6 is drawn in this way only for the sake of clarity.

FIG. 6

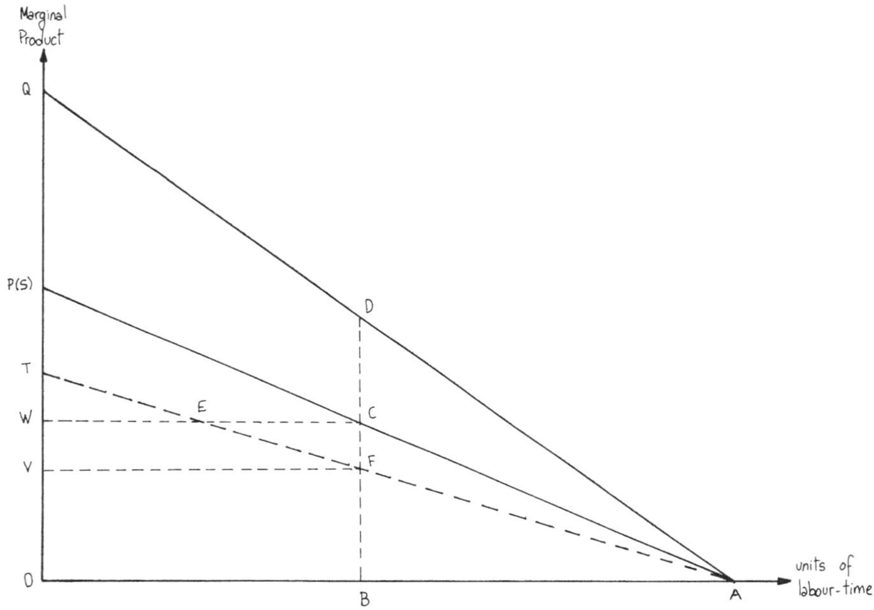

In fig. 6, if labour is paid by time, it produces the area OBCP, OBCW being wages and WCP profits. If labour is instead employed in a sharecropping system, and it is then still willing to supply OB units of labour, production would be OBDQ, and the rent and share will depend on the contract, as we shall see immediately. If sharecroppers impute to their own units of labour-time a cost equivalent to the ruling wage (which in this case is an equilibrium wage, since at that wage, OW, the whole available labour force OB would be employed under the wage-labour system), then the sharecroppers must get a share such that their own marginal return is along the line PA (which could be also called SA). What they get, over the wages they would have got, is precisely the profits that the landowners made, and the landowners get as rent the whole of the extra production (in fig. 6, P(S)CDQ which is greater than WCP(S)). If the part of profits in total production when using wage labour is equal to the increment in production when changing over to sharecropping, then there is no point, under the assumptions, to make the change. If it is lower, then it is worthwhile, which accords with common sense, since an incentive system will be introduced when profits are low and when there is room for workers to work harder.

Perhaps the sharecroppers will impute to their own work an implicit cost OV, lower than OW. If they do so, the landowners may be able to trick them into producing more and getting just the same. The line TA (fig.6) cuts WC so that WE is equal to EC. The sharecroppers' remuneration is now OBFT (equal to OBCW), and the landowners keep their profits and get as extra rent the whole increase in production.

INCREASE IN THE SUPPLY OF LABOUR-TIME

Finally, we could assume (as in fig.7) that under sharecropping the pool of available labour increases, because sharecroppers are ready to use family labour, and/or because they are ready to work longer hours, in the belief that they are, in a sense, working 'for themselves'. It is clearly different to help one's father rather than work for a capitalist.

Here again (considering the triangles OAP and OBQ) we are saying that two different production maxima can be achieved (which was not the case for figs. 3, 4 and 5). One could ask how it is that, if adding units of labour-time beyond OA did not increase production (since A is the point where marginal productivity of labour paid by time is assumed to drop to zero), we feel entitled to draw the line QB. The assumption is that the improvement in the intensity and quality of work, plus the increase in the units of labour-time, will allow a change in methods of production (a change in crops, perhaps).[3]

So, in fig.7 we draw PA and QB (there is no reason why they should be parallel) in such a way that QB is always above PA.

FIG. 7

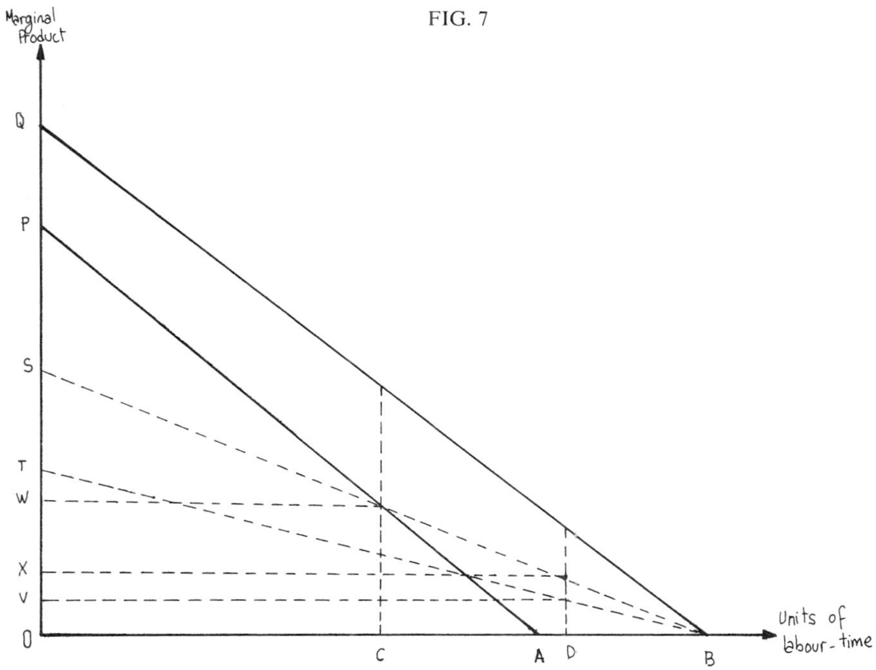

Sharecropping then becomes indeed quite profitable for landowners even if sharecroppers impute a high 'wage' to their own labour-time units, and it becomes even more so if sharecroppers, since their own labour and that of their families have a low opportunity cost in the absence of alternative employment opportunities, decide to work until near point B, say, until point D. The ratios OS/OQ and OT/OQ are two possible terms of the sharecropping contract. The first would correspond to an imputed 'wage' of OW (labour supply is then OC) or to an imputed 'wage' of OX (labour supply, OD). The second ratio corresponds to an imputed 'wage' of OV (labour supply is then also OD). Landowners take much larger rents as 'rentier' landlords than the profits they would make as capitalist entrepreneurs, and in fact, in fig. 7, sharecroppers could indeed try and get better terms than OT/OQ and OS/OQ. The condition for a contract to be established is that both sides make greater incomes than under the wage system.

This paper is not a discussion of Cheung's and Reid's models, but is simply a graphical presentation of points I have made in previous writings. I have considered only work or labour as the variable input, assuming land to be fixed (though in figs. 6 and 7, the intensity of land use is assumed to change). I have not introduced into the analysis risk reduction or diversification, and I have abstracted from other costs, including transaction and managerial costs.

As to the reasons why sharecropping is *not* introduced, I have already remarked on the possible resistance from the workers (since, as in fig. 3, when a given amount of work is done in less time, some workers may then remain unemployed). However, I think the main resistance comes from landowners. And this certainly not because of profit-making motives, but, rather, because (as I have insisted at great length in my previous writings, especially [*Martinez-Alier,* 1968, 1971]) the landowners may fear that by becoming 'rentier', 'absentee', landlords the threat of land reform increases. 'Direct cultivation' with wage labour legitimises to a greater extent their position. There is also the threat from the usual legislation protecting tenants (giving security of tenure and fixing the terms of the contract for long periods, often affording sharecroppers the possibility of becoming cash-tenants with fixed rents). This legislation is part and parcel of the syndrome against 'indirect cultivation'. More could be said on these matters (on the difference between tree crops and annual crops, for instance, though vineyards have often been farmed under a sharecropping system), but I would be repeating myself. This paper adds, then, nothing of substance to what I have said on previous occasions. It is merely a graphical presentation of a *part* of my views on sharecropping (which is a system I do *not* particularly favour). It is clear, in my view, that under quite plausible assumptions sharecropping is more efficient, and more profitable to landowners, than wage labour.

As a final point, I would remark that I have assumed competitive conditions (apart from the 'stickiness' of the wage rate, which cannot drop below a certain level even if marginal productivity under full employment conditions would drop to zero). I have talked about 'landowners' and 'sharecroppers' without implying collusion on either side, i.e. assuming that, even if the farms are large, there is no monopsony or oligopsony of labour, since labour is mobile. This accords with real conditions in most regions of the world since agricultural workers move around in search of employment. The graphs may be taken to refer to what happens in the latifundist sector as a whole or to what happens in any average, representative latifundium. I would also add that this analysis refers to sharecropping proper, and not, certainly, to the labour system in Andean haciendas before the land reforms, nor, for instance, to the *jajmani* system.

NOTES

1. This analysis I first presented at a symposium on latifundist agriculture in the Iberian Peninsula at the Instituto de Economia Agrária of the Gulbenkian Foundation, Oeiras, Portugal, in December 1979. I thank José M. Caballero for his comments.

2. One may think, for instance, of the change from a three yearly rotation (cereals, leguminous crop, fallow land) to a two yearly rotation (cereals, leguminous crop) in Andalusia in the 1920s and 1930s, a change simultaneous with the introduction of sharecropping [*Naredo*, 1977]. One may think also of how labour-intensive crops such as cotton, tobacco, vegetables are often grown by sharecroppers.

3. To give a somewhat fanciful example, let us imagine that the thirty students of a class on rural relations decide to take over a small farm in order to experiment with farming and communal living. One can predict that because of the poor quality of their work, caused by lack of experience, by their urban patterns of work and rest, and so on, they might achieve a certain production, which if we assume the farm to be rather small, will perhaps not increase at all even if one more student adds himself or herself to the commune. If instead of students we get experienced labour, we can assume they will produce the same production in less time (say, twenty workers would do the work of thirty students). And we could further assume that they might use more intensive methods of cultivation, sowing more in time, weeding more effectively (students cannot tell weeds from wheat), also perhaps taking up more difficult crops, or using waste products to feed animals (which the students did not keep because they did not know how to). The production function will shift to the right (as in fig. 6) even without the introduction of any new capital, or at least with very little capital from outside the farm (they raise their own seeds and their own animals, which provide dung). Not only thirty workers might find productive occupations, but also perhaps some extra members of their families (as in fig. 7). If this example makes some sense, then in real life the role of 'students' is performed by common workers paid by time, and the role of 'experienced workers' might be taken by these same workers when being paid under an incentive system. One should also remark that the 'experienced workers' arrangement is not more socially virtuous than the 'students' commune', since now the students would have to be fed by the workers. The question of who gets the benefits from extra work depends on the institutions of property and on the conditions of the labour market. For instance, the experimental re-introduction of sharecropping in the collectivised agricultures of some eastern European countries should be considered in this light. Also, the complaint often expressed against sharecropping, that, as piecework, it causes unemployment, should be kept in mind.

REFERENCES

Cheung, Steven N.S., 1969, *The Theory of Share Tenancy*, Chicago: University of Chicago Press.

Marshall, Alfred, 1920, *Principles of Economics*, Eighth edition, London: Macmillan.

Martínez-Alier, Juan, 1965, 'La actitud de los latifundistas de Cordoba respecto al empleo de los obreros', *Anales de Economia*, October-December.

———, 1968, *La estabilidad del latifundismo*, Paris: Ruedo Iberico.

———, 1971, *Labourers and Landowners in Southern Spain*, London: Allen and Unwin.

———, 1977, *Haciendas, Plantations and Collective Farms*, London: Frank Cass & Co.

Mazumdar, Dipak, 1975, 'The Theory of Sharecropping with Labour Market Dualism', *Economica*, August.

Naredo, J.M., 1977, 'La Crisis de las aparcerias en el secano de la postguerra', *Agricultura y sociedad*, Madrid, no. 3.

Reid, Joseph D., 1976, 'Sharecropping and Agricultural Uncertainty', *Economic Development and Cultural Change*, April.

Sharecropping as an Efficient System: Further Answers to an Old Puzzle

José-María Caballero*

One of the focuses of the economic literature on sharecropping has been the question of efficiency and, intimately associated with this, that of the kind of relations under which landowners prefer to organise the exploitation of their lands.

Sharecropping has been traditionally regarded as inefficient because *ceteris paribus* in equilibrium less inputs would be committed per unit of land than under either wage-labour or fixed-rent farming, output per acre thus being smaller. If this were actually the case, the widespread use of sharecropping relations in a world where land has progressively become more and more scarce with respect to non-land inputs would be puzzling: the market should have got rid of such an uneconomic institution. Instead, we find that, after careful and rational balancing of pros and cons, landowners do often prefer sharecropping to wage-labour relations. How is this to be explained?

Authors have put forward different answers. Some have argued that a combination of perfect competition in land and labour markets and a perfectly observable labour input will render sharecropping Pareto efficient [*Cheung,* 1968; *Reid,* 1976]. Others have stressed the capacity of landowners to force tenants to supply the right amount of labour by using short-term leases and threatening unproductive tenants with eviction [*Johnson,* 1950]. Others contend that imperfections in the labour market which result in a shadow wage rate substantially smaller than the market wage rate, or the combination of large seasonal variations in labour requirements and transaction costs in labour hiring, can make sharecropping more profitable than wage labour [*Martínez Alier,* 1981; *Mazumdar,* 1975]. Finally, other authors have emphasised the advantages of sharecropping in dispersing risk, whether with respect to uncertainty in output performance (or output price) alone or together with uncertainty in the inputs' markets [*Cheung,* 1968; *Newbery,* 1975; *Newbery and Stiglitz,* 1978; *Rao,* 1971; *Reid,* 1976; *Stiglitz,* 1974].

I shall present here an argument based on the distinction between relations of production and relations of distribution. I shall assume perfect competition in all markets and no uncertainty.

Perhaps the crucial difference between sharecropping and wage-labour farming is that whereas in the former the contradiction between landowner and labourer is located in distribution, in the latter it is located in production. Under

**Centre of Latin American Studies and Wolfson College, Cambridge.*

sharecropping the two parties quarrel over the key distribution parameter – the percentage rental. But since both owner and tenant have an interest in making output as large as possible (within the existing opportunity cost restrictions) and the tenant is manager of his own labour, there is no quarrel regarding production and the labour process; on the contrary there is often co-operation. By contrast, under wage labour the labour process acquires a contradictory character: labourers have no interest in the result of their work, from which they are alienated. Design and execution, work and product, labour and management are separated and in opposition here. Labour has to be supervised.

Different locations of the labourer–landowner contradiction produce different types of deviations from efficiency. In sharecropping the contradiction in the distribution of the product results in allocative inefficiency. In order to maximise his income, the rational tenant stops working in his tenancy when the marginal return to the labour there applied equals its opportunity cost elsewhere, which under perfect competition is the market wage rate. Since the marginal return to the tenant falls short of the marginal product of his labour, for he only collects a percentage share of output, the amount of labour allocated to the tenancy will be suboptimal. On the other hand, the wage-labour system is free from allocative inefficiency, since the owner will go on hiring labour until the marginal labour product equals the wage rate, but the contradiction in the labour process gives rise to x-inefficiency. Labourers will tend to shirk. Only by expending money in labour supervision can shirking be stopped.

Thus, under wage labour the right amount of work-days or hours may be hired but there will be a tendency for workers to shirk, i.e. to operate at low-effort rates, if they can get away with it, and they will unless supervision is strict with supervision costs probably high. On the other hand, sharecroppers may put in a suboptimal amount of work-days or hours but will not shirk because they would be deceiving themselves; they work less time but at high effort rates. To the extent that wage labourers have opportunities to shirk, thus reducing effective labour per unit of hired labour, and supervision costs are high, thus increasing labour costs and hence reducing labour hiring, the wage system may finally result in less-effective labour being actually put in than under sharecropping. The allocative efficiency of wage-labour farming may thus be outweighed by its x-inefficiency.

To prove more formally that this may be the case, and to investigate under what set of values of the relevant parameters this would occur, we must compare labour intensity under sharecropping and under a modified, more realistic model of the wage-labour system: one where the tendency of wage labourers to shirk is recognised and supervision costs explicitly introduced. I shall first present this model.

Let us define effective labour, L, as the product of labour-time, t, and the level of labour effort, represented by an effort parameter, θ, which may be assumed to vary between $\bar{\theta}$, the value of the effort parameter when there is no supervision,[1] and 1. Thus

(1) $L = t\,\theta$ where $\bar{\theta} \leqslant \theta \leqslant 1$

To make effective labour a multiplicative function of t and θ is the simplest possible definition. The question remains open of what does θ exactly stand for and how can it be measured. Here θ is intended to indicate a combination of the intensity, intelligence and care put into each time unit worked.[2]

Wage labour must in general be supervised. Supervision costs, S, are assumed to vary with the level of effort the employer wants his hired workers to achieve. Thus

(2) $S = S(\theta)$

The supervision cost curve is assumed to have the familiar shape of cost curves, shown in figure 1, which implies

$S(\bar{\theta}) = 0$,
$S(1) = \infty$,
$S'(\theta) > 0$,
$S''(\theta) < 0$ for $\bar{\theta} \leqslant \theta < \ddot{\theta}$,
$S''(\theta) > 0$ for $\ddot{\theta} < \theta \leqslant 1$.

FIGURE 1

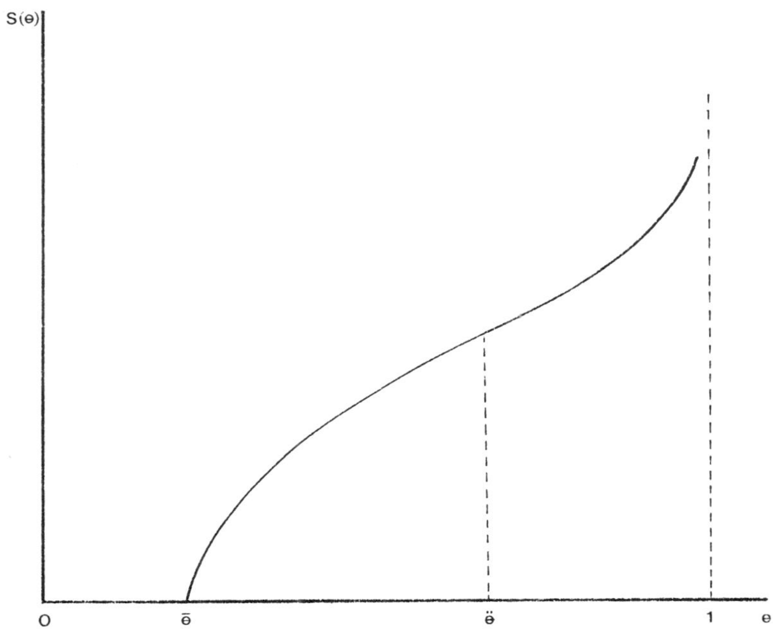

Assuming that only one crop is grown, whose production depends on effective labour and land, T, which is considered fixed, making the crop's price

equal to one, and assuming that there are constant returns to scale in supervision (i.e. once the desired effort level has been decided the supervision cost per unit of labour-time does not vary with the number of labour-time units hired) the problem of the labour hiring landowner is to maximise his profit, P:

$$\max_{\{t,\,\theta\}} P = F(L,T) - (w + S(\theta))\,t,$$

where w is the wage rate and w + S(θ) the labour cost per unit of labour-time, a linear function of θ.

It has been customary in the literature to assume implicitly that wage labourers put in full effort without need of supervision, i.e. $\theta = 1$ and $S = 0$. This is unreal and conceals the contradictory nature of the wage-labour system. The present model seems a better approximation to reality.

The solution of the maximisation problem gives the equilibrium equations

$$(3)\ F'_L = \frac{w + S(\theta)}{\theta} \text{ and}$$

$$(4)\ F'_L = S'(\theta).$$

These equations differ from the customary equilibrium solution in that under the traditional model (4) does not apply and (3) reduces to $F'_L = w$.

To interpret the equilibrium solution let us turn to figures 2 and 3.

The determination of θ^*, the equilibrium level of θ, is shown in figure 2, where $\bar{\theta}$ has been taken as origin. S′ and S(θ)/θ are the marginal and average supervision cost curves. Curve G(θ) is obtained by solving (3) and (4), which gives the equilibrium condition

$$(5)\ w = (S'(\theta) - \frac{S(\theta)}{\theta})\,\theta \equiv G(\theta).$$

According to (5), in equilibrium w is θ times the difference between the marginal and the average supervision cost. In figure 2 equilibrium is reached at B where G(θ) intersects the wage line.

Two features of the equilibrium level of θ are apparent. First, low levels of θ, $\bar{\theta} \leqslant \theta \leqslant \theta'$, are incompatible with positive wage rates. This is a reflection of the assumption of first decreasing and then increasing marginal supervision costs, which in the absence of empirical information is the most reasonable shape to assume for S(θ). If this is actually the case, it is common-sense that equilibrium can only be reached after the marginal exceeds the average supervision cost and the latter begins to rise, for otherwise by increasing the desired level of θ the average supervision cost could be decreased. The second feature is that, for all positive wage rates, the larger the wage rate the bigger the effort level desired by the employer (although not necessarily the larger the amount of effective labour). This can be easily explained. An increase in the wage rate, by raising the cost of labour-time tends to reduce the amount of time units hired, while by lowering the proportion of supervision costs in total labour costs, tends to increase the effort level desired by the employer. In other words, high wage rates, by making effort cheaper relatively to labour-time, induce the substitution of θ for t in the production of L.

FIGURE 2

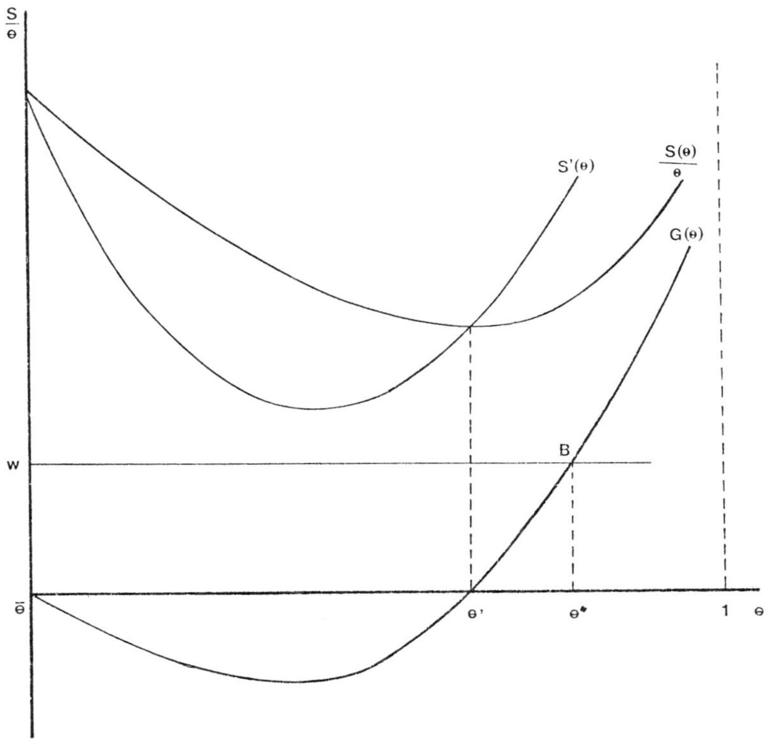

Having determined θ^*, we turn to the determination of t*, the equilibrium level of t. This is shown in figure 3, where for simplicity the marginal product curves are assumed to be straight lines. Line CH is the marginal product of effective labour, identical to the marginal product of labour-time when there is full effort ($\theta = 1$). Under the traditional model, equilibrium would be reached at G, with t_0 labour-time units being employed. BH is the marginal product of labour-time at equilibrium θ, i.e. at θ^*, identical to θ^* times the marginal product of effective labour, according to (1). AJ, parallel to BH, is the same marginal product as before but net of the supervision cost corresponding to θ^*. The difference between OB and OA is the supervision cost $S(\theta^*)$. AJ is the relevant curve for landowners' decision making. According to equation (3), which may be rewritten as

(3a) $w = \theta F'_L - S(\theta)$,

equilibrium is reached at point F, where the net marginal product of labour-time equals the wage rate. The inclusion of supervision costs and effort variability in this model makes the equilibrium labour-time employed, t, smaller than under the traditional model.

FIGURE 3

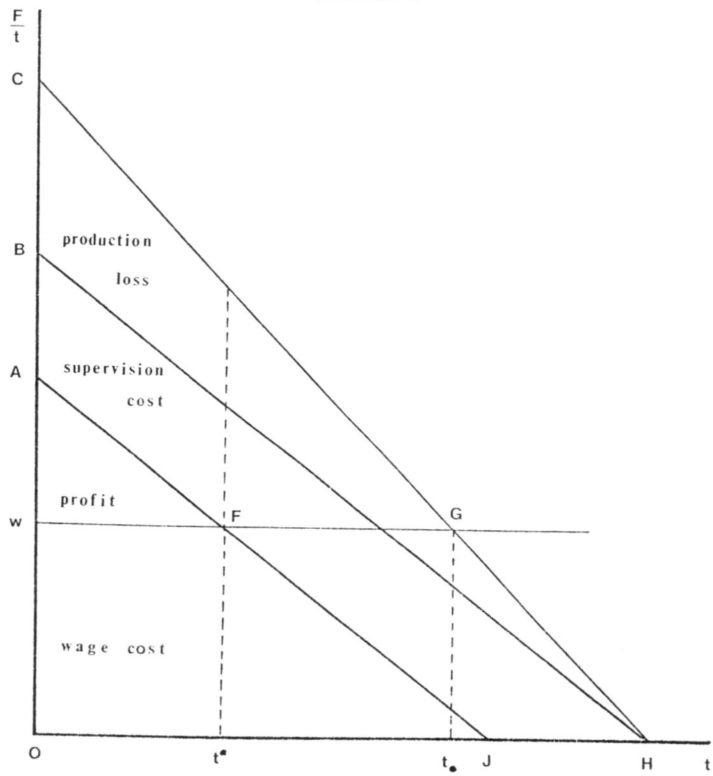

The production loss due to less than full effort level, the supervision cost necessary to sustain the equilibrium level of effort for all labour-time units hired, profits, and wage costs are all explicitly shown in the figure. In addition, it may be pointed out that area AFGC is an approximation to the loss due to the lack of incentive of the wage-labour system – a measure of the waste implicit in the contradictory nature of the capitalist labour process.[3]

We may now enquire under what conditions sharecropping will be more labour intensive than wage labour.

If, in the Marshallian tradition, we assume that there are no binding contractual stipulations concerning the labour input, so that tenants are free to fix the amount of labour put into their tenancies, there will be no supervision and therefore no supervision costs; and since tenants are free to decide upon their labour they will not shirk. We may therefore assume that all units of labour-time they choose to put into their tenancies are at full effort level, i.e. $\theta = 1$.[4] Under these conditions, if we designate by $\bar{F}'_L$ the marginal product of effective labour (or labour-time, identical here) in sharecropping, by r the percentage share in the product accruing to the tenant, and by w the market wage rate which, under perfect competition assumptions, is also the opportunity cost to the labour put into the tenancy, the equilibrium condition in sharecropping is[5]

$$(6)\ \bar{F}'_L = \frac{w}{r}.$$

This condition may be compared with that for wage-labour farming given by (3). Assuming decreasing marginal returns to effective labour, for sharecropping to be more labour intensive than wage labour the marginal product of L must be smaller under the former than under the latter, i.e. $\bar{F}'_L < F'_L$, which substituting in (3) and (6) gives the condition

$$(7)\ \frac{S(\theta)}{w} > \frac{\theta - r}{r}.$$

This inequality is always satisfied if $\theta \leqslant r$. If $\theta > r$ it may or may not be satisfied. It will never be if supervision costs are zero, and it will always be if labour is a free commodity to the employer – a not farfetched assumption if land is abundant and there is some kind of labour bondage. Otherwise, the larger $S(\theta)$ and the smaller w the more likely that (7) is satisfied.

Two propositions follow:

1. For labour intensity to be larger under wage labour than under sharecropping, the necessary but not sufficient condition is that the equilibrium effort level of wage labourers (measured by the effort parameter θ) be larger than the percentage share accruing to the tenant. This condition becomes sufficient if supervision costs are zero, unless labour is a free commodity.

2. For percentage effort levels larger than percentage tenant shares, the larger the supervision costs and the smaller the wage rate the more likely that labour intensity is bigger under sharecropping. The sufficient condition for this is a percentage tenant share equal or larger than the percentage effort level of wage labourers.

To complete this presentation, the labour inputs under wage labour, sharecropping and fixed rent are jointly shown in figure 4.

In the top part, CH and AJ respectively show the ordinary and net marginal product curves, as in figure 3. HI is the marginal return to tenant labour, i.e. CH times r. $t_{w\text{-}1}$, $t_{s\text{-}c}$ and $t_{f\text{-}r}$ are the equilibrium labour-time inputs under wage labour, sharecropping and fixed rent respectively. The latter is also the equilibrium labour input under the traditional wage labour model and under owner-operation. As drawn, $t_{s\text{-}c} > t_{w\text{-}1}$. But this is not necessarily always the case; wages above w' ($= r\ S(\theta^*)/\theta^*\text{-}r$) will make $t_{w\text{-}1}$ to exceed $t_{s\text{-}c}$ (although not necessarily $L_{w\text{-}1}$ to exceed $L_{s\text{-}c}$). If IH lay everywhere above AJ, $t_{s\text{-}c} > t_{w\text{-}1}$ would always be satisfied; and since IH cannot lie everywhere below AJ, as long as there are some positive supervision costs, there must always exist some non-negative wage rate for which $t_{s-c} > t_{w-1}$.

FIGURE 4

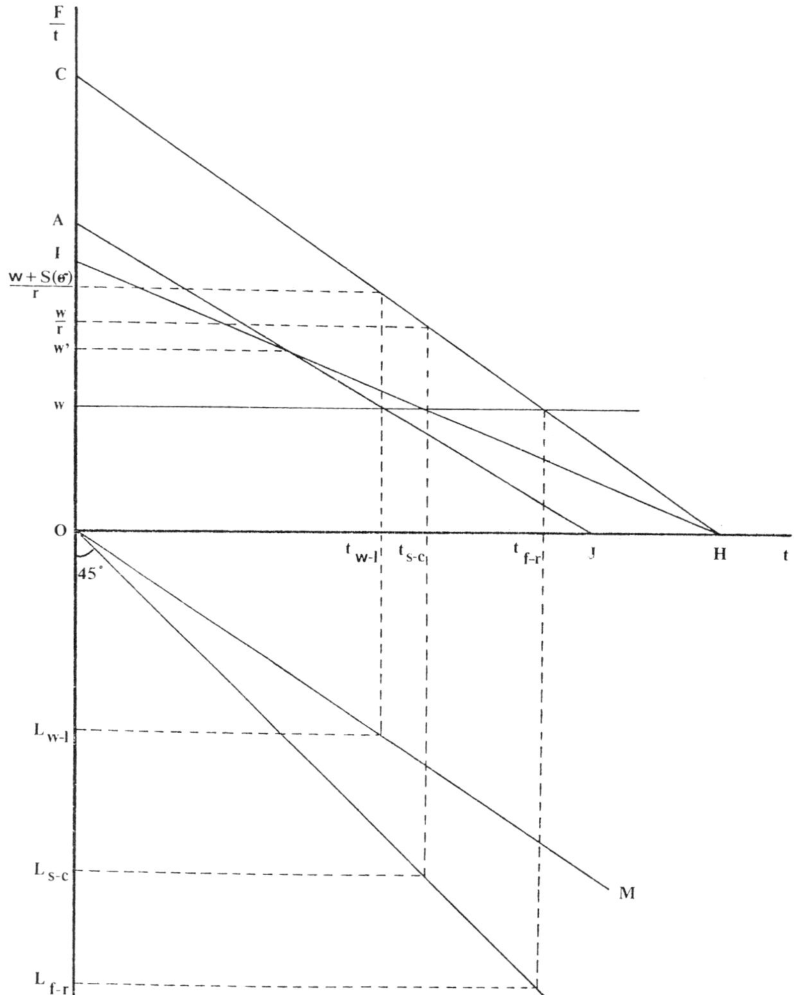

In the bottom part, the equilibrium effective labour inputs under the three systems are shown. Since it is assumed that both fixed-rent and sharecropping tenants work at full effort, L_{s-c} and L_{f-r} are obtained by projecting t_{s-c} and t_{f-r} on the 45 degrees line. On the other hand, L_{w-1} is obtained by projecting t_{s-c} on line OM, whose slope with respect to the Ot axis is θ^*. It is clear that whereas labour intensity under sharecropping may or may not be larger than under wage labour, fixed-rent contracts and owner cultivation always entail, under the stated assumptions, more intensive labour application than wage labour or cropsharing.

If fixed-rent cultivation is more labour intensive than sharecropping – output per acre and hence the surplus per acre potentially appropriated by the landowner being therefore larger – why do landowners often prefer sharecropping to fixed-rent contracts? Fixed-rent leases seem to have all the advantages. Since the tenant is the master of his own labour there is no contradiction in the labour process, and since the distribution is fixed in advance as a lump quantity rather than a percentage, the contradiction between labourer and landowner does not give rise to allocative inefficiencies. It is not immediately apparent, therefore, why fixed-rent leases should not always be chosen by the owners.

The standard answer is based in the advantages of sharecropping in dispersing risk. But an argument can be provided that does not need to assume uncertainty: the pattern of production over time may be known in advance. It hinges upon the natural ups and downs of agricultural production (or the not always so natural vicissitudes of food prices), and the relative bargaining power of landowners and tenants.

In essence, the argument is as follows. If production varies substantially from year to year, there will be a firm resistance from tenants to fixed-rent leases that threaten survival in bad years. Fixed rents could, thus, have a strong ceiling, provided that tenants have sufficient negotiating power. Sharecropping may then be more profitable to the landowner over a cycle of several years, even if output per acre is larger under fixed-rent.[6]

This is illustrated in figure 5, where output and rent over time from a certain amount of land are shown for the two systems.

ABCD represents output performance over time under fixed rent; for simplicity it has been assumed to follow a strictly cyclical pattern. A′B′C′D′ is output under sharecropping, smaller than under fixed rent because of less labour intensity. A″B″C″D″ is the net output accruing to the sharecropper, i.e. A′B′C′D′ times the tenant's percentage share r (=OA″/OA′). OS is the subsistence minimum of the tenant family. The landowner can impose a fixed rent of SR without affecting the tenant's subsistence. If tenants have sufficient negotiating power, this may be the maximum possible fixed rent. Over the cycle of OJ years it gives SRTP revenue to the owner. On the other hand, under sharecropping the owner can collect A′B′C′D′D″C″B″A″ rent over the OJ period without encroaching upon tenant subsistence. As drawn, this is larger than SRTP, and the landowner will therefore prefer the cropsharing system.

The two arguments developed in this paper do not intend to cover all situations which may make sharecropping more labour intensive than wage labour or preferred to fixed rent. In particular, I believe that the most relevant explanations of sharecropping in overpopulated peasant agricultures in the third world

FIGURE 5

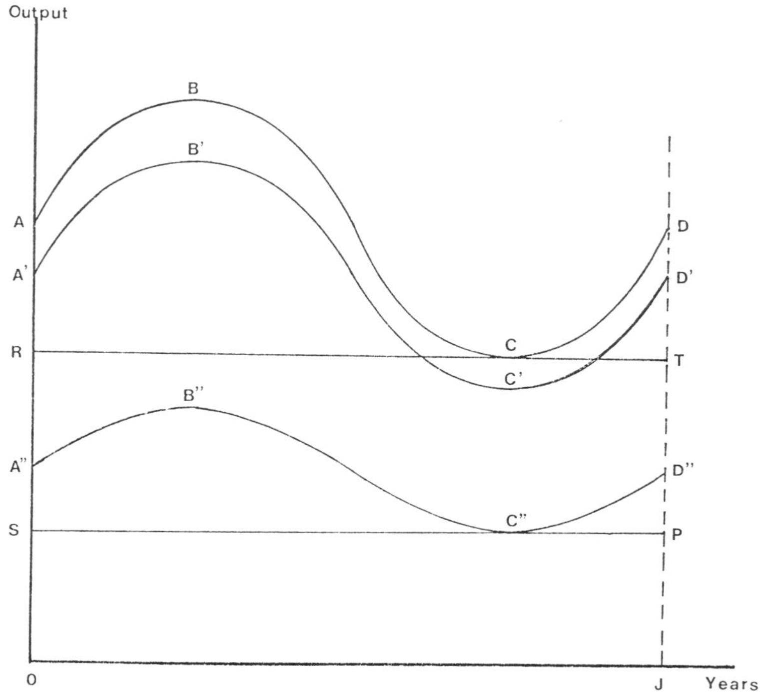

are those which stress labour-market imperfections. However, the argument focusing on how wage labourer and landowner try to overcome the contradiction in the production process, by means of shirking in the former case and by securing labour supervision in the latter, can be usefully applied to many situations. The notion, for instance, that the need to supervise wage labourers may act as a deterrent to capitalist agriculture, could be incorporated into many analyses of the development of capitalism in agriculture. Let me give an example.

It is often found in field studies of peasant villages and country towns that the wealthiest, more entrepreneurial village members, whose urban operations extend to various fields – such as small industry, money-lending, transport and commerce – and are carried out with capitalistic methods and rigour, do nevertheless employ rather traditional sharecropping arrangements in their farms, perhaps to a larger extent than other village members less entrepreneurial in urban activities. The explanation of this apparent paradox may well be that, unable to 'buy' suitable supervision, if they introduce wage-labour relations in their farms, work would have to be supervised by the owner himself, thus reducing the time available for profitable urban activities.

NOTES

1. $\bar{\theta}$ need not be zero since workers may of their own will put in some effort, because of fear of losing the job in the long-run or because they may take some pride in their work or because they may feel responsible to their employer or other reasons.

2. Vanek /1970: 235 *and* 241-2/, who has discussed the effects of effort variation in the context of worker-managed firms, includes intensity and quality, apart from duration, in its definition. Stiglitz /1974: 242/ includes five characteristics: pace and thoroughness, which may be assimilated to Vanek's intensity and quality, efficiency in the allocation of work to specific job tasks, efficiency in job decision making under uncertainty, and inventiveness. However difficult the precise definition of effort may be, it is clear that it refers to a very real aspect of work life. Like the elephant, nobody would fail to recognise it. Measuring θ will of course depend on the precise definition adopted and on the specific nature of the work process being investigated. I do not think a general *a priori* measuring test can be devised.

3. This results from comparing WGC, which is the profit obtained when there is full effort and no supervision costs, i.e. under the traditional model, and WFA, the profit under the more realistic model. For this to be a true measure of the welfare loss we should deduct the satisfaction that wage labourers derive from shirking. If our interest, however, is to compare two alternative modes of production organisation, say wage labour and owner-operation, we may perhaps be justified in assuming that the dissatisfaction experienced from work at less than full effort under the wage-labour mode is equal or even larger than that experienced from the same work time but at full effort level under the owner-operation mode. The worker, presumably, derives pleasure/displeasure not only from the effort he puts into work but also from the mode itself under which this is deployed. If the satisfaction caused by shirking, under wage labour, is cancelled by the satisfaction caused by working for himself, under owner-operation, AFGC will be a true measure of the welfare loss incurred when using the former rather than the latter system.

4. That sharecroppers always work at full effort levels, so that they only adjust labour-time, seems to me the best possible assumption. But, for any equilibrium level of L, L^*, determined by equation (6), any feasible combination of t and θ along the hyperbola $L^* = \theta t$ may be chosen without affecting the results. To compare with the wage-labour system we do not strictly need to assume that sharecroppers always work at full effort. It suffices to assume that sharecroppers' θ is larger than $\bar{\theta}$, the effort level of wage labourers when there is no supervision. We can then normalise making sharecroppers' $\theta = 1$.

5. Let y be the tenant's income, L_1 the labour-time allocated to tenant farming, L_2 that allocated to wage employment, and $\bar{L}$ his total labour availability. His problem is

$$\max_{\{L_1, L_2\}} y = r\,\bar{F}(L_1, T) + w\,L_2$$

s. t.

$$L_1 + L_2 = \bar{L}$$

Forming the Lagrangean, setting the partial derivatives equal to zero and solving we obtain (6).

6. The argument could easily be reversed and turned into an explanation of landowners using fixed rents to 'tie' tenants through their indebtedness in bad years. This would assume that landowners are powerful enough to impose fixed-rent leases that encroach on tenants' minimum subsistence income in bad years.

REFERENCES

Cheung, S.N.S., 1968, *The Theory of Share Tenancy*, Chicago: University of Chicago Press.

Johnson, D.G., 1950, 'Resource Allocation under Share Contracts', *The Journal of Political Economy*, Vol. 58, pp. 111-23.

Martínez-Alier, J., 1981, 'Sharecropping – Some Illustrations'. Paper presented to the Symposium on Latifundist Agriculture in the Iberian Peninsula, Oeiras, Portugal: Instituto de Economia, Agraria, December 1979 (rewritten in 1981). Reprinted in this special issue.

Mazumdar, D., 1975, 'The Theory of Sharecropping with Labour Market Dualism', *Economica*, Vol. 42, pp. 261-71.

Newbery, D.M.G., 1975, 'The Choice of Rental Contracts in Peasant Agriculture', *Agriculture in Development Theory*, edited by L.G. Reynolds, New Haven and London: Yale University Press.

Newbery, D.M.G. and J.E. Stiglitz, 1978, 'Share Cropping, Risk Sharing and the Importance of Imperfect Information', Economic Theory Discussion Paper No. 8, University of Cambridge.

Rao, C.H.H., 1971, 'Uncertainty, Entrepreneurship, and Sharecropping in India', *The Journal of Political Economy*, Vol. 79, pp. 578-95.

Reid, J.D., 1976, 'Sharecropping and Agricultural Uncertainty', *Economic Development and Cultural Change*, Vol. 24, pp. 549-76.

Stiglitz, J.E., 1974, 'Incentives and Risk Sharing in Sharecropping', *Review of Economic Studies*, Vol. 41, pp. 291-355.

Vanek, J., 1970, *The General Theory of Labour-Managed Economies*, Ithaca: Cornell University Press.

C. Concrete Analysis of Sharecropping: Individual Country Studies

Sharecropping and the Plantation Economy in the United States South

Jay R. Mandle*

Plantation economies were constructed in those parts of the western hemisphere where European expansion took the form of the establishing of large-scale farms supplying metropolitan markets with tropical or semi-tropical staples. In a plantation mode of production four attributes are present. First, large-scale export-oriented agriculture dominates the society. Second, the labour-force requirements of that agricultural sector are greater than can be supplied through the functioning of an unrestricted domestic labour market. Extra-market mechanisms, therefore, must be present to provide the required labour to the plantations and these mechanisms are dominant in defining the class relations of the society. Finally, a specific culture both emerges from and reinforces the class relations which are articulated in this way.

Slavery, of course, was the foundation upon which the plantation economy was built. It provided labour to plantations more cheaply and in greater numbers than otherwise would have been the case. Fogel and Engerman estimate for the United States South that even a 50 per cent increase in the income free white farmers were earning on their own land would not have been sufficient to induce them to shift to plantation work. Indeed, with emancipation, even a 100 per cent increase in income was inadequate to attract black workers to gang labour on the plantations. In explanation, Fogel and Engerman cite the resistance which existed to the highly regimented pattern of plantation work and note that 'it was only by applying force that it was possible to get blacks to accept gang labour without having to pay a premium that was in excess of the gains from economies of scale.' They write that 'from the master's viewpoint, the advantage of force, when judiciously applied, was that it produced desired behaviour ... at a lower cost than could have been achieved through financial inducements' [*Fogel and Engerman*, 1974: vol. 1, 237].

With emancipation it was no longer possible for planters to exercise ownership rights with respect to their labour force. Juridical freedom meant that, for the first time, former slaves were able formally to bargain with planters. In principle, therefore, the ending of slavery was a major turning point; a new basis for class relations had been established.

Yet because juridical slavery had ended, it does not follow that, therefore, the trajectory of society had necessarily changed and that the plantation economy had been dismantled. Elsewhere, for example in some parts of the

*The author is Professor of Economics at Temple University. He would like to thank Stanley Engerman Louis Ferleger, Joan D. Mandle, and Isadore Reivich for having read and commented on earlier drafts of this paper.

British West Indies, a 'new system of slavery' had been created in the aftermath of legal emancipation [*Tinker*, 1974]. There, a state-sponsored system of indentured immigration had been developed as an alternative to slavery. Through the use of this system, labour continued to be supplied to sugar plantations in numbers greater than and at costs lower than would have prevailed if an effective domestic labour market had been operative.

The question which arose in the South after the Civil War was the same one that had existed in post-emancipation societies such as British Guiana, Jamaica, and Trinidad. How closely would the new system of agricultural organisation and labour-force mobilisation approximate slavery? Specifically from the point of view of the planters, the issue which arose was to what extent would alternatives to work on the plantations be available to the former slaves. For, if such opportunities were effectively limited, then the freedmen would have to continue, after emancipation as before, to supply themselves to the plantations at relatively low costs. In that situation yet another new system of slavery would have been created in fact if not in law.

Robert Higgs argues that the agricultural system which emerged in the Reconstruction South was far different from slavery. He writes that 'many writers seem to believe that share tenancy was a method of race control imposed on helpless blacks by all-powerful whites . . .' But, Higgs goes on, 'the evidence is almost entirely inconsistent with such an interpretation.' In fact, Higgs concludes, although white efforts to control black labour 'no doubt had some effect on black mobility, in general they failed to achieve their objective of immobilizing the black labour force' [*Higgs*, 1977: 45, 119].

In advancing his argument, however, Higgs inadvertently provides an important clue to its weakness. He writes that with regard to occupations 'blacks had a continuum of preferences in which, other things being the same, land ownership was most preferred, followed by fixed rent tenantry, share rent tenantry and – the least preferred – working for wages.' With regard to the preference for land ownership, Higgs was, of course, correct. But land precisely was the one choice which would have allowed the former slaves to escape dependency on the planters. It was also the one option which was denied to the vast majority of the freedmen. As Higgs euphemistically puts it, 'in the beginning the practical choice was usually between shares and wages' [*Higgs*, 1977: 45]. It was this failure to provide land to the former slaves which was the critical element in the re-establishing of a post-slavery plantation economy in the South.

The demand for land was the dominant theme in black proposals for recreating the South. Vincent Harding writes that 'the search for land, the need to hold on to land was . . . central to the black hope for a new life in America.' This demand was summarised at a national convention of black leaders held in Syracuse in October, 1864. Among the proposals issued was the demand for '. . . a fair share of the public domain, whether acquired by purchase, treaty, confiscation or military conquest.' As Harding correctly notes, this call reflected the fact that 'in an agrarian society . . . freedom meant the right to land – the land they had nurtured' [*Harding*, 1981: 315, 247, 297].

The analytic foundation for this demand and the factor which differentiates land from the other preferences listed by Higgs is that the possession of land provided the opportunity for independence from plantation work. The possession of land would have placed the southern black population in a position comparable to the one occupied by small-scale white farmers before the Civil War: they could have refused to work on the region's cotton plantations. None of the other options possessed this advantage. At the same time, sharecropping represented a means by which plantation agriculture could be provided with a cheap labour force and thus be given a new birth.

The providing of land to black labour would have been a radical act, far beyond the limits of ideological acceptability to Lincoln and his administration, not to mention the government of Andrew Johnson. Even the nearly one million acres of land which came into the possession of the Freedmen's Bureau was insufficient if the task had been to create a black peasantry, since there were roughly one million potential black farmers. In fact, only a small fraction of even this publicly held land came into the possession of the former slaves. What was required, as Dubois noted, was the availability of from 25 to 50 million acres, a scale of land transfer which would have necessitated the dispossession of the South's planter class [*DuBois*, 1968: 691].

Instead of that the only land which came to be owned by blacks in the South was that which was made available by white owners in the market. Higgs attempts to portray this set of market sales in a favourable light by arguing that 'land was simply too abundant to be monopolised by whites'. But what he fails to note is that, according to his own data, the effect of these land sales in the deep South states of Alabama, Georgia, Mississippi, North Carolina and South Carolina was to keep landless between 94 per cent and 96 per cent of black farmers in 1876 [*Higgs*, 1977: 52]. Furthermore, the land which did come to be owned by black farmers typically was inferior to that in the large plantations. Frequently this land was in holdings too small to provide commercial viability and often was offered to reward deferential behaviour rather than entrepreneurial skill [*Mandle*, 1978: 52, 29-30].

Ransom and Sutch note that part of the difficulty that the former slaves experienced in gaining access to land lay in the fact that they, typically, were too poor to buy good land in adequately large holdings. But they emphasise that this was not the main problem. Rather, decisive difficulties lay on the supply side of the market. Ransom and Sutch write that 'the tacit resistance to black landownership among the whites was usually sufficient to prevent such land sales.' When it was not, Ransom and Sutch go on, 'whites who might agree to sell land to blacks were not uncommonly threatened with physical violence.' They cite Whitelaw Reid in this regard:

> In many portions of the Mississippi Valley the feeling against any ownership of the soil by negroes is so strong, that the man who should sell small tracts to them would be in actual personal danger. Every effort will be made to prevent negroes from acquiring lands; even the renting of small tracts to them is held to be unpatriotic and unworthy of a good citizen [*Ransom and Sutch*, 1977: 86].

The failure to gain access to land condemned the former slaves to a weak bargaining position in offering their labour to the plantations. This was so because the failure of land redistribution and inadequacy of land sales meant that the range of employment opportunities available to them was sharply constrained. In the South the difficulty of gaining access to land was joined by the region's slow industrialisation process and racial discrimination in hiring. The result was sharply to limit the number of employment opportunities available aside from plantation work. A migration northwards to industrial employment in that region was forestalled since the jobs which were being created there were claimed by immigrants driven from Europe by that continent's onset of economic growth. Furthermore, the movement of southern blacks into frontier agricultural land, such as that of the Exodusters, was limited by finances and when it did occur was greeted by grievous racial hostility [*Painter*, 1977]. As a result, the range of options available to the southern black labour force was quite circumscribed, a situation which minimised black labour's bargaining strength in the market.

It was sharecropping which emerged in the aftermath of the Civil War to provide the new link between planters and plantation labour. Under the new system gang labour was ended. Individual croppers were assigned to housing provided on specific lots. Initially the terms of sharecropping agreements varied considerably, but within a year or two these terms became standardised with one half of the output produced by a cropper paid to him and one half retained by the plantation owners. A typical contract obliged the cropper 'to work faithfully and dilligently without any unnecessary loss of time, to do all manner of work on said farm as may be directed' by the planter, 'and to be respectful in manners and deportment' to the owner. In exchange the cropper planter agreed to furnish a mule, feed, tools and seed. In addition, the planter agreed to advance provisions to the cropper which were to be paid for from the part of the output due the sharecropper [*Ransom and Sutch*, 1977: 91]

There recently has been an extensive literature which had emphasised the fact that sharecropping was the object of bargaining and market relations, thus implying that it was not exploitative. This argument is made in two dimensions. First, the emergence of the institution of sharecropping itself represented a concession offered by the planters in response to the former slaves' strong antipathy to gang labour. The sharecropping system contained a relaxation of regimented and centralised methods of cultivation compared to gang labour. For that reason its emergence is viewed as a victory for the former slaves, the product of the bargaining relationship between the freedmen and planters. In addition, it has been argued that the terms of the sharecropping contract, too, were subject to negotiation and bargaining. Specifically, Joseph D. Reid Jr. argues that the terms of the share contracts were altered and adjusted in response to changes in opportunity costs, crop prices and yields [*Reid*, 1973]. In short, both the institution of sharecropping and the terms of the sharecropping contract were subject to market influences and thus non-exploitative.

There is no reason to deny the assertion that, at least implicitly, bargaining over sharecropping did occur. It is likely that minor modifications in the terms of sharecropping contracts did occur, though Reid, in making his argument, presents no evidence on its frequency.[1] In addition it certainly is true that the abandonment of gang labour and the establishing of sharecropping arrangements was a victory for the former slaves. Indeed, given their relatively weak bargaining position and the desire of the planters to renew the use of gang labour, its demise is testimony to the depth of feeling which the ex-slaves possessed towards it. It appears that no matter the cost black labour in the South virtually refused to work in the gang system.

But there is nothing in either of these two arguments which undermines the contention that plantation agriculture was re-established after the Civil War. Bargaining did occur, but it did not concern the fundamentals of the relationship between planter and croppers. It was bargaining between unequals. In the absence of land redistribution and without significant numbers of alternative employment opportunities the market strength of the freedmen was far inferior to that of the newly reconstituted planter class. As a result, though the black workers did achieve concessions, those concessions were obtained in a context in which the planters were able to achieve their overriding goals: the attaining of large numbers of workers at low costs and under ubiquitous supervision. Litwack summarises by arguing of even the sharecropping agreements entered into under the auspices of the Freedmen's Bureau:

> Although designed to protect the interests of planters and freedmen alike, the contract in practice gave employers what they had wanted all along – the crucial element of control by which they could bind the ex-slaves for at least a year and compel them to work and maintain proper behaviour [*Litwack*, 1979: 413].

The law in the South further reinforced the subordinate position of the croppers. The way in which this occurred has recently been identified and discussed by Harold D. Woodman. Woodman notes that in southern law after the Civil War a clear distinction was drawn between tenants and sharecroppers. Under state law in the region sharecroppers were considered to be wage labourers. As such, croppers were subjected to managerial supervision and control to a degree far greater than was present in a tenant/landlord relationship. In this connection crop lien legislation gave to landlords a claim for rent and advances which was superior to that of other creditors. Since croppers were only employees, landlords could therefore, if necessary, take control of the crop as it was gathered. Furthermore, these same statutes made it possible for planters to use criminal as well as civil law to control croppers. Southern law with regard to sharecropping in short defined the relationship between cropper and planter in a way which reinforced the dominance of the latter over the former. These laws, writes Woodman, 'allowed planters to exercise a degree of control over their work force far beyond that available elsewhere in the nation' [*Woodman*, 1979: 336].

The result of all of this was that plantation agriculture was re-established after the Civil War in the American South. Sharecropping, in the context of the limited availability for blacks of alternative employment opportunities, allowed southern planters to recreate the essential conditions of plantation farming: the use of cheap and abundant labour under close managerial supervision [*Jones*, 1968: vol. 12, 154-5]. The difference was only that in the post-bellum period compared to the ante-bellum era the organisational basis for plantation agriculture was sharecropping rather than slavery.

Close planter supervision existed on about 80 per cent of farm land under cotton, the South's dominant crop. The content of this supervision is best revealed in a US Department of Agriculture description of a typical work day on a plantation as late as 1925. According to this report a normal work day on a closely supervised plantation started with the ringing of a bell indicating the start of labour for that day. Another bell, usually at about sunset, indicated the end of labour for that day. The land upon which the cropper worked had been assigned to him by the landlord. Supervision occurred during the daily rounds made by the farm manager in which he gave instructions on the details of field work. The tools with which the sharecropper worked were, of course, owned by the planters and were supplied out of a pool of such equipment [*Mandle*, 1978: 48]. Plantation agriculture, in short, was very much alive in the South even after formal slavery had been abolished. As summarised by Edgar T. Thompson, 'the Civil War did not, contrary to an opinion often encountered outside the South, destroy the plantation system ... Legal slavery was abolished but the large landed estate, the physical basis of controlled labour, remained – a fact which alone necessarily ensured a high degree of continuity with the past' [*Thompson*, 1975: 242].

The re-establishing of plantation agriculture in the South doomed that region to a major delay in economic growth. In this retardation the post-bellum South is among the regions referred to when Eugene D. and Elizabeth Fox Genovese assert that 'no slave-country or region crossed the threshold to industrialisation and none adjusted to emancipation so as to launch a new cycle of growth that passed into structural change' [*Genovese*, 1979: 11].

The process of economic development centres on the application of scientific, engineering and technological change to production. Simon Kuznets, the leading empirical student of the development experience, writes that the 'mass application of technological innovation ... constitutes much of the distinctive substance of modern economic growth'. Countries which have become developed, according to Kuznets, are the ones which have managed to take advantage of advancing technology in production [*Kuznets*, 1973: 166, 171]. The hypothesis advanced here accepts this view of development and argues that the structure of plantation agriculture produced a bias against technological innovation. This bias can be seen by examining both the incentive to innovate and the capacity to do so which emerged in a plantation setting.

The extent to which planters were motivated to seek out new methods of production was limited in two ways. Assuming that the planters were profit maximisers, they would tend to use relatively more labour but fewer tools and less equipment than would be the case if labour's price were relatively higher. The typical plantation production function thus was labour intensive. Such a pattern of production, however, militates against technological advance since new technology is often embodied in new capital equipment and tools. The consequence of a regime of cheap labour was to reduce the amount of capital used in production. In doing so capital as a carrier of new technology, too, was minimised.

A second mechanism may also have inhibited technological change in plantation economies. According to Nathan Rosenberg the reluctance of labour in Great Britain to accept factory discipline provided entrepreneurs in the early Industrial Revolution with an incentive to adopt labour-saving technology. According to him the threat of withdrawal of labour and the development of labour scarcities focused the attention of decision-makers on this potential threat to profits. As a result this threat acted as a focusing device and spurred industrialists to change methods of production [*Rosenberg*, 1969].

The plantation economy was an unlikely place to find this kind of mechanism. For one of the foundations of the plantation economy was a close supervisory function exercised by the planter, a function which, in Thompson's words meant that he was 'no mere landlord or farmer; he is also chief of state and head of government.' The means of controlling labour on plantations varied over time, but what was always present was 'the authoritarian character of the institution to a degree exceeding that of the economic enterprises of European and northern United States capitalism' [*Thompson*, 1975: 29, 35].

But not only was the search for productivity improvement constrained in the plantation economy. Even if there had been intense desire for such advances, problems with respect to the availability of such new techniques would have created substantial difficulties. Significantly, however, at least in part these problems of the 'supply' of new technology, too, can be traced to the structure of the plantation economy. In the first place, many of the technological problems associated with tropical or semi-tropical agriculture were difficult. For example, in the harvesting of cotton the developing of a mechanical picker was particularly a problem because the bolls on the cotton ripen unevenly. A mechanical picker was, therefore, required to leave intact and uninjured some bolls, while harvesting others. Similarly in sugar the stalk had to be harvested with little or no damage, often in terrain which constrained the weight of the harvester. Furthermore, the cut in the stalk was required to be clean so as to minimise the loss of the stalk's sucrose content.

While these difficulties presented special challenges, the evidence indicates that they were not decisive in limiting the availability of new techniques; what appears to be even more important than technical difficulties was the only limited extent to which there was an attempt to overcome these problems. Here again we return to the consequences of the structure of the plantation economy. To the extent that the population of such a society was denied alternative opportunities and education, it was not possessed of the skills, opportunities, or incentives to engage in a search for new methods of production. Thus, without even taking into account the question of whether the culture of the planter class militated against scientific advance, the disadvantage of the plantation economy compared to a freer economy with respect to the ability and opportunity to search for and introduce new methods of production is obvious. The very mechanisms by which plantations were rendered commercially viable involved decreasing the proportion of the population motivated and even physically able to undertake the kind of experimentation and trial and error processes necessary to advance production methods. As a result, the admittedly difficult technical problems of harvesting plantation crops could not have been the object of the kind of attention which may well have existed in non-plantation settings.

A second consequence of the slave period relevant to this issue related specifically to the capacities of the planters. These were men who were now, in the aftermath of emancipation, called upon to adopt an entirely new approach to production. Always in the past, masses of labourers, through the institution of slavery, had been the key to the planter's ability to produce profitably. Now under the threat of labour scarcity, one possible response by the planters would be to turn increasingly to science and technology as an answer to their production and profitability problems. To do so, however, was at best difficult and for some planters impossible. Given a long enough period of labour scarcity, some planters, no doubt, would have accomplished this transition. But habits of mind and patterns of life meant that it was a difficult change for all. As a result for most, at least during the early post-bellum years, there was little search for new production methods. Rather the search was for new methods of slavery: to discover new ways by which the traditional pattern of production could be retained.

A variety of empirical measures tends to confirm the hypothesis that plantation agriculture was development-inhibiting. The most direct test of the thesis is a comparison of increases in productivity in plantation-grown cotton with that in other, non-plantation crops. William N. Parker has recently published just such an estimate. It indicates that between the 1850s and the first decade of the twentieth century the productivity increase in cotton was only about a third of that achieved in oats and corn [*Parker*, 1979: 237]. Similarly directly supportive evidence of the thesis concerns increases in per capita income by state. In this series the states in which cotton was most heavily concentrated grew more slowly than did other non-plantation southern states, not to mention the lag they experienced compared to states outside of the region. More indirect indexes of growth such as school attendance and literacy rates and changes in the industrial structure of the labour force also, without exception, tend to support the hypothesis that plantation agriculture inhibited economic growth [*Mandle*, 1978: 58-61].

Implicit in this argument is the view that in the absence of a plantation economy the South might have prospered, or at least grown more rapidly than it did. Parker himself, in 1970, speculated along these lines and concluded that if the slaves had been freed in the late eighteenth century, the South might have rivalled New England after 1860 in economic development instead of lagging behind [*Parker*, 1970: 118]. The assumptions which underlie Parker's analysis are, of course, heroic and as a result his conclusion must be treated with due caution. Nonetheless, this experiment in a hypothetical alternative, too, tends to support the argument that plantations retard development.

It is not absolutely certain that the region would have benefited if it had possessed an alternative agricultural structure to that of plantation farming, though the argument that it would have is plausible. Equally, it is not absolutely certain that with land reform the ex-slaves would have benefited. With land ownership it is possible that all or virtually all of the former slaves might have been commercial failures. If so, they, with the passage of time, would have lost their land either through sales in the market or bankruptcy proceedings. But a more likely possibility is that the pattern which emerged would have been a mixed one. Commercial success would have been present among some of the black farmers, perhaps associated with skill, effort and luck, while failure would also have been present, perhaps associated with the contrary attributes. What might have occurred, in short, is a process of economic differentiation in which, among the black population, like the white population, a class of successful agricultural entrepreneurs might have emerged in concert with a landless agricultural working class.

What can be argued, with some conviction, is that this latter pattern would have been superior to the one which actually came to prevail. For with this kind of differentiation significantly more southern blacks would have experienced rising incomes and wealth than actually did occur. Perhaps equally important would have been the fact that commercial success would not only have been feasible, but, at least in part, would have been the result of the behaviour of the individuals concerned. This stands in contrast to the virtual imposition of poverty which resulted from the functioning of the sharecropping-based plantation economy. Finally, the emergence of a wealthy black agricultural class might have had an impact on the region's politics. Successful black farmers might have been able to challenge for positions of leadership in the region, a challenge which, if successful, might have gone a long way to forestalling the political deprivation which blacks actually did endure in the South during and after Reconstruction.

A programme of land allocation to the former slaves might have been good for them; it might also have been good for the South. As an alternative to the re-imposition of plantation agriculture through the use of sharecropping, land reform might have stimulated regional growth as well as allowing at least some blacks to prosper. To be sure this is all conjectural. But what is certain is that the failure to carry out such a programme doomed both the black population and the South to a century of poverty, the legacy of which we in the United States continue to live with today.

NOTES

1. Ransom and Sutch show that bargaining did not occur over the 50% share, but principally over the number of acres allotted to each sharecropper [*Ransom and Sutch*, 1977: 89].

REFERENCES

DuBois, W.E.B., 1968, *Black Reconstruction in America 1860-1880*, New York: The World Publishing Company.

Fogel, Robert William, and Stanley L. Engerman, 1974, *Time on the Cross: The Economics of American Negro Slavery*, 2 Vols., Boston: Little, Brown and Company.

Genovese, Eugene D. and Elizabeth Fox, 1979, 'The Slave Economies in Political Perspective', *The Journal of American History*, Vol. 66, No. 1.

Harding, Vincent, 1981, *There is a River. The Black Struggle for Freedom in America*, New York: Harcourt Brace Jovanovich.

Higgs, Robert, 1977, *Competition and Coercion, Blacks in the American Economy, 1865-1914*, Cambridge: Cambridge University Press.

Jones, William D., 1968, 'Plantations', in David Sills, ed., *The International Encyclopedia of the Social Sciences*, Vol. 12.

Kuznets, Simon, 1973, 'Modern Economic Growth: Findings and Reflections', in Simon Kuznets, *Population, Capital and Growth*, New York: W.W. Norton & Company.

Litwack, Leon F., 1979, *Been in the Storm So Long. The Aftermath of Slaves*, New York.

Mandle, Jay R., 1978, *The Roots of Black Poverty. The Southern Plantation Economy After the Civil War*, Durham: Duke University Press.

Painter, Nell Irvin, 1977, *Exodusters. Black Migration to Kansas After Reconstruction*, New York: Alfred A. Knopf.

Parker, William, N., 1970, 'Slavery and Southern Economic Development: An Hypothesis and Some Evidence', *Agricultural History*, Vol. XLIV, 1.

__________, 1979, 'Labor Productivity in Cotton Farming: The History of a Research', *Agricultural History*, Vol. 53.

Ransom, Rodger L. amd Sutch, Richard, 1977, *One Kind of Freedom. The Economic Consequences of Emancipation*, Cambridge: Cambridge University Press.

Reid, Joseph D., 1973, 'Sharecropping as an Understandable Market Response – The Post-Bellum South', *The Journal of Economic History,* Vol. XXXIII, 1.

Rosenberg, Nathan, 1969, 'The Direction of Technological Change: Inducement Mechanisms and Focusing Devices', *Economic Development and Cultural Change,* Vol. 18.

Tinker, Hugh, 1974, *A New System of Slavery. The Export of Indian Labour Overseas 1830-1920, London: Oxford University Press.*
06Thompson, Edgar J., 1975, Plantation Societies, Race Relations and the South: The Regimentation of Populations, Durham: Duke University Press.

Woodman, Harold D., 1979, 'Post Civil War Southern Agriculture and the Law', *Agricultural History*, Vol. 53.

The Cycle of Sharecropping and the Consolidation of Small Peasant Ownership in Turkey

Çağlar Keyder*

The prevailing existence of small peasant ownership as the dominant mode of tenure in Turkey invites investigation into the historical genesis of this structure. At one extreme it is possible to argue that the present structure had been more or less homeostatic since the Byzantine Empire, or at least since the classical period of the Ottoman Empire. Alternatively a particular development of legislation may be chosen to constitute the break after which a structure of small property was established. Usually, a theory of linear development is introduced to co-exist uneasily with the conceptual and historical narrative. Depending on the political and theoretical choices of the writer, the linear development may be presented as an ever-growing concentration of holdings, or fragmentation and inability to consolidate holdings due to inheritance practices. Of course, these two tendencies may be brought together and observations accounted for by stating that one or the other was the dominant tendency because of the workings of a factor extraneous to the socio-economic structure.

The interplay between concentration and fragmentation translates, in Turkish history, into a tension between sharecropping and small peasantry. As we shall see, the absence of wage labour meant that concentration – the increase in the scale of production – required sharecropping tenants, whose status as sharecroppers was not necessarily permanent. There are two important observations relating to sharecropping which I will attempt to account for in the argument below. The first is that there are perfectly plausible data indicating the growth of sharecropping practices during certain periods in Anatolian history; and equally plausible data indicating the reverse tendency in the same geographical areas but during different periods. The second observation is that traditional sharecropping has totally disappeared during the period since 1950, and small peasant ownership seems to have been successfully entrenched. Accordingly, we will argue that a cyclical development took place, with the cycle coming to a close in favour of small ownership after the 1950s. As we shall see, the independent variable giving rise to tenurial oscillation was the long cycles of the world economy, and the principal mechanism of transmission was the prices of agricultural produce.

The model of the sharecropping cycle requires a theoretical background deriving from several sources, and I shall first discuss three particular con

**Middle East Technical University, Ankara.*
I wish to express my thanks to V. Brown, S. Pamuk, and the participants in a Durham conference on the rural sociology of the Middle East, for comments and suggestions.

tributions which indirectly constitute major conceptual inputs to the following argument. I will subsequently present the peculiar features of the Ottoman-Turkish social formation that justify the present model. Finally, I will briefly sketch the history of Anatolian landholding to illustrate the historical applicability of such a perspective.

Michel Gutelman, in a book on agrarian structures, discussed the dynamics and the relations of exploitation in what he called the 'parcellary mode of production'.[1] The parcellary mode of production is a term which applies to a structure characterised by a large number of small peasants and a few large owners. Small peasants all work in owner-occupied plots from which a subsistence can barely be extracted. The dominant tendency is the fragmentation of the land because families tend to have several sons, and inheritance customs provide for the dividing up of the land. Thus, at each generational cycle, the land inherited by each offspring falls short of the size required for subsistence. Either, one of the brothers takes over the whole land and is required to pay a compensation to the others, or the family decide to purchase new land from the large landlord in the village to provide a sufficient share for each of the offspring. In both cases, small owners remaining in the village need to pay the price of the amount of land required for subsistence less their share from the inheritance. Gutelman argues that since there are no prior savings, the land is purchased on credit, and the newly setting up family are required to pay the instalments on the principal and the high interest rate on an annual basis. Before the full price of the land is paid back, a new generation comes of age and the cycle recommences. The creditor may be either the landlord who sells the land himself or the usurer. In either case, of course, interest-bearing capital gains importance in the division of the product. Payment for the land, which is regularly subtracted from the output, in fact, constitutes a rent, and thus forms the basis of the relations of exploitation.

Harriet Friedmann's contribution was to point to the importance of the world market as a factor in determining the type of tenure in wheat farming.[2] In a 1978 article, she argued persuasively that the protracted fall in prices during the great depression of the nineteenth century led to a situation where the cost structure of large farms employing wage labour squeezed the agricultural capitalists out of the market. In their stead, family farms were formed and consolidated which had recourse to unpaid and flexibly remunerated family labour. Although it does not seem that these family farms were later subject to a cyclical dynamic following upon the increase of agricultural prices, it is obvious that the world market and the transmission of its tonality through the price mechanism were essential determinants in the increasing importance of family farming. In the model which will be advanced below, I will argue that the cyclical movement of prices determines the pendular motion between independent farms and sharecropping.

Sharecropping is usually perpetuated through a bond that is external to the terms of the tenancy contract. Bhaduri, in an early article, identified the mechanisms of usury as providing that bond.[3] When a debtor–creditor relationship exists between the sharecropper and the landlord, the terms of interest do not allow for the breaking of the tenancy contract and, in addition, they provide the landlord with the additional power to impose his terms. Thus sharecropping continues as long as there is indebtedness, and the landlord's terms ensure that the tenant never disposes of a sufficiently large product to be able to break the cycle of usury. Only if there is a radical change in productivity while the terms of tenancy remain constant (through a new technology, for example) would it be possible for the tenant to retain a sufficient volume of the product (provided that prices did not decline as much as the increase in production) to be able to extricate himself from the bond of usury. Of course, Bhaduri's model of sharecropping has been constructed for the case of the Indian sub-continent, where all land is private property, and, consequently, a landless tenant does not have the choice of expending his labour on land he could reclaim or open up. By contrast, we will be dealing with a situation where access to land is not so restricted and where sharecropping results from a difficulty in continuing with independent farming, rather than from land unavailability.

I shall now turn to the specificity of the Anatolian (Ottoman–Turkish) context in order to argue the cyclical dynamic between sharecropping and independent peasant households.

The history of agrarian structures in Anatolia unfolded against a background of a high land/labour ratio. Anatolian soil was, on the whole, tired and not very fertile; its irrigation was difficult due to an absence of clemently located rivers; as a consequence most of the land was characterised by yields which even in fortunate years of plentiful rainfall could not be considered auspicious. On the other hand, however, there was no shortage of land: reclaimable land at the margin was of the same quality as already cultivated land. Only a very small proportion of land under cultivation was distinguished by higher fertility. Thus, it was always possible to move out of established settlements and start anew with freshly reclaimed land. This is not to suggest that this option was frequently exercised; there are, however, large scale upheavals in Anatolian history – such as the sixteenth-century uprisings – during which entire villages were relocated on new land. For the marginal population who migrated into Anatolia, for newly sedentarised tribes, for peasants driven out of their native villages, the option existed to settle either on freshly reclaimed land or on the outskirts of an already inhabited area.

The population of Anatolia remained stable during the period between the mid-sixteenth and mid-nineteenth centuries. Soil fertility was not conducive to a Malthusian increase in population, nor was there any departure from the traditional technology of a wooden plough plus a pair of oxen. An average household could crop and cultivate an area of around five hectares. This produced a bare subsistence in normal years. If at any time the demographic structure of the family allowed an increase in land cultivated, this situation was sure to be rectified through frequent wars and demands of the Ottoman army which waged campaigns over a vast territory. Given the long-term stability of the population, and the technology, we can easily imagine that agricultural production remained stagnant, except for short-term changes resulting from weather conditions. Labour was the scarce factor in agriculture, and this last statement is perhaps the key to the uniqueness of the Anatolian agrarian structure.

The virtually unlimited availability of land served as an effective barrier against the enserfment of the peasantry, which seemed to possess the ongoing option of setting up independent farms. Furthermore, the fact that private property rights, in the western legal sense, had not been established, implied the preponderant presence of 'state land'. It was 'freehold' that was the exception as all the land that was not actively cultivated, including the marginal land, fell under the rubric of state ownership. But, families or individuals could occupy this land, and, in fact, the system of taxation amounted to a recognition by the state of usufruct rights established through continued cultivation of a plot of agricultural land. Hence, a peasant family could open up and start a farmstead on former state land which, as a result of such possession and cultivation would become taxable land with proper possession rights exercised by the peasant holder.

The Ottoman Empire had taken over and perpetuated an agrarian structure which depended on the presence of a strong centralised state and the absence of serfdom. Under both Byzantine and Ottoman rule a free peasantry and strong state formed the bases of a landholding model capable of defending itself against 'feudalising' tendencies of powerful local lords. To remain strong, the state had to struggle against the centrifugal tendencies of satraps in order to preserve the most stable tax base: the free peasantry.[4]

As is well known, this struggle was not always successful and the free peasantry often succumbed to coercion from locally established state officials. It must be recognised, however, that this local subjugation did not necessarily imply a change in agrarian structure. Often, the change consisted of an increase in the magnitude of the traditional tax. There would be attempts to alter the organisation of production only if alternative markets, crops, or technologies were available and production organised on the basis of peasant holdings was not adequate to reap the gains which might result from a new organisation. As long as the Ottoman state survived, however, such attempts to enserf the peasantry had to overcome legal and institutional obstacles and often active opposition by the bureaucracy. The re-centralisation of power during the first half of the nineteenth century and the forcible dismantling of large estates in the 1830s was the last instance of the workings of this cycle, and the conflict was resolved at the political level in favour of a strong centre. The 1858 land legislation which made the establishment of *çiftliks* difficult represented a culmination of this dimension of re-centralisation.[5]

As has been argued, the immediate consequences of the combination of material conditions and the politico-legal framework (as defined by the impositions of the state) was the unsusceptibility of the peasantry to enserfment, or obversely, the continued existence of an independent peasantry which translated into the absence of a dispossessed peasantry. As is well known, a principal contributing factor to the development of capitalist relations of production in Europe was the dispossession of the peasantry. That is to say, the monopoly over land enjoyed by the lords also meant that the peasantry – when the conditions were ripe – would become 'free' in the sense of not being encumbered by access to or ownership of the means of production. In the Ottoman Empire, however, the peasantry did not become dispossessed until technological change arrived in agriculture proper after the 1950s – and when such dispossession did occur, it came about in a very different way. Hence, there was no possibility for the evolution of capitalist relations of production for the simple reason that the peasantry did not become dispossessed and that consequently wage labour was not available. British capitalists who invested in land in western Anatolia learned this the hard way. Under pressure by the consulates of the Great Powers, the Porte had agreed in 1865 to extend the privilege of private property in land to foreigners. First British and later German investors purchased large tracts with the intention of employing imported technology and native labour in order to produce for export markets. They soon found, however, that there was hardly a supply of labour, and when peasants could be coaxed to accept employment, the wages they had to pay were as high as British levels. Since the soil offered no particular advantages, and a level of technology to make up for the high wages could not be introduced, the would-be plantation owners departed after a few years.[6] Consular reports of the period are replete with their complaints which fail to diagnose their predicament: foreign capitalists had found themselves in a situation akin to the North American case (where wages were high due to the availability of free land), while expecting a situation resembling that of south Asia (where wages were low because the land/labour ratio was low and available land was held under the monopoly of a ruling class of landlords).

The foregoing argument suggests that dispossession of the peasantry as an historical possibility was excluded given the political, legal and material conditions of the Ottoman Empire: thus rendering impossible the development of capitalist agriculture with wage labour. Wage labour, however, is not the only form extending the scale of production. In fact, it is possible for large landlords to increase the scale of production without altering the labour process of the independent peasantry – through sharecropping. Given the discussions proffered above concerning the availability of land at the margin and labour being the scarce factor of production, it can be deduced that a large landlord has to be able to attract families away from their independent farms in order to extend the scale of production. We also know, however, that historically we cannot talk about secular extension of the scale of land controlled by large landlords. That is to say, there must be a mechanism which ensures the resilience of the independent peasantry, the ability to bounce back from sharecropping status. Although data are lacking, it seems that the reduction of independent peasantry to sharecropping status became a widely noted phenomenon only after the increased monetisation of the economy during the nineteenth century. In particular, as the money market extended to villages and as usury came to play an important role in the life of the peasant, sharecropping became a real alternative.

Growing integration into the world market during the nineteenth century provided two different reactions, by the small peasant and by the landlord, which contributed to the emergence of sharecropping. The small peasant, inasmuch as he was marketised, was rendered sensitive to fluctuations in prices. As world market prices moved against his products, he would be hard pressed to find the funds for his necessary purchases. Thus, what used to be a random need for outside support (resulting from crop failure, a wedding or the death of a cow)[7] now became more structured, and dependent on the world market. On the other hand, the large landlord who recognised the opportunity to benefit from growing demand for agricultural products, was in search of additional labour. In a context of relative labour shortage, there had to exist a mechanism acting as a constraint upon the independent peasantry, to transform them into sharecropping tenants.

In the light of the changes which growing integration into the world market brought, let us review the basic premisses of the model. It was argued that land was relatively available, thus excluding monopoly ownership and dispossession of the peasantry; with the added consequence that wage labour was not an extensive development. Also, the specificity of the relations of production in the Ottoman Empire meant that the political authority did not support the attempts by large landlords to concentrate their holdings to the disadvantage of the independent peasantry. We are also making one further assumption: that the degree of marketisation attained implied a sensitivity to world market prices either directly because production was for export, or indirectly because local markets followed the lead of world markets.

The capitalist market is characterised by cyclical behaviour with respect to prices. The periods of prosperity and depression that succeed each other also result in price fluctuations where booms are associated with increasing prices in general and a favourable movement of the terms of trade for agriculture in particular; and depressions imply absolutely and relatively declining prices for agricultural products. These fluctuations of the world market are, of course, independent of individual producers, who only respond to the new economic environment defined by changing prices. The pattern of their response will depend on the prevailing and 'feasible' relations of production. We have already mentioned that H. Friedmann described a fluctuation between capitalist farming employing wage labour and independent petty commodity production based on family labour. This particular fluctuation was determined by the movement of prices before and during the great nineteenth-century depression which implied a return from capitalist wheat farming to family farming because the latter had economic advantages allowing it to withstand the impact of lower prices. We want to address ourselves to the question of whether we can describe a similar, cycle-sensitive behaviour for the case of owner-occupied production versus sharecropping tenancy. The difference would be that while capitalist versus family farming polarity involves different modes of labour organisation, owner-occupied versus sharecropping farming does not. Neither the technology nor the crops change in the transition from the one to the other. Instead, the difference lies in the appropriation of the surplus by the landlord: the rent paid by the tenant to the landlord in excess of the traditional tithe already extracted by the state from the independent peasantry. The argument for the switch between owner-occupied production and sharecropping, therefore, may not rely on the adequacy of the labour organisation relative to the recompensation offered by the world market. In other words, what needs to be explained is neither a change in the mode of subsumption, nor one in the mode of real appropriation.

The long cycles of the world economy translate into changes in relative and absolute prices for agriculture. The assumption of market integration necessitates a degree of price-sensitive behaviour on the part of the peasant producer.[8] This is not to say that independent peasants act in perfectly profit-maximising fashion; there is no doubt that basic subsistence considerations are paramount and accordingly most decisions betray risk-avoiding behaviour. Nevertheless, there will be behaviour changes responding to secular trends in prices, which can be traced to market incentives.

During the upturn of the cycle demand for agricultural goods will increase rapidly, but since the adjustment of supply is much slower, relative prices will move in favour of agricultural producers. At the same time, the price level in absolute terms will be increasing with the consequence of lowering, in real terms, the debt burden of the debtors. Increasing prices are advantageous to debtors as William Jennings Bryan well knew. Thus, in both absolute and relative terms, prices will be providing an incentive for the peasant producer to extend the scale of production: he will want to open up new land, to increase the proportion of commercial crops he produces, to purchase his means of production. To undertake such an extension, the peasant has to resort to borrowing, either from a large landlord or from a usurer in a nearby town. As long as prices continue to increase, the burden of the debt (while probably remaining far above any 'market' rate) will continue to decline; and as long as the terms of trade move in favour of the agricultural producer, the debtor will be able to meet his financial obligations. Thus the decision to expand production, and to increase the degree of marketisation will have been rewarded in the market – for as long as the boom continues.

The cyclical behaviour of the capitalist economy, however, ensures the succession of a boom period with a recession, and a consequent reversal in the fortunes of the agricultural producer. With the depression, prices stop increasing and they may even fall absolutely. At the same time, the terms of trade turn against agriculture, a trend which may last the entire duration of the downturn of the world economy. During such a period, the unfavourable movement of the terms of trade forces the peasant to regress to a lower level of subsistence and retrace his steps back to a lower degree of commercialisation. At the same time, the debts he had accrued during the expansion of production in the previous period, now have to be serviced in fixed money terms, despite the fact that his monetary income has fallen due both to a smaller proportion of marketed output, and to the fall in prices. Hence, it is most probable that the peasant will have to default on his debt, which will force him to sell, at disadvantageous rates, the means of production he had acquired during the previous period.

We must remember that the boom period during which peasant families establish independently on land over which they may have usufruct, also implies a relative shortage of labour for large landlords. Since potential tenant families are able to set up independently and they may borrow from merchants, usurers, or landlords, the supply of labour in the form of sharecropping tenants, is scarce. Thus it is likely that boom periods force large landlords to withdraw temporarily their marginal land out of cultivation. By contrast, during a period of falling prices, when peasant families default on their debts, it is the large landlord who has the wherewithal to purchase the means of production brought to the market by the peasants.

From the point of view of the peasant family defaulting on a debt and thus giving up the option of independent cultivation, parting with the team of oxen and the wooden plough is tantamount to losing their status as independent cultivators. They may continue to enjoy usufruct rights over the land they had been cultivating, but since land is plentiful, and their land is of marginal quality, there would be no way of marketing the fields even if they held proprietary rights. Instead, the peasant family are reduced to a status where they have to become tenants because they cannot pay their debts, or because they have lost the means of cultivating their own land. On the other hand, the large landlord may now expand cultivation to formerly abandoned marginal fields because of the newly available labour. Thus, the proportion of land under independent cultivation and the proportion of independent cultivators in the population decline while the proportion of sharecropping tenants and the ratio of land cultivated through sharecropping increase. The reverse tendency dominates during the boom period: i.e. the proportion of sharecropping declines while production under peasant proprietorship and the proportion of independent cultivators increase. This cycle of reduction into and emancipation from sharecropping tenancy follows the contours of the cyclical development of the world economy, and thus constitutes an illustration of the sensitivity of class relations to prevailing economic conditions. At the same time, inasmuch as sharecropping tenant families continue to undertake production as before – with the same technology and probably producing the same crops – we may consider the shift in status to be no more than the imposition of a new surplus-extracting relationship. That is to say, the independent peasant family, instead of merely paying the traditional tax to the state, under sharecropping tenancy, has to yield a larger proportion of the annual output as rent. In this sense, Gutelman's account of the 'parcellary' mode of production is germane. It is the indebtedness relationship that allows for the additional extraction of the surplus produced by the large landlord. It is also true that the organisation of production remains 'parcellary' in the sense that the 'peasant economy' continues to survive, with the relation of exploitation interfering not during real appropriation, but only after production is completed – in the appropriation of the surplus. From the point of view of the household, the rent in kind is simply another item to be included in economic calculation. Such would not be the case if the transitions were between serfdom with labour-rent and peasant ownership, or between peasant proprietorship and wage labour (as in the case analysed by Friedmann).

According to this model, greater market integration, commercialisation and possibly the expansion of output correspond to the entrenchment of an independent peasantry, while declining commercial opportunity and subsistence orientation imply that sharecropping gains ground against independent peasantry. As long as land is not in the monopoly of large landlords this cycle may continue, but the long-run outcome would have to be in favour of the independent peasantry. If there is at any time a technical innovation which breaks the cycle of indebtedness by allowing the peasant household to capture sufficient surplus, then the balance may permanently tip against the sharecropping alternative. A parallel development may be the greater availability of credit at lower interest rates. In either case, if lower prices during the downturn of the cycle do not force the peasant family to default on their debt, they will be able to avoid selling their means of production and, consequently, could continue independent production. They will thus become part of the category in the middle; between the landlord who rents land and the poor peasantry who are subject to the cycle of sharecropping. These middle peasants, because of the quality of their land, their demographic composition, or because of having accumulated more than a single plough and a team of draft animals, are able to weather the downturn of the cycle. They remain independent cultivators during boom and crisis. Such a perspective indicates that a higher proportion of the peasantry are incorporated into the 'middle' category as a result of technological change or greater credit availability.

On the other hand, with technological change, the large landlord's need for labour will also diminish and he will neither leave his fields uncultivated in the absence of tenant farmers, nor will he exact sharecropping status if he has debtors. In fact, these developments coincided to break the cycle of sharecropping in Anatolian agriculture after the 1950s. Below, I shall suggest a possible application of this model to Anatolian agriculture during the hundred years before 1950, and from 1950 until the end of the post-war boom period.

For the Ottoman Empire as a whole, the standard Kondratieff cycles provide a good indication of the rhythms of agricultural production and incomes.[9] European purchases of Ottoman exports followed the same trend line as changes in total output and volume of world trade which provide the contours of Kondratieff cycles; and the terms of trade for agriculture in general and for exports in particular also behaved accordingly. That is to say both prices and the value of exports (corresponding to agricultural prices and the volume of marketised agricultural output) moved with the cycle. During 1850-1873, and 1896-1926 the trend line rose; for 1873-1896 and 1926-1946 it stagnated or declined. In the latter two periods agricultural prices declined absolutely at the same time as terms of trade were moving against agricultural producers. We may, therefore, expect a development in favour of the independent peasantry during the Victorian boom, and again before and after the First World War, while the relative share of sharecropping tenancy would increase during the great depressions of the nineteenth and twentieth centuries.

The purchasing of already cultivated land from large landlords especially by Greeks and Armenians during the Victorian boom, becomes understandable in this context. Faced with increasing opportunities in the export market, first Christian minorities then foreign passport holders made an attempt to establish commercially oriented family farms, mostly in the Aegean littoral. It was in the 1850s and 1860s that the production of labour-intensive (therefore family scale) crops such as raisins and figs increased.[10]

We may conjecture that exporting merchants played an important role during this wave of the extension of small cultivators. Since credit was made available on the basis of future output, especially Greek peasants would have borrowed in order to shift or extend cultivation. Ultimately the origin of these funds was European financial markets; and the collapse of the funds market after 1873 (as reflected in the cessation of credit inflows into the Ottoman economy in general) also implied a shortage of credit for peasant growers. The increase in lands under peasant ownership stopped, and did not receive a similar boost until the next boom period.

The major occurrence of the next boom period, from the point of view of the Anatolian peasantry, was the construction of the German-financed Anatolian Railroad. The railroad opened up large tracts of previously uncultivated land and helped immigrants and volunteers to settle on it.[11] Such new settlements were in the plains of the interior, climatically suited to the cultivation of the traditional Anatolian wheat. Especially in the case of immigrants, land was distributed evenly among the families, contributing to a large-scale peasantisation of the interior. For earlier settlements, the railroad served to consolidate the independent peasantry by extending the market. In fact, during this period around the turn of the century, Anatolian growers began to account for an increasing proportion of Istanbul's grain needs (traditionally a market in which imported grain dominated) and the suppliers were peasants depending exclusively on family labour.

There was another development later in the period which was even more significant. Greeks and Armenians had increased the land under their control during the previous depression with consequent resentment and complaints on the part of the Moslem population. This sentiment, however, only became vocal during the boom. Then, in a succession of purges, Christian minorities were driven out or eliminated, with the land reverting mostly to the control of the Moslem small peasantry in the same locality. The final purge of this nature occurred during and after the Turkish war of independence. Between 1922 and

1924, an estimated two million Greeks (more than two-thirds of rural origin) left Anatolia and their lands were taken over by the Moslem population.[12] It seems that although the mode of appropriation of these lands varied from one locality to another, one concern was constant: those who had lost family members in the wars (i.e. widows, orphans) and those whom the community judged to be needy, always received some land.

In fact, at the end of the war period, emigrations and the perishing of a large proportion of the able-bodied male population raised the land/labour ratio to a new high. The new government attempted to respond to this situation by passing a law which required surviving males to till the land of the widows, orphans, and the disabled. Government agencies were also required to contribute labour from among their employees. At the same time, however, commercial opportunities were increasing, and especially after 1925, when new Turkish banks entered the scene, credit for export-oriented agriculture was plentiful. A larger source of credit for the peasantry was merchant's credit, or funds advanced to growers on the basis of future contracts. This last source of credit expanded, as did trading activity in general, and by 1927 observers were commenting that the commercialised peasantry had overextended themselves by having borrowed too much.[13] It may safely be stated that neither the pre-war nor the post-war period was conducive to the extension of sharecropping.

The first signs of the coming crisis were felt during 1926, after which agricultural prices began to decline and the defaulting of many borrowers resulted in the seizure of their land and animals by banks and usurers. An additional difficulty for the Anatolian peasantry was that the traditional tithe which was proportional to physical output had now been replaced by fixed money taxes assessed on the value of land and number of animals owned. These taxes suddenly became weighty burdens when the value of output declined, and it was common for the peasant to have to sell all his belongings to be able to pay his taxes. In the absence of urban employment, finding a sharecropping position was probably considered desirable. In fact, throughout the crisis period, there was an extension of large ownership, sharecropping, and economic dependency relationships between the small peasantry and the landlords. Much of the 'village literature' of Turkey is a legacy of this period when the *agha* (the large landlord with poor tenants who are bound to him through ties of usury) had at his disposal all the economic and political means to exploit the small peasantry. Agricultural output stagnated in the absence of commercial opportunity; and marketised output declined absolutely. Although, after 1936, the volume of production and area under cultivation began to increase at around the same rate as the growth of population (and the labour force in agriculture since there was no urbanisation during those years) a primarily subsistence-oriented production pattern became more prevalent. The internal terms of trade served to penalise the farmers as the government sought to transfer a surplus from agriculture into industrial investment. During the war years military mobilisation and confiscation of farm animals created further difficulties that were exacerbated by the forced levy on the output. Thus, the entire period between 1929 and 1945 may be considered one of agricultural stagnation, and even regression from the perspective of agriculture's integration with national and world markets.

It was not until the end of the Second World War that major transformations began to occur which eventually fixed the predominantly peasant-proprietor character of the agrarian structure. Peace had signalled the establishment of American hegemony over the 'free world', and Turkey was in an especially privileged position to take advantage of this change. The European Recovery Programme designed to reconstruct the economy of Europe was extended to cover Turkey and to develop its economy along the envisaged division of labour in the world market. American experts preparing reports on the Turkish economy diagnosed a shortage of capital and lack of infrastructure blocking the development of agriculture. Until the mid-1950s external funds were directed predominantly towards the rural sector: aids and grants from the US contributed greatly to the fulfillment of technological change in Turkish agriculture as the world economy as a whole was expanding under American hegemony.[14]

The bulk of these funds was used to import tractors, such that the number of tractors increased from 1,750 in 1948 to 40,000 in 1955. A not inconsiderable sum was allotted to the building of a road network, designed especially to integrate the agricultural areas with production potential to the main cities. This development, of course, served to provide every peasant with easy access to urban markets. With the indication that this production potential in agriculture could be unleashed, the political opposition to the one-party regime that had ruled Turkey since independence took the form of a resentment by the agrarian population of the difficult decades of the 1930s and 1940s. In effect, 1950 witnessed the replacement of the Republican People's Party by the new Democrat Party, which intelligently identified itself with the new dynamism in agriculture, and based its campaign on a platform of 'enriching the peasant'. Thus a political dimension was added to the economic boom such that state policies (and credit policies of the semi-official banks which reflected them) during the decade of the 1950s quite consciously favoured the agricultural sector, and had the impact of increasing the incentive for the extension of peasant farming.

The economic boom, especially after the euphoric years of the early 1950s, gradually became an urban phenomenon as well and this development opened the gates of the big urban centres to rural migrants. The phenomenon of the shantytown entered the Turkish scene after 1950. From our point of view, availability of urban labour markets to rural migrants signifies a development through which a new variable acts as a safety valve to introduce an alternative source of income to the peasant household. Through this mechanism many peasant families found the wherewithal to sustain petty production and to remain as owners of property despite economically unviable situations.

The most important development from the point of view of the consolidation of peasant property, however, was the opening up of vast tracts of land for cultivation: between 1948 and 1955 land under-cultivation increased by 50 per cent. This growth was well in line with the model we have advanced above, except that it was much more rapid and more extensive compared to earlier booms. The reason, of course, lay in the availability of tractors and newly constructed roads. The area reclaimed was state land, previously designated as 'meadows and pastures'; however, it was not only the rich landlords who could purchase tractors that took the opportunity to increase their holdings. First of all, tractors were not simply sold on the market: in most cases they were purchased on credit. In a 1952 survey, 93 per cent of households owning agricultural machinery were found to have financed their purchases by credit amounting to 60 per cent of their outlay.[15] Political patronage mechanisms played as much role as economic ability in the allocation of credit. Besides, credit had become more readily available as the Agricultural Bank pursued a new policy under the new government and increased its loans tenfold between 1948 and 1958. The geographical distribution of the tractors indicated that their incidence was rather low in regions most characterised by landlord regimes, while regions of market-oriented small farming received a disproportionately large share. Secondly, the reclamation of land was not carried out unilaterally by politically powerful tractor owners. The village community frequently became involved in the decision-making process, and the reclaiming process itself led to a chain of events whereby other peasants benefited indirectly. To illustrate the chain of events I shall piece together the scattered evidence in order to obtain two alternative and representative scenarios of land reclamation. The first variant applies to villages with a relative preponderance of sharecropping tenants. In those villages the distribution of land was uneven due to the low number of 'middle peasants' who survived the depression with their holdings and draught animals intact. The majority of the peasants were poor and were bound to a richer landlord by ties of usury and/or sharecropping, with no likelihood of escaping their situation until the reversal of the economic conjuncture. When the economic conjucture was reversed and the landlord (or landlords) in the village reacted by purchasing a tractor, the new technology was used to replace the sharecroppers or to take over the better land cultivated by the poor peasants. In other words, the landlord put his tractor to use primarily as a means of ploughing and cropping his own property or the land he had recently seized under some legal pretext. The response to the loss of tenant status by the sharecroppers, or to the loss of better quality land by poor peasants, was, either collectively or under the authority of the village government, to take action to reclaim land from what until then had served as the village commons. Thus, the state-owned marginal land which had not been under cultivation until then came under the possession of former sharecroppers who had to borrow funds in order to invest in land preparation and in draught animals. It is probable that some of these funds served to pay the former landlord for the use of his tractor during the reclamation process. Given the expanding supply of credit mentioned above, the establishment of owner-occupied farms was not constrained through a scarcity of funds. Furthermore, the uncertainty associated with occupying state land was soon rectified as de facto possessed land was formally distributed with proper titles. 'Land distribution commissions' established in accordance with the diluted land reform act of 1946, travelled from village to village to formalise this transfer of newly opened state land to the least propertied peasants. Through this scheme more than 400,000 families were granted more than 3m hectares of land.[16]

The second type of opening up occurred in villages with a more even distribution of land and a greater proportion of 'middle peasants'. Such 'middle peasants' with political connections, or sometimes merely entrepreneurial spirit, could obtain bank credit towards the purchase of a tractor. Most commonly two or more peasant households joined together to make the investment, and then agitated at the level of the village government to reclaim land from the state-owned commons. Once again the occupation of state land was eventually formalised through cadastral ratification, but since there were relatively fewer sharecroppers driven out (i.e. fewer landless peasants) most of the village residents received a share of the new land. It is also probable that the tractor owners received a better share as compensation against the use of their machines in the reclaiming of land. Contrary to more simplistic accounts, rich peasants who purchased tractors could not unilaterally seize the commons land.

In both models, the extension of small peasant production was the dominant tendency. The number of owner-occupied farms increased from 2.3m in 1950 to 2.5m in 1952 and to 3.1m in 1963.[17] Yet this development was effected fairly painlessly, that is to say, without necessarily requiring heavy borrowing. In most cases tractors became available for hire, reducing the dependence on oxen, and the price of oxen did not increase since tractors effectively substituted for animal power. With the easier availability of credit, borrowing of funds for extension of production in the usury market declined. Migration to the city also reduced the pressure on loanable funds (and on land) by removing from agriculture those peasants who would have borrowed to establish their own operation, and began to constitute a new source of funds as remittances were sent back to relatives in the village.

From the very first years of tractor use, there was a developing rental market allowing small owners to cultivate their land and obtain a return on it even if sufficient labour was not available due to demographic cycles or to employment elsewhere. An especially important dimension of mechanisation in this regard was the introduction of mechanical harvesters, mostly in the form of combines. Given the size and intensity of cultivation, harvesters always served more than the needs of a single farmer. A single harvester usually served to harvest the grain in several villages, depending on roads and the terrain. By 1955 there were 5,615 combines in Turkey and a lively rental market for their services, which drastically reduced the peak labour demand.[18] Due to the existence of this rental market, a small owner could now supplement his agricultural income through seasonal work in an urban area while continuing to cultivate his fields. Because of this availability of non-agricultural sources of income, he would be less vulnerable to the cycle of agricultural earnings.

The introduction of tractors and other labour-saving machinery constituted the requisite technological change which broke the cycle of sharecropping. Not only did tractors and harvesters decrease the demand for labour by large landlords, thus eliminating one of the constituent factors in sharecropping but they also allowed for an unprecedented extension of cultivation and output to serve to consolidate small peasant property. Given the definition of the 'middle peasant' we proffered above (as the independent owner who disposed of sufficient accumulation to escape the cycle of indebtedness and sharecropping) a large proportion of previously poor peasants gradually entered the ranks of the middle peasantry during the 1950s.

These developments of the 1950s continued into the 1960s. Additionally, fertilisers and other agricultural inputs were increasingly subsidised by the government, making the adoption of new technology easy for small owners. In fact the widespread availability of tractors and combines in rental markets in large part severed the link between scale and use of technology. There was a very small difference, if any, between small and large owners on similar land, as regards the productivity of land and labour.

The reclaiming of land and its distribution; political favouring of the agricultural sector; the spread of mercantile and communications networks and of new technology contributed to the consolidation of peasant ownership. Through formalisation of distribution of new land and a liberal credit policy the state contributed to this consolidation. This consolidation, in turn, compelled the state to be responsive to the needs of small farmers. Because peasants did not borrow heavily in usury markets they were far less vulnerable to the impact of falling prices, and the integration of the national labour market (later with Europe as well) allowed access to an alternative source of revenue.

After the 1950s' transformation of Turkish agriculture relevant questions no longer concerned the sharecropping–small ownership dynamic; rather given the continued relative absence of capitalist farming, differentiation and typology of transformations undergone by the small peasantry constituted the more important puzzles.

NOTES

1. M. Gutelman, *Structures et Réformes Agraires, Instruments pour L'Analyse,* Chapter 4, 'Le mode de production parcellaire', Paris, 1974.
2. H. Friedmann, 'World Market, State and Family Farm: Social Bases of Household Production in the Era of Wage Labour', *Comparative Studies in Society and History*, vol. 20, no. 4, (1978).
3. A. Bhaduri, 'A Study in Agricultural Backwardness under Semi-Feudalism', *Economic Journal*, vol. LXXIII, no. 329, (1973).
4. I have treated this theme in two earlier articles: 'The Dissolution of the Asiatic Mode of Production', *Economy and Society*, vol. 5, no. 2, (1976); and 'Agenda for Ottoman History', *Review*, vol. 1, no. 1, (1977), with H. Islamoglu.
5. 'Çiftlik' is the term applied to commercial estates which began to emerge in the seventeenth

century, and catered to the growing market in Europe. These involved, especially in the Balkan lands, a radical change in the status of the peasantry, who were reduced to virtual serfdom through the new organisation of production. In the narrower sense, 'çiftlik' referred to a single individual registering the lands of an entire village in his own name. It was this latter practice that the 1858 law tried to combat. For earlier çiftliks see T. Stoianovich, 'Land Tenure and Related Sectors of the Balkan Economy, 1600-1800', *Journal of Economic History*, vol. XIII, no. 4, (1953); for the 1858 legislation see Ö.L. Barkan, 'Türk Toprak Hukuku Tarihinde Tazimat ve 1274 (1858) Tarihli Arazi Kanunnamesi', in his *Türkiye'de Toprak Meselesi*, Istanbul, 1980.

6. O. Kurmuş, *Emperyalizmin Türkiye'ye Girişi*, Istanbul, 1974, pp. 99-115. Ş. Pamuk brought this point to my attention.

7. Cf. K. Marx, *Capital* III, New York, 1968, p. 599. 'The mere death of his cow may render the small peasant incapable of renewing his reproduction on its former scale. He then falls into the clutches of the usurer, and once in the usurer's power, he can never extricate himself.' I have treated credit and usury in an earlier article, 'Credit and Peripheral Structuration: Turkey in the 1920's', *Review*, vol. III, no. 4, (1980).

8. I am not ignoring the extensive literature on the market behaviour of small peasantry. My view is that even if the risk-taking behaviour of the peasant reveals considerable inertia, there *is* a degree of sensitivity to market proces. Cf. J. Scott, *The Moral Economy of the Peasant*, New Haven, 1976.

9. See Ş. Pamuk, 'Kapitalist Dünya Ekonomisi ve Osmanli Diş Ticaretinde Uzun Dönemli Dalgalanmalar, 1830-1913', *METU Studies in Development*, 1979-1980 Special Issue.

10. See F. Rougon, *Smyrne,* Paris, 1892, Chapitre 3.

11. Ş. Pamuk, 'Commodity Production for Export and Changing Relations of Production in Ottoman Agriculture, 1840-1913', forthcoming in *The Ottoman Empire in the World Economy*, eds. S. Faroqhi and H. Islamoğlu-Inan.

12. See my *The Definition of a Peripheral Economy: Turkey 1923-1929*, Cambridge, 1981, Chapter 2.

13. Ibid., Chapter 5.

14. M. Singer, *The Economic Advance of Turkey*, Ankara, 1977. This book includes a useful overview of the agricultural transformation during the 1950s, and contains most of the official figures.

15. *Türkiye'de Zirai Makinalaşma* (results of a survey conducted by the Ankara University, Faculty of Political Science), Ankara, 1954, pp. 119-120.

16. DIE (State Statistical Institute) *Zirai Istatistik Özetleri, 1940-1960*, Ankara, 1965.

17. *1950 Ziraat Sayimi Neticeleri,* Ankara, 1956; and *1963 Genel Tarim Sayimi Ornekleme Sonuçlari*, Ankara, 1965.

18. DIE, *Zirai Istatistik Özetleri*, op. cit.

Tuscan Sharecropping in United Italy: The Myth of Class Collaboration Destroyed

Desmond Gill*

THE MEZZADRIA SYSTEM

When Italy was unified (1861) Tuscan agriculture was dominated by sharecropping. This is true of all provinces except that of Grosseto in the south west, which insofar as it was cultivated at all was more or less capitalist.[1] It will be easier to understand the impact that the development of capitalism had on the system if we first make a necessarily cursory examination of it in its pure, or 'classical' form.[2]

It is clear that some kind of sharecropping existed in Italy at the time of the Roman Empire [*Rouveroux*, n.d.: 17-24]. But the earliest evidence of Tuscan *mezzadria* seems to be that contained in contracts dated 818 AD and 821 AD [*Imberciadori*, 1951: 38]. The fact that only two contracts have been discovered for the ninth century should not, of course, lead us to think that the system was necessarily uncommon at that time. But it was not yet dominant.[3]

From about 1000 AD the demographic increase which affected Tuscany as much as the rest of Europe provided a need for more food. Land reclamation took place to meet this, particularly in the Arno Valley and on the Serchio plain. The existing infrastructure of small farms was extended and deepened during this period. *Mezzadria* contracts found increasing favour. When the Black Death struck the region in 1348 causing stasis, *mezzadria* had become by far the most common farming method.

Sharecropping was suited to the social, political and economic conditions of the day. The newly rich merchants of the communes were attracted by the modest but secure investments this provided for surplus commercial profits, and their share of the agricultural produce was consumed by their enormous households. Since the sharecropper had a direct interest in productivity, the landowner's ignorance of farming techniques was held to be of little importance. The necessary land market was created largely by the freeing of the serfs in the latter half of the fourteenth century. This deprived the magnates of their traditional labour force. Soon the magnates too, as well as religious institutions which owned vast tracts of land, recognised the advantages of returns in kind. For his part, the *mezzadro* had a holding which did not require any outlay of capital from him, and like the landowners he wanted a variety of subsistence crops rather than cash crops.

People controlling a number of farms soon began to group them together in

* Centre for the Advanced Study of Italian Society, University of Reading.

The author wishes to thank the following people for their help: Dr. Paul Corner, Director of the Centre; Deborah Gill; Martin Minogue, School of Administrative Studies, University of Manchester; Dr. Rex Taylor, Medical Sociology Unit, Aberdeen.

estates, with a central processing point consisting of vinification equipment, olive presses and the like. It was here that business between the two parties was conducted, and the position of the sharecropper's account communicated to him. He was not allowed to keep a copy of this himself.

The landowner's control over the system was ensured by the contractual term that the labour capacity of the *mezzadro* and his family should match the labour requirements of the holding. Conversely, the amount of land should be just adequate for what the landowner considered to be the family's needs. Because the family operated a limited amount of land, they could not amass enough saleable surplus to achieve any degree of independence. At best they could make enough cash to pay their 50 per cent share of the modest running expenses. These conditions gave the landowner the right to control the details of the sharecropper's private life. He gave or withheld permission for members of the family to get married, checked their physical condition and moral conduct, and could order the partial dispersal of the family or the adoption of living-in help, in order to ensure that the family's labour input equalled but did not exceed the requirements of the farm.

The structure of the sharecropping family was a patrilocal, (single) multiple family household.[4] In order to meet the labour requirements of the holding, a family was often extended into a *frèrèche*.[5] This last arrangement was usually temporary: if the family became too big for the farm one or more brothers could be expelled with his wife and children. The other method of adapting the family's working capacity to the requirements of the farm was to take on a living-in hand called a *garzone*, who could on occasion be completely integrated into the family through marriage.

The paterfamilias was the *capoccia*, whose word was rarely challenged. He governed the entire family, often including his own father, had the latter passed his active prime. The father would have been *capoccia* in his turn, and passed the position down to his son when he himself saw fit, not when the rest of the family thought it was time for the succession to take place. If the 'natural' successor, the eldest son, was felt to be incapable of managing the responsibilities which the position entailed, another member of the family could accede, either by appointment or by election by the 'enfranchised' members of the family. The landowner was entitled to vet the appointment of a new *capoccia*, who therefore depended on him for his supremacy. The head of family was to some extent, then, the landowner's agent. Families of thirty were by no means uncommon.

The other key family figure was the *massaia*. She handled a small cash allowance granted to her by the *capoccia*, for the purchase of goods not produced on the farm, which in the case of the better-off families included luxuries. She was the wife of the *capoccia*, or if the wife was not considered by him to be competent, or if she did not exist, another woman in the family. In the *frèrèches*, the *capoccia* was one brother and the *massaia* the wife of one of the others. This was considered a fair decentralisation of power.

The landowner could evict the *mezzadro* without notice or explanation. The *mezzadro* committed a criminal offence if he left the farm of his own accord, since to do constituted breach of contract. The landowner, then, had virtually absolute control over the sharecropper. He normally placed this control in the hands of a salaried bailiff, who was not a trained agronomist but served as an agent charged with the task of making sure that the sharecroppers did not retain more than their authorised share of the product. His function was considered necessary because the landowners tended to be absentee, spending most of their time off the estate engaged in a profession or commerce. This system, predominant in Tuscany by the eve of the Renaissance, continued well into the period of United Italy, when the advent of capitalism had certain effects on it which will now be described.

CAPITALISM: ITS CONSEQUENCES FOR TUSCAN AGRICULTURE

Capitalism had both technical and political effects on Tuscan agriculture. During the years of Hapsburg physiocracy, landowners in Tuscany had enjoyed low taxes. The new nation, as part of its attempt to participate in western capitalism, wanted an industry-led economy, and in order to encourage this imposed the burden of taxation in large measure on agriculture. Mazzini, dealing with Tuscany as part of the Jacini enquiry into agricultural conditions throughout the kingdom, points out that although there had been an overall rise in prices during the past half century, almost two thirds of the gain which this brought to those engaged in agriculture was cancelled out by the progressive increase in taxation since unification [1881: 367-8]. He lays great stress on this, returning to it several times, and sums up his feelings [1881: 423] by saying that although individuals were probably not being ruined by taxation, agriculture itself was suffering enormously, because the people responsible were withholding investment from it and giving in to the more tempting promises of industry and commerce. They wanted to sell their lands and leases but were unable to find buyers.

Furthermore, at this point, around 1880, the price rises were more or less at an end, but taxes were not reduced significantly. For the four years 1858-61 (state, province and commune taxes inclusive), on 100 lire of estimated taxable revenue the figure had remained at 27.72, the highest it had been in recent decades (except for a single year, 1849, when it had been 27.74). On unification the tax immediately began a steep rise. It increased every year, reaching 64.19 in 1878 [*Mazzini*, 1881: Appendix 26]. Admittedly we must set against this Mazzini's calculation [1881: 368] that the estimated revenue was roughly half the real revenue, since the estimates were based on the old Hapsburg land register,[6] but it remains true that the position of people involved in agriculture was deteriorating. Jacini's statement in the final report to his commission's findings sums up the tax situation: he points out that on 1,000,000 lire of income 300,000 were paid in direct taxes alone, not to mention the innumerable indirect taxes, such as the grist tax [*Radi*, 1962: 30].

Italy's attempt to achieve an industrial takeoff involved the adoption of import tariffs. Like the fiscal policy, these were designed to nurture infant industry. We must assess the effects thay had on Tuscan agriculture.

Wheatgrowers managed to get a high protective tariff. They needed one. The development of steam shipping and the opening up of the fertile, easily worked heartlands of North America brought a flood of cheap, good quality grain on to the European market. This exacerbated the effects of Russian and Turkish competition, which had been forcing Italian domestic prices down for some time. In 1878 wheat fetched 38.11 lire per quintal, which fell steadily to a low of 22 in 1888. Imports during the 1870s had varied little, remaining around 300,000 tons, but in the 1880s they rose sharply to peak out at over a million tons in 1887. The barrier against this onslaught was raised from a basically fiscal 1.4 lire per quintal imposed as part of the 1878 general tariff, through 3 in 1887, 5 the following year, to 7.5 in 1885, which was the highest wheat tariff in Europe.

This tariff was certainly an advantage to wheatgrowers in the Po valley and the south, but how did it affect Tuscan wheatsellers? Wheat was not an unimportant crop: the sharecroppers grew it because it was the basis of their diet. But other factors also have to be considered. Firstly, the Tuscans had already begun to reduce their wheat acreage [*Mori*, 1955: 47-48]. Secondly, of course, no more than half of the wheat produced by sharecroppers was available for marketing. Thirdly, yields were poor, although nowhere in Italy was output impressive. Britain at this time produced an average of thirty-two quintals per hectare, other countries over twenty. Jacini calculated Italy's average to be about eleven. The highest productivity was in the province of Milan with fifteen quintals. In the south (with a few exceptions such as the province of Foggia, which produced 14.5) it was generally 6-9 quintals. Mazzini [1881: 258-302] gives productivity tables for a number of farms scattered all over Tuscany in zones of varied terrain. Examining the wheat yields of the sharecropping farms in those tables, we find that the average is 6.4 quintals per hectare. The range of farms makes it certain that this was typical of Tuscan sharecropping at the time, and so we now conclude that productivity there was among the lowest in Italy. Therefore, the only agricultural product which received a massive protective tariff was one of limited importance in Tuscan accounting.

Italy's protectionist policy involved her in a commercial war with France. This had important implications for Tuscan wine production. Wine was unprotected, and Tuscany faced a problem similar to that of growers in the south. In Tuscany, as in the south, wine was one of the few sectors which had shown a little capitalist dynamic. In 1868 Tuscany had produced 1,500,000 hectolitres, five per cent of the national product. Only Lombardy and Sardinia had produced less in that year. The average for the quinquennium 1889-93 was 3,146,917 hectolitres, an output exceeded only by Sicily and Apulia [*Pinzani*, 1963:86]. Mori [1955: 488n.] publishes the following table:

Tuscan wine exported to France

1881-86 (average p.a.)	1,500,000 hl
1887	2,800,000 hl

First 2 months 1888	1,000,000 hl
1889	1,439,000 hl
1890	936,000 hl

A fact to be borne in mind is that the tariff war with France began only at the end of the first two months of 1888. While the figure for those two months is probably inflated because France, in anticipation of the tariffs, was stocking up, the subsequent fall-off in exports is clear from this table. Although new outlets were to be found later, there is no doubt that this sector, the only one in which performace had improved, was damaged by the loss of its French market. This should not be exaggerated: winegrowers in the south were ruined at this time, wholly dependent as many of them were on trading with France, but in Tuscany the growth had been contained, not only by the conformation of the terrain and the general reluctance to modernise on the part of a large proportion of landowners, but also by the sharecropping system itself, which inhibited the development of monoculture. Nevertheless, the detrimental effects took place at a time when the region could not afford further difficulties.

Protectionism, then, while its effects should not be overemphasised, increased the problems of a Tuscan agriculture already in crisis, and did not offer aid in specific sections to cushion the blow.

The answer to the setbacks resulting from the economic policies of the new nation could have been to adopt a dynamic capitalist approach, as the Po valley farmers did with such success. But in Tuscany this course was not taken, for a number of reasons. First, with the best will in the world there was a limit to what the landowners could do in this direction, because of the natural conditions of the region. Mazzini describes them in the first chapter of his report. Well over half of Tuscany is hilly, rainfall and river flow are seasonal, and the soil is seldom good. Cold winds are frequent, and tend to occur in spring, when growth is tender, or in autumn, when fruit is ripening. Winter frosts can damage olive trees so badly as to make them unproductive for several years. There are violent hailstorms capable of wiping out entire grape or olive harvests in ten minutes. The irony is clear: because of the terrain the only products suitable for modernisation, wine and olives, are precisely those at greatest risk from the weather. Tuscany is simply not suited to high-cost agriculture.

Perhaps the most important consideration limiting the degree to which agriculture in the region was modernised was political. The landowners had clear examples in areas outside Tuscany of the power of an organised proletariat and the dangers of class division. In the Po valley the first big strike of wage labourers took place just before the 1882 corn harvests [*Radi*, 1962: 64]. The consensus of opinion among Tuscan landowners was that the sharecropping system represented the best way of avoiding such problems.[7] They argued, and many of them believed, that the system was a partnership between capital and labour: since both parties stood to gain from this partnership in equal proportions, there was no class conflict inherent in *mezzadria*, the argument ran. In spite of contrary advice from many well-qualified observers, such as Jacini himself in his final report, the social order was retained. Mazzini

[1881: 456] gives the proportions of the various categories of rural workers at the time of his report. Over half were sharecroppers, and less than a quarter were wage labourers. The rest owned the land they worked or controlled it through rent or lease. Twenty years later, according to the 1901 census, 53.2 per cent were sharecroppers and 24.7 per cent were wage labourers. Since the census included the province of Grosseto, which Mazzini did not, it would seem that if anything there was an increase in the proportion of sharecroppers during the period under consideration, since Grosseto province had a higher percentage of wage labourers than the rest of Tuscany. To maintain sharecropping was against the national trend.

The landowners, then, by and large refused to be drawn into the capitalist mode of production. There were limited exceptions. The case of wine production has already been referred to, and some industrial crops were introduced as well, such as tobacco and sugar beet.[8] The limitations were imposed partly by the terrain and climate, but also by the landowners' fear of the social consequences of capitalisation. But the majority who did not modernise could not hope to compete with the new market forces, and the attractions of agriculture for them diminished rapidly. They did not try to renew it even according to its own logic, and so it soon reached the point of collapse.

In Chapter 20 Mazzini describes the farming techniques almost universally practised. The sharecropper thought only of his family's short-term requirements, not of the long-term effect of his cropping patterns. Wheat was often grown on the same patch for two years or more. When it did form part of a rotation the alternating crop was generally another cereal, such as maize. Rarely were pulse crops introduced, which, being nitrogen fixing, would have benefited future cereals.[9] Fertilising was connected to livestock rearing. For the period 1850-80 records show an overall fall in livestock numbers [*Mori*, 1967: 152]. There was some expansion between 1875 and 1879, but after that there was stasis [*Pinzani,* 1963: 88]. In spite of this increase the stocking rate was low. Mazzini explains that nowhere was there more than one cow per hectare, and usually there was only one per two or three hectares. Generally these animals provided all the fertiliser that was used. It was stored in the open, as a rule, and therefore subject to leaching. Only in a few areas, such as around Lucca, was it supplemented with bought-in materials, such as butcher's remains, and the more thoughtful farmers used spent grapes, the dregs of olive oil, and the family's own natural wastes.[10]

Implements were primitive. Iron ploughshares were to be found, but so were wooden ones, and the spade was widely used.[11] A depository of threshing machines in Florence, a hundred years after the machine had been invented, was unknown to most people and ignored by virtually everybody.

The soil had been exhausted by thoughtless techniques, past investments had been overexploited, even on some of the estates where a degree of modernisation had taken place,[12] and fresh injections of money could not in the circumstances be expected. In the late 1880s powdery mildew struck the vines [*Radi*, 1962: 25]. Before they could recover, phylloxera, which by wiping out so many French vines in the 1870s had given Tuscany a chance for partial development, had worked its way down Italy, reaching Chianti in the mid-1890s [*Pinzani*, 1963: 87]. Thus, the one sector which had shown itself to be capable of at least some economic dynamism was gravely damaged, and at the end of the century the region's agriculture was in a dire state indeed.

The final aspect of the effects of the advent of capitalism which ought to be mentioned here was the result of the limited acceptance of capitalist methods discussed above and stressed by Snowden (see notes 8 and 9). These developments had considerable implications for the politicisation of the sharecroppers: they were subject to a degree of proletarianisation which took various forms.

The opportunities for specialisation in Tuscany were greater than in the other areas where sharecropping was still the dominant tenurial form, Umbria and the Marches, because in Tuscany large, centralised property was more common. The emphasis began to shift slightly from the individual work of the sharecropper to the group activity of the estate. Olive presses and vinification equipment were installed at the central processing point or, where they already existed, were replaced with models giving a bigger output. This development is quite clear in a number of contracts. For example, the one of 1848 on the *Fattoria dei Piani* in the Bisenzio valley, has a clause simply obliging the sharecropper to consign to the landlord's house the stipulated share of goods. Contracts drawn up later in the century usually have clauses pertaining to the use by the sharecropper of the estate's central facilities and payment for that use. The significance of this is that the sharecroppers were beginning to develop a common interest and a collective relationship to the landowner. This was new.

Sharecropping also began to involve a certain amount of paid labour done for the landowner. The contracts for the 1880s and 1890s on the *Fattoria dei Piani* stipulate that 'the sharecropper shall provide all the labour necessary for pressing olives and handling the oil ... under the direction of olive-press operatives employed at the owner's expense'. Limited though it was, the new mechanisation represented the acquisition of new means of production, of which the landowner retained control, and the use of specialised operatives to work them.

Finally, a clear deterioration in the financial conditions of the sharecroppers, of which more details are given below, caused an increasing number of them to take holdings which would not absorb all their time. Thus they could have a certain subsistence quantum and a free house, and seek as much paid labour as possible outside the farm, often in roadbuilding and other public works. They saw that although they had greater security than the wage labourers, the hourly rate for their labour was smaller, since they worked a longer day and every day of the year to gain not much more than the wage earners. This development brought them into contact not only with each other but also with the proletariat, such as it was in the Tuscany of the time.[13]

THE WORSENING CONDITIONS OF THE SHARECROPPERS

Given the disadvantages which sharecropping suffered in a capitalist context, and given that capitalism held limited prospects of success anyway, the logic of the situation might be that the landlords would make their sharecroppers shoulder the increasingly heavy burdens as much as possible. This is what some people say they did (for example Mori [1955: 489] and Radi [1962: 71]), by toughening up the contracts, inserting more severe clauses. Certainly the landowners had a lot of power to change the clauses. The *Codice Civile* of 1865 had, in articles 1651-2, given them the freedom to insert individual clauses and to evict without explanation. The contracts of the *Fattoria dei Piani*, dated from 1848 to 1915, all contain clauses taking full advantage of this freedom, and stating that the sharecropper could only appeal against his eviction after he had vacated his holding. But caution should be exercised in drawing general conclusions about the landowners' use of these powers. It would be surprising if they did not make the sharecroppers pay for the growing difficulties in some areas and in some degree. But it was doubtful whether this response was systematic. Standard contracts printed for the 1880s were sometimes still in use on the eve of the First World War. More original study of the contractual relationship could possibly reveal that the landowners, rather than being increasingly exploitative, were inert, partly because they were absentee in so many cases, and partly because they did not want to risk inflaming an already dangerous situation.[14]

It did not require active aggression on their part to reduce the sharecroppers to a position of intolerable misery: other factors were sufficient for this. The phylloxera which occurred around 1893 affected them badly because it meant a drastic reduction in the volume of one of the main crops which they converted into cash to pay their share of the running expenses or to repay debts. The town population controlled births, but the traditional Catholic sharecropping family did not, so there were more mouths to feed. Although properly speaking the size of the farm should have been related to the size of the family, many families outgrew theirs. Glowing reports of how the holdings were kept as spick and span as gardens, so beloved of the nostalgic, do not take account of the fact that this was possible only because there were too many hands with no productive work to do. Industrial development was minimal, so there was no outlet for surplus labour.

Local authorities (controlled by the landowners) increasingly discriminated against sharecroppers, often requiring more family tax from them than from professional and propertied families. As one prominent Tuscan landowner, who had been Minister of Agriculture, Industry and Commerce, Count Francesco Guicciardini, observed to the *Accademia dei Georgofili*, at the town hall the sharecropper was treated as though he were a well-off citizen. State policies, too, placed him at a disadvantage. The government's refusal to adopt a progressive income tax, and its preference for indirect taxes, made it difficult for him to buy the things he could not produce himself, such as salt and sugar, and so he consumed practically none of these things. He would have been unaffected by income tax, had it been imposed.

These, together with sharply declining harvests, were the main factors affecting all agricultural workers, and the evidence that life had become intolerable for many began to appear round the turn of the century. In 1898, the year after a disastrous wheat harvest, 54 demonstrations, often involving violence, were recorded between 6 and 13 May. The records of the Military Court at Florence tell us that sharecroppers were involved, albeit in small numbers. It should be remembered that although the sharecroppers represented only a tiny proportion of the demonstrators, they were members of a social group traditionally unused to demonstrating. The number of accusations of stealing made against sharecroppers also increased. The Marquis A. Carega Bertolini knew that this was happening on his farms, but did nothing to stop it because he realised that his sharecroppers could not survive without this supplement [*Radi*, 1962: 41].[15]

Perhaps the clearest indication that the sharecroppers, who had lived for centuries in apparent, possibly genuine, near-contentment, now found their lot intolerable, is the appearance of emigration. This was previously unknown in Tuscany. Data are a little confused. Pinzani [1963: 92] gives the following figures:

1890	*1897*
6,000	10,397

wheareas Mori's figures [1955: 491] are:

1898	*1901*
4,626	15,218

The apparent contradictions in the two sets of figures need not trouble us too much. In both cases the picture is clear of an increase around the turn of the century. Particularly relevant to our present purpose is the fact that sharecroppers were included among these emigrants, and represented an ever-growing proportion. On 1 and 8 January 1898 *La Martinella* published letters commenting on the emigration of entire sharecropping families. The writers were stupefied by the novelty of this phenomenon. Thus, while there is some evidence to suggest that by the turn of the century the sharecroppers were already protesting with their feet, it was to be some time before they learned that there were other means of protest.

PROBLEMS OF POLITICAL ORGANISATION. COMPLIANCE AND THE BEGINNINGS OF PROTEST

The most immediately obvious difficulty in the way of a concerted response was the individualism of the sharecropper and the lack of contact with other members of his own 'class', who certainly constituted a 'sack of potatoes', in Marx's celebrated phrase in *The Eighteenth Brumaire of Louis Napoleon*. Theirs was a traditional family life on isolated farms. Meetings were infrequent. They would help each other out sometimes, but only in busy periods when there was little leisure for discussion. There was some contact, however, during the ritual stroll through the village or outside the main estate buildings after Mass on Sundays. It is interesting to note that after the sporadic uprisings of the previous year, the landowners managed in 1907 to have a law passed which took away the sharecropper's traditional right not to work on Sundays [*Mori*, 1955: 493].

The relationship to the landowner was an important obstacle. To all appearances the sharecropper was not an employee but a farmer with all the dignity of independence. His cropping patterns were not usually interfered with much. He organised his day's work at his own discretion. When landowner and sharecropper met it was normally on a personal basis, often at the latter's own house. From a functional point of view the purpose of the visit was that the landowner could check whether the family was capable of running the holding properly, but since in most cases the landowner knew little about farming, the visit took on a different character. His interest in the family was quite often spontaneous, with the children, for example. He would be greeted with the traditional courtesy of 'Buongiorno signore! Buongiorno vossia!' and be given raw eggs to drink. The atmosphere was almost that of an extended family, with the owner as the revered patriarch, dispensing advice on all sorts of problems, not only agricultural. The relationship certainly involved more than the cash nexus existing between employer and wage labourer.

Spontaneous organisation, then, was unlikely. Organisation under the hegemony of a political party presented problems. The socialists had been active among the wage labourers of other areas for some time. In Tuscany they concentrated on the miners and what few urban workers there were. They had little to do with agricultural workers until the mid-1880s when their newspapers began to raise the question of involvement in the Tuscan countryside. The wage labourers were fairly promising material, although the fact that they were almost all casual workers and could not find jobs for more than half the year presented obvious organisational difficulties. But the sharecroppers constituted the real problem. The socialists had not yet developed their ideology enough to incorporate this question. They were still trapped within a crude Marxism which could not accept that peasants formed a class. It was still to be some time, in their view, before the peasants would either fall into the proletariat or be absorbed into the bourgeoisie. The trouble with this attitude was that the sharecroppers did not know that they were supposed to do one of these two things, and never fully grasped the inevitability of the process. Consequently they remained, in spite of the limited degree of proletarianisation discussed above, decidedly individualist. It was only after the experience of fascism that the Italian left acknowledged that the peasants, including the sharecroppers, must be considered a permanent fixture, and took account of this in their evolving theory. This should be understood if we are to explain the Italian Communist Party's 'political success and its stature as the moral leader of Italian politics'.[16]

The doctrine of the early socialists could lead them into either of two possible stances with regard to the problem of the sharecroppers. They could, as many advised, simply not bother with them for the time being and let political evolution do the spadework. Alternatively they could lend history a helping hand by hurrying along the proletarianisation process. Many of them believed this to be at a more advanced stage than it was in reality. That is, they overestimated the effects of the partial capitalisation discussed above and believed that sharecropping was disappearing, which was hardly the case. The second alternative was presented to the party's 1897 congress by Bissolati, Samoggia and Rocca Pilo, at the suggestion of their leading agrarian theorist in Tuscany, Doctor Vacirca. Congress adopted the general proposal, and it was decided to form sharecropper leagues, the main objective of which would be to force the landowners to abandon sharecropping in favour of wage structure.

As we can see with hindsight, this was a tactical error. The effects of the disastrous 1897 harvests had not come fully home to the sharecroppers, and they were still better off than the wage labourers, very few of whom had fixed jobs anyway. So it would have been surprising had they taken kindly to the suggestion that they become proletarian. This explains to a large extent the socialists' lack of success in the Tuscan countryside in the 1890s. Not only was the strategy ill-conceived. When the socialists' loyalty was tested they had to make a choice, and the inevitability of that choice was exposed. For example, from about 1896 the production of Italian straw hats came under competition from the Far East. During that and the following year there were violent uprisings by the platters in the towns of the province of Florence [*Pinzani*, 1963: 89]. The platters demanded that the women of sharecropping families be banned from doing part-time work for the straw industry. By their support for this demand the socialists alienated themselves from the sharecroppers, who saw them as taking away a source of supplementary income.

Socialism as then conceived was not taking root in Tuscany, and party strategy had to be rethought. That matters were less straightforward than had at first been thought was acknowledged in a pamphlet of 1901 by Eugenio Ciacchi, *Ai contadini*, in which the author wrote: 'To begin this improvement in the conditions of rural workers we must exercise a great deal of tact, prudence and care.' The substance of the pamphlet centred on the idea of co-operatives, on the model of the successful *Cooperative di Consumo Agricolo e delle Fratellanze* initiated by Camillo Prampolini in Emilia-Romagna [*Ragionieri*, 1955: 455]. The immediate aim of these organisations was to free the sharecroppers from financial dependence on the landowners by obviating the need for loans from them. The co-operatives would provide for the sharecropper's needs at reasonable prices, having bought up stock from those who had a surplus. Profits would be distributed among members for whatever use they could put them to. The need for solidarity was stressed: if a sharecropper was dismissed unfairly others would refuse to replace him and do all they could to find him another holding. At the same time a programme of preliminary work was undertaken, consisting of general discussion and propaganda, in the socialist newspapers and also in the local markets, where the sharecroppers occasionally met, and on the estates.

The next stage was to translate the growing consciousness of the sharecroppers into powerful action. This took place at the very beginning of this century under the loose general auspices of the *Federazione Nazionale dei Lavoratori della Terra (Federterra)*, though in very limited areas, the precise geographical location of which is surely worthy of our attention.

There was hardly any development at this time north of a line drawn through Cecina and Arezzo. The very first organisations were those in Castelnuovo in Val di Cecina, Pomarance and Radicondoli [*Ragionieri,* 1955: 460]. These villages are in the Colline Metallifere: there must have been some contact with miners. Close behind came the leagues at the southern end of the Val di Chiana, in the villages of Sarteano, Chiusi and Chianciano. As is pointed out by Procacci [1970: 132], these villages are close to the railway line which runs northwards from Rome, along the Umbria-Lazio border, and through the middle of the province of Arezzo (itself a socialist stronghold). A branch line goes to Terni in Umbria, where in 1894 the government had established a massive steelworks. All the early political activity of the sharecroppers began at the railway line and spread along it and outwards from it.

The first rural strike in central Italy was in February-March 1901, just south of the Tuscan border, at the railway line in a sharecropping area of northern Lazio. This small strike petered out with no effect. In 1902 protest began in earnest, in the same area and also around Terni. It spread northwards along the main line and eastwards along the steelworks branch. It was in Umbria that cattle were first brought along and being left unfed and unwatered they added their loud cries to the voices of the protesters. This provided a graphic illustration of what an extended farm strike could involve.

The first strike in Tuscany, on 7 April, was called by a rich twenty-one-year-old, Aldo Mieli, supported by an equally rich friend. They had been elected to the Chianciano local council in September 1901, largely on the farmworker vote [*Ragionieri,* 1955: 460]. In February 1902 they were the leaders of a league consisting of 130 of the commune's 150 sharecropping families. Mieli had asked the mayor to help him persuade the landowners to make certain concessions. The mayor's flat refusal to co-operate was presumably not unconnected with the fact that he owned a large part of the land his council administered. The main concessions requested were as follows. First, although the constitutional terms of the agreement provided for the division of the product on a fifty-fifty basis, it was claimed that in some cases the landowners were taking more than half. An end was required to this abuse. Second, the sharecroppers wanted relief from the running expenses which had increased beyond their means since unification, and consisted in the main of higher land tax, the application of sulphur and copper sulphate to vines, and machine threshing, which by now had become fairly common in their area. Third, the more servile obligations of the sharecropper, such as digging ditches for no payment, the gifts of eggs, hams and so on, as well as the traditional 'courtesy' of allowing the landowner the first choice of grapes, were felt to be part of an outdated tradition. The owners had no interest in making these concessions, but when the strike was called they agreed to participate in a conciliation committee. On the committee Mieli, being a landowner himself, sat on their side of the table. They talked through the night, and in the morning were able to announce that all the demands were to be met, not in full but in very large measure.

The Chianciano proprietors had been taken by surprise, but when their startled colleagues in nearby Chiusi heard what had happened they met and decided their negotiating policy in advance of the anticipated strike. When this took place, on 11 May, it led to concessions far less damaging to the landowners than those exacted by Mieli's group.

The defeat of the movement took place at Sarteano, a few kilometres away. There the proprietors offered a few 'spontaneous' concessions, dealing with individual sharecroppers. But the latter stood firm and demanded a collective deal for the commune and bigger concessions. The landowners refused these, and when the strike came on 18 May they had the police ready. Many arrests were made, and the next day a large number of letters of eviction were issued. Apparent stability returned to the countryside for a few years.

The 1902 strikes were spasmodic, and limited to a small part of the province of Siena. Other provinces were entirely untouched [*Procacci*, 1970: 138]. It is possible to criticise the organisation of the strikers. For example, the farms from which colleagues had been evicted after the Sarteano episode could have been left empty. At Chianciano, when it began to rain the demonstrators rushed back home to cover the hayricks.

But it was early days yet. What is significant in the accounts of the early league formations is the surprisingly positive attitude the sharecroppers had to them, in spite of centuries of compliance and ignorance. They were working under socialist hegemony, in general terms, but were by no means being led by the nose. On the contrary, they themselves often took the initiative. At Radicondoli two hundred of them asked the socialist leader Vittorio Meoni how to form a league, rather than sitting passively waiting for a visit from a socialist activist. The league at Pomarance was begun and run entirely by sharecroppers.

The countryside remained calm until 1906. This time the centre of activity was a different part of Tuscany (though still near the railway line), Rignano in the eastern part of the province of Florence. The beginning of the protest has been traced to a single estate, the *Fattoria di Torre a Cona*, where all the sharecroppers struck [*Radi*, 1962: 106]. Neighbouring communes followed suit, though strike action was restricted to half a dozen of them. Although the demands varied a little from one commune to another, they were much along the lines of those which had been made in the Sienese villages four years earlier. It was not a very determined effort, and cooled off without the claims being met. In one area, Sesto Fiorentino, the strikers were more stubborn, and the police were called in. There were arrests, and the strike was broken.

The 1906 protest did not match that of 1902, in size, vigour or success, and it seemed that the movement had petered out completely. Improvements to the agreements had been modest and affected only a few areas. For further developments to take place the situation required a catalyst. It found one in the Great War.

THE WAR AND ITS AFTERMATH: THE OFFENSIVE OF THE SHARECROPPERS

The war raised the consciousness, ambitions and hopes of the sharecroppers.

They came into contact with urban and northern agricultural workers, and with intellectuals. Demands that they, along with other rural workers, be given land after the war were increasingly insistent, but the authorities were unenthusiastic about this proposal until November 1917. Farmworkers suffered especially badly during the war. They were nearly all in the infantry, who incurred greater casualties than other sections of the armed forces. 64 per cent of the war orphans were the children of rural workers, 30 per cent those of non-agricultural workers; only 3 per cent had industrialist or merchant fathers and 2.7 per cent professionals [*Fano*, 1975: 475]. The appalling disaster of Caporetto reduced failing morale to zero, and the generals feared that their remaining cannon fodder might desert en masse. It was suddenly felt in high places that the promise of land was a good idea after all. The slogan 'Land to the peasants, who are fighting for it!' became official, the carrot which led more peasants to their deaths, and was withdrawn after the slaughter ceased.

The sharecroppers came back to farms which had been neglected during their enforced absence, to face a worsening economic situation and deteriorating living standards. Agricultural prices were rising steeply, but this was of limited benefit to them, since it was only the landowner who could benefit from the rises. Therefore the sharecropper's position in the partnership deteriorated in comparative terms, and as other prices were rising too, it also deteriorated in real terms. It is important to note that the sharecropper's position deteriorated in comparison with that of the wage labourer, who obtained wage increases which kept up with and in some cases even went ahead of inflation in 1919-20. Those sharecroppers who had to employ labour in order to meet their contractual requirements had to pay more for it (and there were more of these, because most of those who had died had been the adult males, and many of the survivors were mutilated).

At this point, then, the sharecroppers were highly inflammable material, and the time was ripe for an increase in political organisation, and a hardening of the expected reaction. It was at this point that the Catholics showed themselves as a political force to be reckoned with in rural Tuscany. For the unaligned sharecroppers, of whom there were many, especially in the northern half of the region, they were an alternative to the socialists. The fact that there were two groups leading the sharecroppers constituted a split which undoubtedly weakened the movement. At the same time, competition for support gave it extra impetus. The Catholics had grown in influence during the war. They entered the war much weaker than the socialists, but emerged from it just as strong [*Radi*, 1962: 153]. They controlled much of northern Tuscany, for example San Casciano in Val di Pesa, Sesto Fiorentino, Fiesole, the district of Pistoia, Regello, the whole of the Mugello, and other zones.[17]

In northern Italy the *biennio rosso*, the two Red years 1919-20, was seen by participants as the conclusive battle between the bourgeoisie and the proletariat predicted in the Communist Manifesto. In rural Tuscany the political line-up was more complex. There were three groups. The Catholics appealed to the individualist land-hunger of the sharecroppers.[18] Their 'White' leagues grew rapidly during 1919 under the umbrella of the Popular Party. Their main aim was to involve the sharecropper more in management, and eventually convert his contract into rental, lease or freehold. The socialists, perhaps encouraged by the success of the Russian revolution, had allowed their policy emphasis to shift away from piecemeal, pragmatic improvements and reverted to the old dogmatic aim of proletarianising the whole of the region, with eventual collectivisation in mind. Their support was organised in 'Red' leagues. Their main power bases were in southern Tuscany, but they penetrated into northern areas too, while there was little compensatory White presence in the south. Finally, the landowners had the *Associazione Agraria Toscana* (AAT), formed in January 1919 in anticipation of trouble. While it claimed not to have a class basis (technically it was open to sharecropper membership) its real purpose, rather than its vaunted concern for the objective interests of agriculture, was revealed in the words of its statute, quoted by Degl'Innocenti [1975: 108]: 'vigorous opposition to unjustified attacks on landed property'. Those attacks had been made by socialists, but forthcoming White attacks were to be considered equally unjustified, even more so. The AAT became very influential. Its 1,264 founder members grew to 5,324 by November 1920, who controlled a third of all Tuscan agricultural property [*Cantagalli*, 1971: 598].

After the war the first real negotiations with the AAT were made by the Whites. On 22 May 1919, a thousand sharecroppers in the commune of Prato in the Bisenzio valley, an area completely under Catholic control, formed a local federation of leagues. They made demands to the landowners for various economic improvements, and also asked that their holdings be administered 'with the reciprocal agreement of the parties' [*Degl'Innocenti*, 1975: 108]. As the owners did not reply, they turned to the AAT, who agreed the economic clauses but refused the one which would have given the sharecropper a say in management.

It was the same story when the Florentine Provincial Federation of Leagues, formed on 1 June, approached the AAT in July. The resulting *Concordato di Firenze* contained clauses of an economic nature, setting out the apportionment of expenditure between the two parties, abolished the unpaid work which the sharecropper had traditionally done, and recognised the Federation. But the Association did not budge an inch over the question of management: Article 13 stated that 'The proprietor administers the holding in the common interest of the parties'. The Florence section of the AAT accepted the agreement on the condition that no contract should contain a single clause that was not of a purely economic nature [*Radi,* 1962: 187]. As the year wore on agreements of this kind were reached in various parts of the region, improving the sharecropper's economic position but excluding him from management.

In practice, the landowners did not feel obliged to apply the concessions, which were modest, in substance no different from those made to the Red leagues back in 1902. One landowner, Pier Francesco Serragli, addressing the Georgofili late in 1919, said that many of his colleagues were still failing to meet their agreed obligations to the sharecroppers, taking advantage of their weakness. Furthermore, the White leagues did not fight for the application of the agreements. The only effective weapon would have been the strike. The government had a non-intervention policy over political demonstrations, and it was unlikely that the army or the police would have broken any strikes. But the

Catholics would not resort to these tactics: they believed in 'the solidarity of classes, collaboration between capital and labour'. The strikes which began in September 1919 for the application of agreed economic clauses were led by socialists [*Radi*, 1962: 187]. When four thousand White supporters staged a strike around Prato on 24 February 1920, their leagues had them back at work by the end of the month [*Degl'Innocenti,* 1975: 109].

Pressure grew from both Reds and Whites for a uniform regional contract. The AAT agreed to begin talks on 18 May 1920. When the contract was finally drawn up on 6 August the Whites had been left out of the negotiations completely. Why?

The Reds had made considerable strides in their areas, particularly in the provinces of Arezzo and Siena. Their demands were purely economic: they did not threaten the landowners' control, as the Whites did. As long as they did not demand the abolition of sharecropping the owners could tolerate them. The Reds saw their role as that of preparing for a capitalist revolution, which would in turn release the sharecroppers into the proletariat, and into their hands. It was not in their interest, as they conceived it, any more than it was in the interests of the landowners, to allow the estates to be broken up. In order to force the AAT to negotiate they called an all-out strike of their members. Alone they would probably not have been strong enough, but enough Whites, in defiance of their leagues, joined them to make the owners give way. Having temporary control of much of their rivals' support, they were in a position to insist that the Whites be excluded from the discussions. Separate talks with the Catholics were mooted, but they presented conditions which laid so much emphasis on the sharecropper's right to management and to first choice should the owner decide to let or sell his farm, that the AAT felt unable to negotiate at all on this basis, and refused to go ahead with the talks, continuing with the socialists. Now that the sharecropper movement was firmly under way, both proprietors and Reds feared the same thing, an end to unified control of the means of production. The Catholic movement in rural Tuscany suffered a serious setback thanks to a strange alliance of landowners and socialists.

The *Patto delle Leghe Rosse*, or 'Red Pact', as the 1920 regional contract came to be known, was not a revolutionary document. It provided for full wage-labour payments for jobs done by the sharecropper for the owner, instead of the existing average of 75 per cent, and shifted more of the administrative expenses on to the owner's shoulders. But the produce was, quite simply, to be divided in half. The socialists were content for the time being to reduce the landowner's returns in order to encourage him to change to more productive, capitalist methods. A particular gain was the requirement that a 'fair reason' be provided for evictions. The three 'fair reasons' were (Article 21):

(a) criminal conviction of the sharecropper;
(b) the takeover of the farm for the owner's own use;
(c) transformation of the system of administration to capitalist methods.

The Catholics had been pressing for a fair reason for evictions. In the 'White formulae' presented to the AAT on 1 July one of the demands had been that in the case of an eviction 'the proprietor must demonstrate the fair reason in court'. These formulae were thrown out, but the socialists not only won the principle, they also managed to stipulate what the 'fair reasons' were to be, including (c).

So the socialists had made decent advances, and the proprietors had retained control over their land, even though the loss of the right to evict arbitrarily had weakened that control to some extent. But the Whites had been humiliated. They had seen some of their sharecroppers working with the socialists. They had seen the socialists take over one of their most important ideas, that of the 'fair reason', and adapt it to socialist political ends. The Catholic leagues, in a despairing and ill-fated attempt to rescue themselves from the oblivion to which they were soon to be consigned, denounced the new regional contract on the grounds that i) it did not provide the sharecropper with the first option should the owner decide to let or sell the farm, and ii) it did not provide that the sharecropper could, if he wished, take a tenancy of the farm, even against the owner's wishes. It should be noted that the leagues were acting independently of their Tuscan leaders, who took a more moderate stance [*Degl'Innocenti*, 1975: 110].

The decisive defeat of the White leagues was now under way. On 16 October 1920 [*Radi*, 1962: 202] they persuaded the AAT to add to the Red Pact a clause stating that if the majority of sharecroppers in a particular zone, or all those on a single estate, wish to convert their contracts to rental, then the matter should at least be considered. Then the AAT changed its mind. On 29 October the Whites drew up a contract with the local Agricultural Association in the Mugello, which as well as the usual economic clauses included provisions (Articles 3, 4 and 5) that

(a) the administration of the holding is in the hands of the proprietor, who conducts it in collaboration and harmony with the sharecropper;
(b) a proprietor who wishes to sell his farm must first offer it to the sharecropper;
(c) a proprietor who wishes to let his farm must favour the sharecropper.

The local Association agreed, but was pressurised by the regional Association into striking out all except the economic clauses, which were if anything less demanding on the landowners than those of the Red Pact! The final mortal agony of the Catholic leagues made them appear dangerous. It expressed itself in direct action. There was sporadic preliminary activity. Whites refused to share their wine and to work at the olive presses. Some even sold their cattle and pocketed the proceeds. But the police soon put a stop to that.

The co-ordinated White assault on the Red Pact began on 15 November around Prato [*Cantagalli*, 1971: 598], and immediately spread throughout the areas under Catholic control. In the Mugello some violence occurred. The main tactic was the rural equivalent of a factory occupation. The bailiff and his assistants were ejected from the estate buildings, white flags were raised on the strawstacks and hencoops, and the sharecroppers announced that they no longer recognised the landowners, and would not until their new conditions were met. It was this activity that seemed to represent to the landowner a threat of real revolution. It was not the expected bolshevik revolution, which had already been compromised in Tuscany. The Whites, not the Reds, posed an immediate threat to the control which the landowners had over the region, and it was this activity which first brought the fascist squads out of the towns and into the Tuscan countryside.

THE ADVENT OF FASCISM

This paper cannot discuss the fortunes of *mezzadria* under the fascist regime, since this did not become such until after the period under discussion. The fascists who appeared in the Tuscan countryside at the culmination of the events described here, to put an end to the social and political advances made by the sharecroppers, were not a coherent political party with a carefully planned programme for post-war national recovery. They were the *squadristi*, a motley collection of malcontents – former *arditi* (crack assault troops), well-born men of private means, small businessmen, and sometimes mere adventurers.[19] They were variously inspired by an impassioned nationalism, a burning desire for individualist upward social advancement, and the blind fear that the Russian revolution could spread to their own country.

At first the fascists met no real resistance. In San Casciano in Val di Pesa, and at Signa, small squads went around in lorries, having a word with the sharecroppers who were occupying the estates, firing their pistols into the air from time to time, which was enough to make the occupants vacate the properties. Francesco Giunta, a leading Florentine fascist, took part in these sorties. When he heard that in San Piero a Sieve, in the Mugello, a dozen or so sharecroppers had taken over the property of Countess Cambray Digny, he decided to offer the services of his squad at once. Giunta was very fond of the Countess. San Piero a Sieve was his home village, and he had spent much time at her villa in his youth. By the time he arrived there on this occasion, with his squad of twelve, led by Luigi Zamboni, Bruno Frullini, Manfredo Chiostri and Italo Capanni, her estate had already been re-occupied for her by the *carabinieri*, but a large crowd of sharecroppers had gathered in the main square of nearby Scarperia and gone in procession to take over the property of Senator Marquis Gerino Gerini, the *Fattoria della Casaccia*. The squad felt it their duty to help. On their way to the estate they passed the house of one of its *mezzadri*, seventy-two-year-old White Sitrialli. He was on the threshing floor outside the house, not at the estate. The family took refuge inside the house and threw the bolts. The fascists fired at the door, a bullet passed through, and the old man died instantly.[20]

CONCLUSIONS

The significance of the death of Sitrialli lies in the fact that it is the first example of genuine squad violence, rather than the sabre rattling which preceded it, in rural Tuscany, and in the fact that it was directed against a White family. It was only later, in the early months of 1921, that the squads began to make regular sorties into Red strongholds.[21] The uprisings which immediately followed the signing of the Red Pact were instigated by Whites, not by Reds, and it was the Whites who first incurred the wrath of the squads.

The fear of revolution was even more confused in Tuscany than it was elsewhere. Even in the Po valley, the first rural area of Italy to suffer fascist violence, the hopes and fears of revolution were unrealistic [*Corner*, 1975: 104-5]. But in Tuscany the dictatorship of the proletariat was an almost totally imaginary threat, which depended on what was going on in other regions more than on anything the Tuscan socialists were actually doing. The strikes they staged in early 1921 were for a better application of the Red Pact, that was all. But the example of Red activity in the province of Bologna was there to frighten the Tuscan landlords. There the socialists had a virtual monopoly of the leagues. Since 1915 they had had an annually renegotiated agreement with the landlords. But early in 1920 they sent a draft contract to the proprietors, declaring themselves unprepared to negotiate and determined to impose its terms unilaterally. Not only was up to 70 per cent of some of the crops to go to the *mezzadro*, but the document heaped huge expenses on to the shoulders of the landlords (Art. 39): all land taxes, a variety of insurance premiums, including health insurance and pension contributions, all costs of fertilisers and disease control chemicals. Rather than talk, the socialists took to direct action at once. Mobile Red patrols burnt crops, attacked landlords and destroyed stacks. In certain cases, fields were left untended, then declared uncultivated, enabling the socialists to occupy them according to the terms of the Visocchi decree.[22] In May 1920 the landowners asked the Minister of Agriculture Falcioni to issue a decree enabling the police to recover their land for them. This he did, but the government had a non-intervention policy as far as political demonstrations were concerned, and did nothing else about the problem. The strikes continued, with the countryside feeling abandoned by the authorities, another reason why the landowners allowed the squads to behave as they did.

But in Tuscany the socialist demands were not nearly so devastating. There one of the most important factors affecting the reaction of the landowners was an imagined social threat, an apparently grave danger to the structure of society. The revolutionary government in Russia was now an established reality. The Italian Socialist Party, at headquarters, remained fairly moderate, and could not accept the twenty-one conditions for continued membership of the Third International which Lenin communicated to them in October 1919. In particular, they could not accept Condition 17, and call themselves a communist party, and Condition 7, which demanded the expulsion of Turati, Modigliani and the other reformists. But at the January 1921 Conference no less than one third of the delegates accepted these conditions and split. Italy now had her own Communist Party. Furthermore, there were the occupations of the northern city factories in 1920. The factories had been re-occupied on 25 September, but the fear of similar events remained. This terrifying pre-revolutionary atmosphere in late 1920 was suddenly extended to the Bologna countryside as a result of the activities of the local socialists.

Certainly, when the Red Strikes were resumed in southern Tuscany, fascism was there to respond with violence. The fact that their demands were moderate made no difference: the squads made no distinction between the Tuscan socialists and their brothers to the north. From then on the clashes developed their own momentum. Strikes caused fascist reprisals, which in turn caused the strikers to take up arms, so that, as Chiurco puts it, the squads were compelled to punish the delinquents who had killed so many patriotic comrades. The fascists at this time were not assiduous students of the *mezzadria* contract, they were not trying to resist the demands of peasants for better working conditions and economic improvements. They were conducting a political crusade, and any rural disturbance at all was viewed as bolshevism. But, as I have argued, the only real threat to the social status quo in Tuscany came from the Catholics. The conservative elements in that society had identified Catholic activity with the Russian revolution, in what Radi calls [1962: 187] an adroit act of political prestidigitation. *La Nazione*, the moderate conservative newspaper of Florence, a city surrounded by some of the strongest White camps in Tuscany, and very few important Red ones, welcomed the formation of the *Associazione Agraria Toscana* in an article entitled 'Against Leninist propaganda in the countryside' [*Degl'Innocenti*, 1975: 108], published in April 1919. In some areas Catholic activity was even confusedly referred to as 'White bolshevism' [*Snowden*, 1972: 283]. The maximalist stance taken by the Catholics in their rejection of the Red Pact was the flashpoint of reaction in Tuscany, and helps us to understand the contradiction that in an area where the socialists were moderate, reaction was so violent and the destruction of workers' organisations so systematic.

The history of the *mezzadria* movement during the period we have examined, then, is one of class conflict more than it is one of economic struggle. A clash of economic interests forms an important part of the history, of course, and the significance of this has not been minimised in this paper. But it is important to stress that the rural workers' movement was resisted from beginning to end according to social and political criteria, in order that the status quo should not be disturbed. *Mezzadria*, an attempt to couple capital and labour in a collaborative effort, was the basis of Tuscan society. It dominated the countryside, and the overriding importance of agriculture in the region's structure explains the interest which the city *fasci* took in its welfare. It was in order to preserve the social structure that capitalism was resisted in the first place. When the area was engulfed, partially but inevitably, in a unified market economy, the course which would have made the most economic sense would have been to take the advice offered by Jacini and other experts and abandon sharecropping in favour of more rational methods, such as specialised cropping plans suited to local conditions and flexible enough to meet the demands of a changing capitalist market. This would have been justifiable in spite of the limitations imposed by the terrain. The Georgofili, the first people to be faced with the problem, were perfectly well aware of these considerations, but they made a conscious decision to retain *mezzadria* for the sake of social stability, or, more exactly, so that they could retain control of their lands. The objectives of the two left-wing parties seemed to threaten the social status quo, one in the long term and the other immediately. The gradualist socialists of the area wanted, through a progressive toughening up of the economic aspect of the partnership in favour of the *mezzadro*, to squeeze the landlords so hard that they would be forced to adopt more productive, capitalist methods. This, according to the doctrine of the socialists of the time, would release the sharecroppers into the proletariat, which, unified at full strength behind them, would enable them to collectivise the land. The Catholics wanted the *mezzadro* to have direct control over his individual farm, either as a tenant or as a small proprietor. They saw this as class collaboration, and so it was in a sense: it would certainly have produced a deeply conservative society and provided an effective discouragement to the advance of socialism. But as far as the landowners were concerned, this would have involved the surrender of their control just as certainly as if their estates had been collectivised. Furthermore, the climax of the history was the violent suppression of all the leagues. As far as the Whites were concerned, this began at the moment when they went all out to achieve greater control for the individual working farmer, through participation in management and the option of taking a rental and first choice in farm purchases. As far as the socialists were concerned, as long as they did not seem to be threatening the system itself, or at least did not seem to be on the point of overturning it, their demands could be negotiated and met in large measure. Only when the activities of the Red leagues of Bologna caused a real fear of bolshevism did the establishment resort to violence even in their dealings with the socialists. At this point, in the atmosphere of terror and hatred on all sides, distinctions became blurred and all league members were viewed generically as revolutionaries.

The foregoing history is about economic interests in the sense that an important advantage of gaining control was the possibility of making more money. Indeed the fascists, once in power, reduced the returns of the *mezzadro* and increased those of the landowner. But the fact that the latter was prepared to accept the inevitably lower returns of *mezzadria* production suggests that it was felt preferable to sacrifice economic aims to social ones.[23] The squads in Tuscany were the result of a determination to see through to the bitter end the implementation of the decision taken by the landowners of the 1870s and 1880s.

NOTES

1. See, however, Pratt [1980]. He finds that there was some sharecropping in Grosseto province, and his analysis shows that it was administered in much the same way as in the rest of Tuscany.
2. *Mezzadria* was the 'classical' form of sharecropping. It was so called because the division of the product and that of working expenses were *ad medium*, at least according to the terms of the agreement. The farmer working under such terms was called a *mezzadro*.
3. Jacquart [1970] and Jones [1968] give detailed accounts of post-feudal landholding patterns and relations of production.

4. In a patrilocal (single) multiple family household the married sons and their families live in the same house with their father and under his authority. See Poni [1978: 202].

5. A *frèrèche* consists of cohabiting families of married brothers, one of whom is head of household. See Poni [ibid.].

6. Tax was estimated on the use for which the land was registered, not on the use to which it was put, nor on the income it allegedly produced.

7. The *Accademia dei Georgofili* was an association of landowners and agronomists. A cross-section of their opinions on the usefulness or otherwise of *mezzadria*, from both a technical and a social point of view, can be read in 'Reale Accademia dei Georgofili, 1934-5'.

8. Snowden [1979: 150ff] tells us in some detail in what areas this modernisation took place.

9. As always, there were exceptions worthy of note. See again, Snowden, [1979: 150-51] for examples of improved cultivation techniques, such as four-course rotations, ploughing, and early trials of chemical fertilising.

10. It should be remembered, though, that around Lucca rentals, leases and small freeholds predominated over sharecropping. The more thoughtful general use of fertiliser may have been due to the fact that the farmers did not have the sharecroppers' dependence on the landowner for technical advice, which frequently was not given, or was ill-informed.

11. A discussion of the social implications of digging and ploughing techniques would require a separate article. Expressing the matter crudely, we can say that the owner preferred the sharecropper to dig the land, since this worked the soil more deeply than the ploughs then available, and so raised productivity. The sharecropper objected to the spade on the grounds that its use involved more labour than that of the plough. His reliance on the landowner to help acquire a plough was therefore an important factor in delaying innovation. See Poni [1963] for a lucid and detailed discussion of this question. Poni is writing about the province of Bologna, but in their broad lines the techniques and the controversies attached to them were similar in Tuscany.

12. The 'past investments' had been those stimulated by the innovatory work done in particular by Baron Ricasoli with one of his bailiffs, Testaferrata (as well as the Pucci family, Cambray Digny and others), in an attempt to compete in the infant capitalist system. Such investments included the planting of the Ricasoli mixture of grape varieties for Chianti, which is still in use today, both in Chianti and in much of the rest of Tuscany. Other technical improvements included Ricasoli's new rotation, which involved the introduction of nitrogen-fixing legumes. But the spread of these improvements ceased when the financial policies of United Italy removed the stimulus to agricultural investment. In any case, all modernisation had taken place within the limits imposed by sharecropping, and monoculture was almost universally eschewed. See Biagioli [1970].

13. For a fuller discussion of this point see Mazzini [1881: 477-8]; Mori [1957: 155]; Sereni [1968: 324]; Pinzani [1963: 90-94].

14. The *Fattoria dei Piani* contracts used from 1880 to 19[illegible]5 show no appreciable change. They contain certain minor manuscript alterations, almost always the same. For example, at one point the women of sharecropping families were forbidden to take work as wet-nurses, and then the restriction was lifted. Nor do the contracts of the second half of the nineteenth century appended to Mori's article [1955] show any particular hardening. Mori's own observation is that the striking thing about the contracts for the whole of the century is their lack of change.

15. Many landowners around this time expressed deep concern at the plight in which their sharecroppers found themselves. If we interpret this phenomenon according to the theory of surplus extraction, we could say that the proprietors feared for their share of the product. Nevertheless, it does seem that an element of disinterested benevolence was involved. A thoughtful, non-doctrinaire examination of this trend would be welcome.

16. See Davidson [1981: 42].

17. For an alternative assessment of the Catholics' importance, see Snowden [1975: Introduction and Appendix], where they are presented as a small, locally contained force tolerated, even encouraged by the Church as a bulwark against bolshevism.

18. For the moment we shall not go into details about the vexed question of the precise political function of the Catholics in the area. Catholic political groups were basically an expression of the Church's fear of bolshevism. The Church did not recognise the new Italian state, and on 10 September 1874 the Holy Penitentiary issued a *non expedit* advising Catholics not to vote in parliamentary elections. In 1877 this directive became a *non licet*. Attempts during the last decade of the nineteenth century on the part of successive liberal governments failed to secure an easing of it, because no government was prepared to give Rome back to the Pope. At all events, the *non expedit* was widely disregarded by rank and file Catholics, who increasingly exercised their right to vote in order to block the socialist advance, to diminishing Vatican disapproval. Pius X's encyclical, *Il Fermo Proposito*, of 11 June 1905, involved a partial relaxation of the *non expedit*, authorising bishops to allow their flock to vote if they considered such a move to be in the interests of the Church. In progressive relaxations, other situations were listed as justifying Catholic voting. Formal political organisation under the Catholic banner was not possible, however, until the *non expedit* was revoked in 1918 by Benedict XV. See Seton-Watson [1967: 59, 61, 94, 219, 223-4, 226-7, 233, 274-5, 278, 279, 388, 515 and 549; *Istituto della Enciclopedia Italiana*: under *non expedit*].

19. For a description of the social composition of the early fascist squads see Snowden [1972: 150ff].

20. The story is told by Cantagalli [1971: 600-2].

21. See Chiurco's accounts in [1923: 29-31] and [1929: *passim.*]. It goes almost without saying that these accounts should be read with an eye open to the exaggerations they contain.

22. This was a ministerial decree issued on 2 November 1919, granting ex-combatants the right to have uncultivated land assigned to them.

23. Corner [1975: 527] points out that sharecropping can sometimes be adopted by capitalist entrepreneurs as a protection against falling prices: it guarantees a minimum return, and that crops do not cost more to produce than they are worth on the open market. He notes that the fascist census of 1931 indicates an increase in the number of sharecroppers in Italy over the previous decade from 30% to 33% (of agricultural workers), which seems to constitute a partial explanation of the reaction to the agricultural crisis of the late 1920s. But, as Corner also notes, that census was designed to demonstrate the success of the fascist policy of 'peasantisation' or 'deproletarianisation'. Even to work a vegetable garden in one's spare time qualified one to be classified as a sharecropper for official propaganda purposes. This *caveat* aside, we should ask whether the view that fascism encouraged sharecropping merely for reasons of social harmony is as accurate as has generally been supposed.

REFERENCES

Biagioli, G., 1970, *Agrarian Changes in Nineteenth-century Italy: the enterprise of a Tuscan landlord, Bettino Ricasoli,* Reading: Reading University, Institute of Agricultural History Research Paper no. 1.

Cantagalli, R., 1971, 'La prima scissione del fascio fiorentino', in Unione Regionale delle Provincie Toscane, *La Toscana nel regime fascista*, 2 vols., Florence.

Chiurco, G.A., 1923, *Fascismo senese*, Siena.

——————, 1929, *Storia della rivoluzione fascista*, 5 vols., Florence.

Corner, P.R., 1975, 'Considerazioni sull'agricoltura capitalistica durante il fascismo', *Quaderni storici*, 29/30, May-December.

Davidson, A., 1981, 'Italian Communism and the Experience of Fascism', *The Italianist*, Vol. 1, No. 1, Reading: University of Reading and Dublin: University College.

Degl'Innocenti, R.M., 1975, 'Cenni sulle lotte agrarie nel pratese nel primo dopoguerra', in Comitato Regionale Toscano per il XXX della Resistenza e della Liberazione, *Mondo contadino e resistenza*, Foiano della Chiana.

Fano, E., 1975, 'Problemi e vicende dell'agricoltura tra le due guerre', *Quaderni storici*, 29/30, May-December.

Giorgetti, G., 1974, *Contadini e proprietari nell'Italia moderna*, Turin.

Imberciadori, I., 1951, *Mezzadria classica toscana*, Florence.

Istituto della Enciclopedia Italiana, 1956-61, *Dizionario enciclopedico italiano*, Rome.

Jacquart, J., 1970, 'Histoire de la campagne toscane', J. Meyriat (ed.), *Tradition et changement en Toscane*, Paris.

Jones, P.J., 1968, 'From Manor to Mezzadria: a Tuscan Case-Study in the Medieval Origins of Modern Agrarian Society', N. Rubinstein (ed.), *Florentine Studies*, Florence.

Mazzini, C.M., 1881, *La Toscana agricola*, Rome.

Mori, G., 1955, 'La mezzadria in Toscana alla fine del XIX secolo', *Movimento operaio*, Vol. VII.

__________, 1957, *La Valdelsa dal 1848 al 1900*, Milan.

Pinzani, C., 1963, *La crisi politica di fine secolo in Toscana*, Florence.

Poni, C., 1963, *Gli aratri e l'economia bolognese dal XVII al XIX secolo*, Bologna.

__________, 1978, 'Family and "podere" in Emilia-Romagna', *The Journal of Italian History*, Vol. 1, No. 2.

Pratt, J., 1980, *The Mezzadria*, Euromed Working Paper no. 9, Amsterdam.

Procacci, G., 1970, *Le lotte di classe in Italia all'inizio del secolo XX*, Rome.

Radi, L., 1962, *I mezzadri: le lotte contadine nell'Italia centrale*, Rome.

Ragionieri, E., 1955, 'La questione delle leghe e i primi scioperi dei mezzadri in Italia', *Movimento operaio*, Vol. VII.

Reale Accademia dei Georgofili, 1934-5, *La mezzadria negli scritti dei Georgofili*, 2 vols., Florence.

Rouveroux, P., n.d., *Le métayage. Ce qu'il faut en savoir*, Paris.

Sereni, E., 1968, *Il capitalismo nelle campagne 1860-1900*, 2nd ed., Turin.

Seton-Watson, C., 1967, *Italy from Liberalism to Fascism, 1870-1925*, London.

Snowden, F.M., 1972, 'On the Social Origins of Agrarian Fascism in Italy', *European Journal of Sociology*, Vol. XIII.

__________, 1975, 'The Social Origins of Fascism in Tuscany', unpublished M. Phil. thesis, St. Antony's College, Oxford.

__________, 1979, 'From Sharecropper to Proletarian: The Background to Fascism in Rural Tuscany, 1880-1920', in J. Davis (ed.), *Gramsci and Italy's Passive Revolution*, London.

The Introduction of Free Labour on São Paulo Coffee Plantations

Verena Stolcke* and Michael M. Hall**

INTRODUCTION

As a country with abundant land and a relatively scarce population, Brazil confronted special problems in the creation of a labour force. Until the 1850s, slaves made up the bulk of the workforce needed by large-scale export agriculture. By the mid-nineteenth century, however, as slavery came under increasing attack, at least some São Paulo coffee planters began experimenting with free labour. The abolition of the slave trade in 1850, moreover, coincided with the penetration and rapid expansion of coffee in western São Paulo, due to the decline of the other main coffee-growing region – the Paraíba Valley – and in response to the growing international demand for coffee. The introduction of free labour in São Paulo agriculture is, in effect, an instance of the creation of a free labour force in a situation of extensive agricultural development under conditions of potentially scarce labour supply. 'Our soil offers unlimited wealth, but we lack labour,' as one contemporary put the crucial obstacle to continued agricultural development which São Paulo planters faced in the second half of the nineteenth century.[1] The most forward-looking coffee planters clearly saw after 1850 that a way of replacing slave labour, or at least supplementing it, would have to be found in the fairly near future in order to provide the sizeable number of workers required for this very labour-intensive crop. Slavery in fact continued until 1888, but it was precisely the increasing debate over the labour question, and the experimentation in the following decades by São Paulo planters, which eventually made possible a relatively non-violent transition to free labour.

The creation of a labour force is never an exclusively demographic question. In the absence of a readily available local reserve of labour, São Paulo planters resorted to the use of immigrant workers. However, their experience with slaves had made them acutely aware of the need for effective forms of labour control. Thus, the issue planters increasingly faced throughout the second half of the nineteenth century was not only that of finding a new source of labour to replace their slaves, but also of how to organise and control free labour efficiently. The development and organisation of a free labour force for the São Paulo coffee plantations was both an economic and a political process. The dynamics of the situation derived from both the economic rationale of the planters and from the bargaining power available to labourers to resist the planters' impositions.

The objective of this article is to understand how São Paulo planters

**Universitat Autonòma de Barcelona.*
***Universidade Estadual de Campinas.*

eventually solved their labour problems. The effective supply of labour at any particular moment is importantly affected by the sanctions and incentives available to employers to enforce their power over labour to extract profit. In the São Paulo case, the absence of an established labour market decisively determined the planters' choices of labour systems and their evolution. The early difficulties with free labour have repeatedly been attributed to the relative unprofitability of immigrant workers in comparison to slaves,[2] yet what induced planters to introduce free labour in the first place was their increasing awareness that slavery was doomed. The more interesting initial question, then, is why planters first chose sharecropping as the labour system under which free labour was to be introduced, rather than a straight wage system or some other arrangement.

The absence of an established labour market not only decisively determined the planters' choice of labour systems but also their fortunes with them. The success of the labour systems introduced was not only determined by cost factors (in the narrow sense of the cost of obtaining immigrant labour), nor even by the planters' ideology (their supposed 'backwardness' or, alternatively, their exemplary entrepreneurial spirit), but importantly by the struggle between planters' actions and workers' responses as shaped by the economic and political circumstances in which they found themselves. As we will attempt to show, it is this interrelationship between systems of labour exploitation and patterns of labour resistance which explains the successive transformations of the forms of labour contracting adopted.

THE SHARECROPPING CONTRACT

In 1847, Senator Vergueiro, the owner of a large estate near the town of Limeira, in the province of São Paulo, became the first planter to introduce immigrant labour for work in coffee production.[3] Vergueiro, as his son later put it, had foreseen that the end of slavery was only a matter of time.[4] Initially the immigrants seem to have been offered two kinds of contract: a sharecropping and a labour leasing [*locação de serviços*] contract, but they opted for the former.[5]

According to the sharecropping contract, the planter financed the immigrants' transportation from their country of origin to the port of Santos, advanced the cost of transport from Santos to the plantation, as well as the foodstuffs and tools the immigrants needed until they could pay for them with the proceeds from their first crops. The planter assigned the workers the number of coffee trees they could tend, harvest and process, and allotted them a piece of land on which to grow their own food crops. In addition, the immigrants were given a house, apparently free of charge. Their remuneration consisted of half the net profit from coffee and from the food crops. The labourers were obliged to repay the expenses incurred by the planter on their behalf with at least half of their yearly returns from coffee. The initial contract did not specify the length of its duration, but stated the amount of the debt owed by the immigrant on account of his transportation costs and other advances. For any amount outstanding after two years, the labourer was to be charged interest, which was to be the case with other advances after one year. Finally, the immigrants could not legally move off the plantation until they had repaid their debts. Should they do so, they incurred a substantial fine. Work was organised and supervised by the planter or his administrator. The labourers were explicitly required to conduct themselves in a peaceful manner.[6]

The planters thus transferred all the expenses of obtaining immigrant labour to the labourers themselves. Consequently, such workers started out already burdened by a substantial debt. It was generally expected that a diligent labourer would take an average of four years to pay off his debt.[7]

In the early 1850s, a number of planters, impressed by the apparent success of the Vergueiro experiment and concerned with the effects of the recent termination of the slave trade, approached Vergueiro and Co. to obtain immigrant labourers for themselves. At this point, contracts became more onerous for the immigrants. Not only did Vergueiro and Co. begin to charge a sizeable commission, to be debted to the labourers, but interest on debts was now charged from the date of arrival, sometimes at a rate of 12 per cent rather than the previous six per cent.[8] Moreover, from the early 1850s onwards the whole immigrant family was held liable for the debt, which was the way found by planters to attempt to protect themselves against losses in the case of the death of the family head.[9] Finally, while free labourers had initially been expected to process the coffee they harvested, by 1856 they were gradually relieved of this task, and instead were charged a fixed fee per unit of coffee harvested.[10]

By 1855 there were about 3,500 immigrant labourers working on 30 plantations in the province of São Paulo.[11] In most cases free labour existed side by side with slavery, although planters early on established a certain technical division of labour. All tasks beyond coffee cultivation and harvesting, and which were said to require constant supervision, or were inappropriate for sharecropping, continued to be executed by slaves. Such tasks included preparing the soil for various crops, planting new coffee, planting yearly crops for the plantation's consumption and, increasingly, the processing of coffee.[12]

Almost simultaneously, a further measure directly related to the labour problem was introduced. In 1850 a Land Law was passed which consolidated private property rights and was intended to prevent immigrants from becoming landowners through simply settling on public lands.[13] The existence of vast expanses of unoccupied territory posed, in fact, one of the serious obstacles to employing the scattered national population of free workers, who never formed a significant part of the plantation work force in the nineteenth century.[14]

Since the recruitment of immigrant labour required an initial investment by the planters, one of their constant concerns was to guarantee this capital investment.[15] However, for the immigrants, the initial debt weighed increasingly heavily on their income. In practice, their returns from coffee cultivation turned out to be markedly lower than the stipulated 50 per cent of net profit. The consequences were far-reaching.

The first serious sign of the mounting discontent among immigrant labourers came in mid-1856, when a group of Swiss workers revolted on the Nova Olinda plantation near Ubatuba. The trouble apparently started when cattle invaded their food plots, and disagreement over indemnity led to the intervention of the police. Eventually, when the Swiss consul visited the plantation, the central issue became the general conditions of the immigrants: quality of food plots, fulfillment of contracts, housing, etc. Planters typically attributed the events to instigation by subversive elements. The conflict was finally resolved when the consul promised the immigrants that they would be transferred to a government colony, that part of their debts would be forgiven, and that no interest would have to be paid.[16] This conflict already reveals what were to become some of the crucial points of contention between planters and workers.

The most important revolt, however, began in December, 1856, among Swiss and German labourers on Senator Vergueiro's model plantation Ibicaba. The presence of a Swiss school teacher, Thomas Davatz, contributed to converting the latent resentment of the immigrants into an organised movement of protest against what they felt were grave irregularities in the fulfillment of their contracts. They did not question the terms of the contract as such but protested against 'the calculation of returns from coffee produced, the charging of the commission, the unfavourable exchange rate used to convert their debts into local currency, the charging of transport from Santos to the plantation, and the strange division of profits from the sale of coffee.'[17] What seems to have sparked the revolt was disillusionment with the results of the 1855 harvest which, contrary to their hopes, did not allow them to reduce their debts. As Davatz later wrote, what he and his compatriots demanded was no more than just treatment.[18] The revolt ended when Davatz was expelled and some of the other leaders left the plantation. One measure of the apprehension which the revolt produced among the planters was the accusation levelled against Davatz that not only was he in alliance with extraneous elements, but that he harboured communist and other terrible plots.[19] Many planters appear to have been genuinely terrified that the revolt would not only spread to free labourers on other plantations (a link with the earlier Ubatuba conflict was repeatedly suggested),[20] but worst of all might incite the slaves.

Much has been made of this revolt and of the frauds committed by planters which supposedly triggered it off. These are generally used to explain the failure of the sharecropping system under which the first immigrants were contracted, and the alleged decline of interest among planters in the use of free labour.[21] It seems, however, that these irregularities (fraudulent measures, dishonest calculation of profits, etc.) only added to the growing disillusionment of immigrants with their living and working conditions. The planters, for their part, acted under a serious misconception. They had not counted on the resources available to the immigrants in resisting what they considered unjust contractual conditions and impositions. The introduction of free labour required a capital outlay whose amortisation demanded a level of exploitation which the planters were unable to enforce. The main obstacle was not the threat of possible reprisals from the immigrants' national governments, since protection was reluctant and limited at best,[22] but rather, under the circumstances, the sharecropping contract did not adequately address the problem of creating a reliable labour force.

Why then did São Paulo planters adopt the sharecropping system? It has long been maintained that sharecropping is less efficient than wage labour, the usual argument being that since sharecroppers receive only a part of the product they will stop work earlier than a wage labourer does.[23] More recently, the prevalence of sharecropping has been attributed to its greater efficiency in risk dispersion.[24] The São Paulo planters' option for sharecropping has been interpreted in similar terms. Holloway suggested that uncertain coffee yields and prices induced planters to forego some of the potential income which was exclusively theirs under slavery for it prevented the frightening possibility that wages might absorb more than the income from the crop.[25] This argument obscures the essence of sharecropping, which is its particularly exploitative character. As Reid has demonstrated for the post-bellum American South, rather than uncertain outcome, what in fact explains the adoption of sharecropping are the distinctive incentive features it contains, in contrast to wage labour.[26] Sharecropping in a situation of scarce labour is in fact more efficient than wage labour. It is a form of the use of labour similar to a carefully negotiated piecework system. Both are forms of incentive wage systems, a way of securing extra effort from labour, of making labourers work harder and better for only a small increase in total remuneration over that of wage labourers. Remuneration in the form of a proportion of the product constitutes an incentive for the labourer to intensify his effort since it is on the amount produced that his returns will depend. He will cultivate with greater care, again because part of the result will accrue to himself. In addition, the supervision required will be insignificant, since control of work is exercised by the labourer himself.[27]

In the case of São Paulo, because of the absence of a local supply of workers, labour costs were at least initially high. Moreover, coffee is a very labour-intensive crop. Because of the incentive element characteristic of sharecropping it could be expected that sharecroppers would tend more coffee trees per worker than would wage labourers. Consequently, fewer workers would be required, and initial investment would be lower. Sharecroppers are usually contracted in family units. Thus, sharecropping also allows the landowner to benefit from the use of the sharecroppers' family labour.[28] Planters had always opposed recruiting single men since it was argued that immigrant families were less prone to abandon the plantations. This may be so, but equally important was surely the fact that the immigrants' families constituted a cheap labour reserve. A sharecropper will usually accept a division of the product that will not fully cover the potential market price of family labour, which would otherwise remain under- or unemployed. Planters, in fact, sometimes prohibited immigrants and their families from working outside the plantation.[29] The planter thus obtained this additional labour at a cost below that which he would have had to pay were he to contract it on the market as wage labour. Since labour needs during the harvest were about one-fifth greater than during cultivation,[30] the workers' wives and children could satisfactorily cover this additional demand.

Labourers were also assigned a subsistence plot, which was a further way of reducing unit labour costs. These plots were usually granted on marginal lands not appropriate for coffee, or on virgin land later planted in coffee. Workers had no effective possession of the land. Ideally, they were expected to produce strictly what was needed for their own subsistence, in this way further reducing the cost of their own reproduction.[31]

By reducing unit labour costs in comparison with wage labour, sharecropping must have initially appeared to the planters as the most appropriate substitute for slave labour. The incentive element must have seemed a satisfactory substitute for the coercion which made slaves work. The question was not merely to fill the potential gap in labour supply, but to do so in a profitable way.

Immigrants, however, were free workers. As sharecroppers, they were in principle free to decide on labour intensity and the allocation of labour. Their diligence and productivity in coffee thus depended on their own appraisal of returns.

Planters and immigration agents sought to create the illusion that immigrants would quickly be able to repay their debts and acquire their own land. In practice, however, immigrants usually had to wait for at least two years before receiving significant returns for their efforts. The share to which they were entitled from the first harvest took almost another year to be paid because of delays in the marketing of coffee; but, since the contracts stipulated that half of the workers' annual earnings from coffee were to be withheld to cover their debts, and in the meantime they had accumulated new debts from further advances, only in the third year could they expect to receive much cash.[32] It is hardly surprising that the immigrants grew increasingly discontented.

The 1856 revolt remained an isolated event.[33] The majority of the immigrants reacted in a less dramatic but at the same time more insidious manner by systematically restricting output in coffee cultivation. Planters soon grew concerned about the immigrants' low productivity in coffee. As late as 1870, an emissary of the Imperial government noted that 'most plantations are not yet in a condition to receive free labour even under the sharecropping system, mainly when workers already start out burdened with a debt ... The transportation costs are excessive and consequently the labourers' share is insignificant ... The delay in the sale of coffee forces the workers to wait over eight months for payment ... As a consequence, the workers generally tend a reduced number of coffee trees, preferring to plant food crops to supply their homes and cover their needs, in addition to obtaining immediate benefit. Yet it is evident that this system cannot be advantageous for the planter, whose main interest is coffee.'[34]

The initial debt, even without arbitrary additional difficulties created by the planters, discouraged any effort by the workers in coffee cultivation beyond what was strictly necessary. As another observer remarked: 'They [the labourers] abandon the coffee trees, which as a result do not produce but deteriorate, and the planter is deprived not only of his share in the product but also of that of the immigrant which is the only source of amortisation [of the debt] he has.'[35]

The contract left open the number of coffee trees to be tended by a family and the size of their subsistence plot. Both of these elements were initially left to the decision of the labourers themselves. This made it possible that, as immigrants became uninterested in repaying their debts within the expected time, they increasingly diverted their labour to food crops whose returns accrued to them directly and immediately. Although many observers remarked on the alleged laziness and lack of interest of the immigrants,[36] in fact what happened was an alternative allocation of labour to food crops, rather than an absolute underuse of labour capacity.[37]

Most of the immigrants in the early 1850s were members of the rural or urban poor who were driven by the severe economic crisis in Central Europe to abandon their home country, in many cases as a matter of sheer survival. What they probably initially hoped for was to make a secure living.[38] Since the conditions they encountered in São Paulo made it almost impossible for them to obtain a profit from work in coffee, they preferred to dedicate a significant part of their efforts to food crops. As a result, productivity in coffee was low. As Carvalho de Moraes quite appropriately observed, 'the planters were at the mercy of the *colonos*'.[39]

The planters' power to control labour and enforce a satisfactory level of productivity in coffee cultivation was limited by the circumstances under which free labour had been introduced, that is, in the absence of a local reserve of labour and under arrangements which required that immigrants repay their costs of passage and initial settlement to their employers. Planters presumably thought that the incentive element contained in sharecropping would effectively replace market forces in reducing wage costs. However, the initial debt cancelled out the incentive element, and the planters lacked any effective means of forcing their workers to produce coffee. The threat of dismissal, which is the usual form of persuasion used by employers to enforce labour contracts, was hardly practical since it would have meant the partial or total loss of the planters' investment. Although it was true that the immigrants could not legally abandon the plantation until they had paid off their debts, neither could the planters make them work beyond what the labourers themselves were willing to do. Even the use of state power, as in the case of the Ibicaba revolt, was of little avail. The leaders of the revolt were expelled, but those who remained did not work any harder. The use of extra-contractual means to amortise their investments had backfired: 'it is incontestable that the planters, when the labourers would not work, could not enforce the fulfillment of their obligations and thus suffered harm from the mistreatment of their coffee trees, the reduction of the harvest, and the total or partial loss of their advances.'[40]

THE LABOUR-LEASING CONTRACT

By the late 1850s, planters faced a dilemma. Economic incentives had failed to produce the expected results: a reasonable level of productivity and amortisation of debt within the expected time. After 1857 the sharecropping system was gradually abandoned in São Paulo. Free labour, however, by no means disappeared. While the number of immigrants engaged in coffee cultivation did not increase during the next two decades, it declined only slowly. Reports reaching Europe on the immigrants' hardships eventually persuaded both the Swiss and Prussian governments to take severe measures which practically halted emigration from the two countries to São Paulo.[41] Nevertheless, in 1870 it was estimated that approximately 3,000 free labourers – some of them Brazilians – still worked on the plantations, a decline of about 500 since 1860, and the debate over possible solutions to the labour problem had not abated.[42]

Many planters continued to explore alternative labour systems and to devise institutional safeguards which they hoped would allow a more effective enforcement of contracts. In order to deal with the related problems of control of productivity and amortisation of debt, planters initially resorted to contract changes. Sharecropping was gradually replaced by a labour-leasing contract [*locação de serviços*]. Instead of a share of the value of production, labourers were henceforth paid a pre-established piecerate for each measure of coffee produced. It was argued that by thus reducing uncertainty over income and eliminating long delays in payment, labourers would feel encouraged to apply themselves with greater diligence to coffee cultivation.[43] In addition, the clause that surplus food production be shared with the planter was generally dropped, and increasingly the size of the food plot was fixed and/or let against a rent in an attempt to discourage immigrants from diverting labour to food crops.[44] Moreover, free labourers no longer participated in coffee processing, either directly or through a fee. This task reverted to slaves until the 1880s, when it was then carried out by wage labour. Significantly, it was this part of coffee production, together with transportation, which were rapidly mechanised in the early 1870s as slave labour became increasingly problematic.

Labour-saving innovations, however, were not introduced in coffee cultivation. Mechanisation of the harvest was not technically feasible, and the mechanisation of weeding by the use of a cultivator would have severely upset labour demand throughout the agricultural year. Mechanised weeding would have either produced idle labour during the cultivation period, or a shortage for the harvest in a situation of general scarcity of free labour.[45]

The labour-leasing contract continued the wage incentive system, but it still could not assure an adequate level of productivity because it did not resolve the basic problem of debt as a disincentive to increased productivity in coffee. Planters, in fact, felt they also needed additional legal powers to protect themselves against the non-fulfillment of contracts and specifically the non-payment of debt.[46]

Some attempts had been made earlier to prosecute immigrants for non-payments of debts. However, the law regulating sharecropping was largely ineffective. It allowed planters only to rescind the contract or demand indemnity for damages; the former implied loss of the immigrants' debt, and the latter increased the debt without, however, providing means which obliged labourers to work to pay for it.[47]

Occasionally, planters also tried to apply to sharecropping a law of 1837 regulating labour-leasing contracts. According to this law, any labourer who, having been dismissed, did not pay his outstanding debt, could be jailed and condemned to public works until he had paid up. In case of abandonment of the plantation, he was to be arrested immediately and not released until his debts had been paid.[48] However, the applicability of this law to sharecropping had proved uncertain.[49] Thus, a further reason for planters to prefer the labour-leasing contract was surely their desire to avail themselves of the more severe penal sanctions contained in the 1837 labour-leasing law.

However, there is little evidence that even the 1837 law was ever widely applied. The contracts had not usually stipulated a fixed period for the amortisation of debt and, in any case, the use of the law to press for repayment did not secure the planters' primary objective, which was to achieve greater productivity of labour in coffee. As long as the immigrants were willing to remain on the plantation, there was little the planters could do to make them work, short of outright coercion, which they knew could produce untoward results.

Productivity in coffee continued to be low.[50] A survey of families resident on the plantation Martyrios in 1869, the property of Senator Francisco Antonio de Souza Queiroz, was published by Carvalho de Moraes. Labourers were contracted under the labour-leasing system. The survey contains data on size and composition of the families and number of coffee trees tended by each. With these data it is possible to calculate the number of trees cultivated per family and per labourer by the consumer/worker ratio of the family.

NUMBER OF COFFEE TREES TENDED PER FAMILY AND INDIVIDUAL LABOURER BY CONSUMER/WORKER RATIO OF THE FAMILY: PLANTATION MARTYRIOS, 1869

c/w ratio	1.0-1.4		1.5-1.9		2.0 and over	
	per family	per worker	per family	per worker	per family	per worker
trees tended	2,109	566	2,071	709	1,940	813
number of families	11		7		5	

Source: Carvalho de Moraes, op. cit., appendix 17. The fact that with increasingly more favourable consumer/worker ratios (1.0-1.4) the number of trees tended per family increases somewhat is due to the fact that those families generally contained a larger absolute number of workers.

The average number of coffee trees cultivated by one labourer varied from 566 to 813 trees. Moreover, the larger the productive capacity of the family (families with a c/w ratio between 1.0 and 1.4), the smaller the number of trees tended by each labourer in it. The overall production intensity per labourer in coffee was low, especially when compared to the usual average of 2,000 to 2,500 trees tended from the 1890s onward by workers who were also producing food crops at the same time. According to the 1869 statistics, those families which had the comparatively largest productive capacity, exerted themselves least in coffee cultivation.

In addition, the quality of work appears to have deteriorated under the labour-leasing contract. Such a system of remuneration not only affects labour intensity, but may also have consequences for the care with which tasks are executed. In general, piecework is not used for many agricultural tasks because the quality of work suffers. There is evidence that under the labour-leasing contract, while the immigrants were keen on harvesting as much coffee as possible, they tended to be negligent in the weedings or abandoned outright part of the trees once they had been harvested.[51]

In effect, neither contract changes nor the use of more severe labour laws succeeded in creating a satisfactory labour force because these measures failed to resolve the basic problem: the initial investment planters had to make to introduce free labour. Heavily burdened by their initial debts, immigrant labourers continued to work little in coffee and were troublesome.

THE COLONATO

Those planters who still worked their plantations with free labourers by the late 1860s gradually introduced further adjustments in the labour contract.[52] While contracts initially had contained fines for abandoning the plantation before the worker's debt had been repayed, they now increasingly stipulated fines for the non-execution of weedings. As one planter noted, 'for us the only advantage of slaves is in discipline, and once a planter wants to renounce a little control and patiently bears the faults of the *colonos*, he will gradually succeed, by means of the fines contained in the contracts, in making all *colonos* submit to regular work.'[53] Food plots were now regularly assigned in proportion to the number of trees tended by the family. Finally, some planters began to introduce a new form of remuneration, a mixed task and piecerate system, the *colonato*, an arrangement which was to prevail on the coffee plantations from the 1880s until the 1950s. Under this system, coffee weedings were paid at a fixed annual rate per thousand trees tended and the harvest at a piecerate.

By paying a separate stipulated rate for weedings, a sort of fixed-minimum wage,[54] which guaranteed the labourers a stable income independent of coffee yields, it could be expected that workers would not neglect the coffee groves outside of the harvest season. In addition, since part of the labourers' remuneration under the new contract depended directly on the number of trees tended and no longer on their yields, it could be assumed that they would feel encouraged to cultivate a larger number of trees. By maintaining the piecerate system for the harvest, however, labour costs could still be adapted to annual fluctuations in yields. Moreover, unit labour costs could be lowered through intensification of effort on the part of the immigrants' family at the time when labour demand was greatest.[55]

By the late 1860s, however, there was a growing awareness among planters that contract adjustments and the existing labour laws were insufficient in themselves to assure a disciplined and profitable free labour force. The decade of the 1870s began amid predictions of an impending labour crisis, both in terms of future supply and of labour control. Planters attempted to deal with the situation in several ways. Many continued to introduce free labour on a limited scale, though privately sponsored schemes had lost most of their appeal.[56] Moreover, planters' faith in this system of immigration was further shaken by a new outbreak of labour conflict on some plantations run with free labour.[57] At the same time abolitionist agitation was growing, especially after the 1868 Manifesto of the Liberal Party which called for an end to slavery. Pressure thus mounted for a comprehensive solution to the labour problem.

Nevertheless, coffee production during the 1870s expanded to almost twice what it had been in the previous decade.[58] In fact, planters were still able to postpone until 1888 the effective end of slavery. They temporarily solved the anticipated labour shortage by generally rationalising coffee production. They managed to disarm the abolitionists with a very limited concession: the Rio Branco law of 1871 which declared that children born thereafter of slave mothers were to be free.[59] Planters also continued to purchase substantial numbers of slaves from the Rio de Janeiro region and from northern Brazil.[60] The great expansion of railroads, almost all located in the coffee-growing areas, which took place in the 1870s also helped postpone an acute labour shortage. The availability of railroad transport allowed planters to reassign to other tasks the slaves they have previously had to use in transporting their crop to market.[61] Moreover, by lowering the cost of transportation and reducing the damage the crop suffered en route, the railroad helped compensate the planters to some extent for possible rises in the cost of slave labour.

The introduction of labour-saving machinery in the coffee industry was another important element in forestalling the crisis.[62] José Vergueiro perhaps best exemplified the spirit shown by the more astute planters when he remarked in 1874 that 'saving labour is the principal objective which we must always have in mind since, if time is money, the saving of labour is also.'[63]

Coffee production on the plantations run by slave labour, still the large majority, became generally more rational. As one São Paulo official put it, not only had the introduction of machinery 'profitably replaced a good part of the work force', but 'the direction of the workers has become more intelligent, and the division of labour has been put into practice.'[64]

Although São Paulo planters managed to cope with their labour needs quite successfully during the period, they did not by any means abandon the question of how to organise the supply of free labour.[65] The issue of negotiating the transition from slavery to free labour became increasingly urgent. It was the old problem: Brazil 'possesses the best climate in the world, almost all the precious metals and a prodigiously fertile soil, but lacks population and for that reason is poor.'[66]

At the 1878 Agricultural Congress, called by the government to evaluate the general state of agriculture, one group of planters opposed large-scale immigration as a solution to the labour problem because of its costs to themselves or to the country. They demanded, instead, laws to combat the alleged aversion of the local population to work. They sought means of disciplining *agregados* and of forcing work from *ingênuos*, as well as provisions to tighten the 1837 law regulating labour-leasing contracts.[67] In the end this position was to be overruled by those planters who believed that reliance on former slaves after Abolition, or on the available local population, to cover their labour needs would be highly problematic, and who saw in subsidised mass immigration the only possible solution. But before this latter group finally succeeded in forcing the state to assume full responsibility for mass immigration, the government made a last attempt to relegate this task to the planters themselves by granting them some additional legal powers to control labour. In response to those planters who were pressing for more stringent labour laws, further legal provisions were passed to discipline the increasingly unruly free labourers. In 1879, the 1837 labour-leasing law was replaced by a new regulation covering both labour-leasing and sharecropping contracts and providing prison sentences not only for abandoning the plantation without just cause but also for strikes and incitement of others to strike through threats or the use of violence.[68]

This law was surely not least the result of renewed outbreaks of labour unrest on some plantations. Almost simultaneously with the Agricultural Congress, the Tyrolean labourers on the plantation Salto Grande in Amparo, owned by Joaquim Bonifacio do Amaral, declared a much publicised strike.[69] The labourers began their movement at the onset of the coffee harvest in protest over a number of abuses and omissions of which they felt they had been the victims.[70] It was feared that the strike might spread to other estates. Although the planter made a certain number of concessions, such as raising the piecerate from 500 reis per alqueire of coffee harvested to 600 reis, and agreeing to charge interest on outstanding debt only after two years, the labourers could not be persuaded to resume work. The owner then stopped providing foodstuffs, a well-known means of forcing labourers back to work, and at the same time had the leaders prosecuted and condemned to prison sentences for non-fulfillment of contract under the 1837 law.[71] These events seem to have convinced even such a staunch defender of privately sponsored immigration as Amaral of the urgent need for subsidised mass immigration, although he still insisted that more severe laws to enforce contracts were equally necessary.[72]

Other São Paulo planters, in contrast, became increasingly emphatic about the need for the liberalisation of labour contracts and legal reforms as indispensable prerequisites for a successful programme of government-sponsored immigration, which they regarded as the only solution to the labour problem. The relatively low coffee prices of the early 1880s, and the difficulties planters faced in securing credit, acted as further deterrents to privately sponsored immigration, since they left many planters unable or unwilling to advance passage money to immigrants.[73] On the other hand, the 1879 law was largely ineffective and even counter-productive. As Antonio Prado, one of the most prominent São Paulo planters noted, immigrants in jail were neither repaying the planters' loans nor harvesting their coffee, and in addition the law only served to discredit Brazilian colonisation in Europe. Prado further observed that the law would soon become unnecessary in any case, at least in São Paulo, since the Assembly of that province had approved a measure in 1884 providing free passage for immigrants who went into agriculture.[74]

Planters by this point were generally aware that 'it is impossible to have low salaries, without violence, if there are few workers and many people who wish to employ them.'[75] The way out of this dilemma, and the essence of the São Paulo immigration system, was explained by another member of the Chamber of Deputies shortly after Abolition: 'It is evident', he said, 'that we need labourers . . . in order to increase the competition among them and in that way salaries will be lowered by means of the law of supply and demand.'[76]

As long as coffee plantations could be run predominantly with slave labour, it was difficult to secure state-subsidised immigration on a large scale. However, by the early 1880s Abolition clearly had become unavoidable, and a comprehensive solution had to be found. The São Paulo planters, whose power in the government had been increasing steadily, then finally succeeded in imposing their solution to the labour problem. Although more than two years and several important modifications of the law providing for subsidised immigration were required before the system functioned satisfactorily, Prado was essentially correct in affirming that the problem had been solved by the mid-1880s. After 1884, the state, instead of coercing labourers directly, sought to achieve the same objective – cheap and obedient labour for the plantations – by flooding the labour market with subsidised immigrants.

By 1886 the provincial government had found an effective way to provide complete subsidies for immigrants, and the effect was almost immediate. By May of 1887, some 60,000 to 70,000 immigrants, by now predominantly Italian, had already been placed in the agricultural establishments of São Paulo.[77] This figure exceeds the estimated 50,000 slaves who were being used on São Paulo coffee plantations in 1885.[78]

Immigration policy remained essentially unchanged until the First World War. Between 1884 and 1914, some 900,000 immigrants arrived in São Paulo, mostly as cheap labour for the coffee plantations. The immigration programme allowed São Paulo planters not only to abolish slavery with only moderate inconvenience but the scheme – aided by at least initially high coffee prices – created the conditions for sustained expansion of coffee production. Between 1888 and 1902 the number of coffee trees planted in São Paulo increased from 221 million to 685 million.[79]

Subsidised mass immigration and the creation of an effective capitalist labour market seemed to solve both of the difficulties planters had previously encountered with free labour – the debt and labour discipline. Since planters no longer had to advance passage money to immigrants and were assured of an abundance of fieldhands by the state, the 1879 law fell into disuse and was finally repealed in 1890 because 'among the economic measures most strongly demanded by the present state of the country is the need to populate it, since public wealth develops in direct proportion to the increase in population.' In order to attract sufficient numbers of immigrants the 'shameful' law of 1879 needed to be abolished. Contracts henceforth should be 'acts of pure convention based on mutual agreement, which elevate the labourer to the category of partner in the contract.'[80] The 1879 law, however, had not only served to enforce contract fulfillment, but also prohibited strike action. Among other liberal reforms of the period was one replacing the 1879 provisions with a decree on so-called 'crimes against the freedom of work' which penalised the inciting of either labourers or employers to increase or reduce work or wages through the use of threats or violence.[81]

Contract enforcement and labour discipline were henceforth to depend on the mechanism of the market. Planters did indeed employ coercion and violence under certain circumstances to keep labourers on the plantations and extract profit, but in general they preferred to deal with the problem of keeping down labour costs by increasing supply. Extra-economic coercion, which was at times substantial, served essentially to improve the planters' bargaining position in the labour market.

Even after the 1880s, planters regularly claimed that there was a shortage of agricultural labour in São Paulo. There are, however, a number of indications that these laments were for the most part debating points to press for continued mass immigration so as to assure abundant labour at the low wages the planters wanted to pay. For one, the two alternative sources of labour, the freedmen and so-called national labour were never tapped in any substantial way before the First World War, when subsidised European immigration became impratical. Both these groups were largely ignored by the planters even during alleged labour shortages.[82]

Moreover, agricultural wages varied little between 1884 and 1914. At the end of the century, with falling coffee prices, there was even a decline, followed between 1902 and 1910 by a moderate rise due to the relatively small number of immigrants entering São Paulo and the large number of departures. This was countered, however, by an increase in immigration in 1912 and 1913 so that by 1914 money wages were back to about the same level they had been in the mid-1880s. In the same period real income of the labourers seems to have declined significantly.[83]

Even at the beginning of this century, when complaints over labour shortages became particularly vehement,[84] there does not seem to have been an absolute scarcity. Such propaganda was to some extent a reaction by planters to a 1902 Italian ban on subsidised immigration – which was evaded in part and offset by the introduction of Portuguese and Spanish labourers. However, planters in the older coffee regions did face difficulties in competing for labour with newer regions at a time of declining coffee prices, and at the wage levels they were willing to offer.[85] An indication of the dubious nature of the planters' complaints is the fact that neither was the 1903 suggestion to cut 20 per cent of the coffee trees so as to increase labour supply ever heeded nor, as far as we know, did coffee ever go unharvested for lack of labour.[86] Interestingly, it was in this context that a return to sharecropping was proposed, which turned out to be unfeasible, however, since labourers, 'fearing a further decline in prices, and having work with fixed results, reject sharecropping'.[87] Instead, planters in the older areas increasingly allowed food crops to be grown between the rows of coffee in order to attract labour.[88]

In effect, the massive influx of immigrants covered labour needs quite satisfactorily before 1914. Not only did subsidised mass immigration drastically reduce initial labour costs, but planters were now required to make only minor advances for food and agricultural tools. In addition, the disappearance of the burden of the initial heavy debt, coupled with stricter regulation of work and fines for non-execution of tasks, now possible to enforce because of the abundant labour supply, produced a marked increase in labour productivity in coffee. By the late 1880s the average number of coffee trees tended by an adult man ranged between 2,000 and 3,000, while women usually cultivated half that number.[89]

Planters continued contracting immigrant labour under the mixed task and piecerate system coupled with subsistence production. The contract was now annual. Even with an abundant labour supply, the *colonato* system continued to be more profitable than straight wage labour for a number of reasons. Food prices were persistently high since foodstuffs were to a large extent imported as a consequence of the planters' almost exclusive interest in coffee.[90] As the president of the province noted in 1887: 'Coffee gives the best returns . . . it would be an error to disdain it in order to cultivate something else.'[91]

Food plots acted as an incentive for the labourers, at the same time reducing the cost of the reproduction of labour. Moreover, since food crops were grown during the coffee cultivation season when labour demands were comparatively lower, the planters could make full use of the immigrants' family labour throughout the whole year. As one observer noted in 1908, 'what really enables the immigrants to make ends meet are the crops they have the right to raise on their own account.'[92] Bonardelli in 1916 estimated that food crops grown by labourers made up one third of their income.[93] The right to grow food crops rather than constituting a payment in kind or a means of fixing labour, was primarily a way of extracting a labour rent in addition to the surplus produced in coffee cultivation.[94] The combination of cash crops with subsistence cultivation was potentially problematic for economic development in general, and for the planters themselves – as late as 1922 planters stressed the need to restrict food crops since 'the landowner cannot permit that labourers plant extensively in cereals when labour is lacking for coffee cultivation.'[95] However, if effectively controlled, and as long as coffee was highly profitable, the arrangement was clearly advantageous for the planters.

A final way the planters devised to reduce unit labour costs was their growing preference, not only for families, but for large families, i.e. units containing at least three workers.[96] A large number of workers in a family made it possible to depress wage costs further. The more workers per consumer in a family, the lower the cost of reproduction of each individual worker and consequently the lower the task-rate could be.[97]

There seems little doubt, therefore, that for much of the period between 1884 and 1914 coffee planting was a profitable undertaking. Even after 1896 when coffee prices started to decline on account of over-production, reasonably well-run plantations were still returning a profit.[98]

This was possible, however, through an increase in the rate of exploitation as well as various frauds.[99] Most Italian observers at the time agreed on the extremely poor working and living conditions of immigrant labourers.[100] Planters, for their part, usually claimed that 'the crisis' at the turn of the century had forced them to treat their labourers the way they did.[101]

The planters' power increased in direct proportions to the growing powerlessness of the labourers. Almost all plantations had their band of *capangas*, armed guards in charge of carrying out the will of the planter and of controlling among other things the entrance and departure of labourers. Planters occasionally attempted to retain their labourers by such means as artificially delaying the end of the agricultural year when the final financial accounts were settled, thus making it unlikely that the labourer 'could find work on another plantation where cultivation for the next crops had already begun'.[102] Nevertheless, the mobility of labour was generally considerable. Some 40 to 60 per cent of the labourers left their plantation each year.[103] Planters exercised extensive control over their labourers, in itself an expression of their abiding fear of indiscipline or worse. Observers sometimes remarked on the *colonos*' 'submissiveness, respectful and docile' behaviour, although this was often more apparent than real.[104]

The obstacles to more than sporadic individual resistance were immense, and planters did what they could to suppress any expression of discontent. For example, all societies or associations of labourers were prohibited.[105] Yet despite severe control, not only were individual labourers often turbulent and sometimes violent, but strikes occurred with some frequency.[106]

Labourers acting individually or nearly so quite regularly carried out acts of violence against planters and overseers. One of the most famous instances, though by no means atypical, occurred in 1901 when Francisco Augusto Almeida Prado, a planter belonging to the prominent Prado family, was so careless as to stroll through his coffee fields one day without the protection of his bodyguards. Several of his labourers took advantage of the situation and murdered him, riddling his body with knife wounds, and chopping it into pieces with their hatchets and hoes. Diogo Salles, the brother of the President of the Republic, was murdered in 1900 by a labourer for attempting to rape the murderer's sister. This was an individual motive, but it is as much indicative of the extreme forms exploitation took, as of the underlying tensions. Other incidents arose from somewhat broader issues. In the Prado murder, the immediate cause was said to be the punishment he had imposed on his labourers for refusing to put out a fire on the estate of one of his relatives.[107]

Collective action was also not infrequent. The first strikes after Abolition occurred in the early 1890s, and the Italian-language daily *Fanfulla* registers several dozen by 1913; from that year to 1930, the *Patronato Agrícola* (a state agency) and the labour press provide references to over a hundred strikes on the coffee plantations.[108] Most such strikes were limited to the boundaries of one plantation, although in 1911 about 1,000 labourers struck on half a dozen plantations in the area of Brangança for twenty days and secured a slight increase in pay. The following year, labourers on more than a dozen plantations in the Ribeirão Preto area went on strike and also secured a small wage rise. The largest strike of the period took place in the same area in 1913. It mobilised between 10,000 and 15,000 labourers, but resulted in a total defeat. Strikes were generally carried out in an effort to secure an increase in wages, or in opposition to such measures as the non-payment of wages, attempts to reduce salaries, heavy and arbitrary discipline and fines, or restrictions on the planting of food crops.[109]

From the beginning, the relations between planters and free labourers were fraught with severe and generally rather explicit tensions. Even had the planters desired to establish paternalistic relations of personal dependence – and the point remains unclear – a number of elements limited the ready use of such arrangements to obscure exploitation and discourage resistance. Not only were São Paulo plantations large, but the work force was both new and foreign-born, thus depriving the planters of many of the traditional sanctions – religious and otherwise – which rural ruling classes have often enjoyed.

As long as sharecropping prevailed, struggle between labourers and planters centred around the share in net profit from coffee. Exceptionally, labourers revolted, but more generally they resisted by withdrawing labour from coffee cultivation. As long as food growing was not subject to systematic restraints, the labourers were able, without jeopardising their own survival, to deprive the planters of part of their labour power. This does not mean, however, that labourers reduced their expectations. What they did was to divert a significant part of their productive capacity to food growing, beyond what they required for subsistence. Though essentially individual, these actions were decisive in determining the readjustment of the labour system.

The transition from sharecropping to the mixed task and piecerate system, the *colonato*, was a process of increasingly systematic exploitation of labour, aided after the mid-1880s by the massive importation of immigrants. Once a capitalist labour market had been created, planting rights restricted to the basic minimum, and increasingly severe labour discipline instituted, conditions effectively disappeared for individual struggle at the level of work against what were considered unsatisfactory returns. By the turn of the century immigrant labourers constituted a homogeneous mass, subject to more or less uniformly harsh conditions. Under normal circumstances they had the possibility of exploiting regional differences to their advantage, or of abandoning São Paulo agriculture altogether either for work in Brazilian cities or in other countries. Labour mobility among plantations and outright departures were considerable. But for those who remained in agriculture, low wages or any additional exactions by planters, such as the non-payment of wages, the prohibition of food growing, or a reduction in wages, could trigger off collective action in the form of strikes. Contradictorily, by increasing labour discipline, planters created the very conditions for collective – and thus potentially much more damaging – action by the immigrants.

CONCLUSION

This is not the place to attempt a survey of the vast and growing literature on the history of Brazilian agriculture.[110] Our contention, however, is that most of these studies have tended to apply macro-models to the plantation labour systems examined in this article rather than posing the crucial question of why planters opted for the various arrangements in the first place. Insufficient attention has thus been paid to the decisive conditions under which free labour was introduced, to the dynamics of the labour systems themselves, and to the reasons for their transformation.

In part, recent debates have continued a long dispute about the supposed feudal, or at least non-capitalist, nature of Brazilian agriculture.[111] The implicit point of reference for both the 'feudal' and 'capitalist' interpretations is straight wage labour as the characteristically capitalist form of surplus extraction. Those who endorsed the feudal thesis emphasised the distinctive features of the plantation labour systems as proof of the non-capitalist nature of Brazilian agriculture.[112] Those who supported the capitalist thesis generally took as their point of reference the larger economic system into which Brazilian agriculture was integrated and argued for the underlying identity between the specific labour systems used and wage labour. While such authors explained the prevalence of these systems by their greater productivity in contrast to wage labour, they still did not account for their transformations.

More recently, yet other interpretations have been proposed. From the perspective of capital accumulation it has been suggested that labour relations in Brazilian agriculture have benefited the accumulation of capital in industry through relatively low prices for agricultural products, and were thus functional for the system.[113] This view has been contested by the suggestion that low agricultural productivity, on the contrary, has set clear limits to accumulation: 'The relatively low prices of agricultural products imply, in reality, relatively high "social costs" and a relatively low "social productivity".'[114] The articulation of the two sectors is seen as contradictory, a contradiction produced by the relatively low level of productivity in agriculture on account of what are seen as predominantly pre-capitalist or backward relations of production. Productivity in agriculture is indeed lower, in money terms, than in industry. But again why planters have for so long opted for labour systems of low productivity rather than introducing more productive innovations is not explained. It seems reasonable that planters should prefer labour-intensive methods of production rather than increasing fixed capital, if the cost of the latter was higher than that of labour. But this posits the question of why the cost of labour was so low for so long. Low agricultural wages imply a restriction in the consumption of wage goods, but then why should planters be concerned with that, rather than with their own levels of profit? Such an attitude has important political consequences generating social tensions, but then capitalist development is hardly a harmonious process. Nor is its dynamic determined by the needs of capital accumulation as a kind of *deus ex machina* but, as we have attempted to show in the specific case of the coffee plantations, by class struggle.

These are macro-analyses. There are also a number of studies specifically treating the first experiments with free labour, most of which centre on the failure of the sharecropping system. It has been variously argued that what condemned the early sharecropping experiments was in the last instance the planters' own backwardness. It is suggested that for free labour to be comparatively as profitable as slave labour – the point of reference is slave labour – planters resorted to forms of coercion which were unacceptable to the free labourers who reacted by rebelling. The implicit conclusion is that planters did not know where their true interests lay. They revealed themselves incapable of appreciating the requirements of a contractual relationship, an incapacity derived from their long tradition as slaveholders.[115] By using coercion rather than economic incentives, they undermined the labour system they themselves had adopted in order to replace the slaves.

These interpretations contain a mixture of elements from both the feudal and capitalist theses. The early sharecropping system is implicitly interpreted as capitalist, but its failure is generally attributed to the planters' traditional ideology which led them to treat free labourers as if they were slaves.

We have argued that São Paulo planters introduced free labour to replace or add to their slave force because the most forward-looking of them were aware by mid-century that some substitute for slave labour eventually had to be found. Under these circumstances, they were willing to experiment with new forms of labour, whose productivity they initially did not know. However, only a minority of São Paulo planters opted for free labour at the time, and the end of slavery was not immediate. Thus there was no acute shortage of slaves, and planters could continue to staff their plantations with slave labour, whose productivity was familiar to them, in case free labour did not respond to their expectations. Not only those who have studied the early experiments with free labour, but also the planters themselves repeatedly attributed the difficulties with free labour to its relatively low productivity in comparison to slavery.[116] It may be true that slaves were more profitable than free labourers, but to go back to slavery was no permanent solution. Thus, to explain the failure of the sharecropping system by its comparatively lower productivity would be reasonable only if it were not for the impending abolition of slavery. The spectre of abolition left planters, in fact, with only two alternatives: either to find a satisfactory substitute form of free labour, or to abandon coffee cultivation altogether.

The planters who initially took it upon themselves to find some replacement for slave labour were the first of a remarkable group of agricultural and commercial entrepreneurs. Their decision to reorganise production using free labour is in itself symptomatic. They viewed labour as a cost and adopted a labour system, sharecropping, which they thought would assure them a cost per unit of output lower than that of straight wage labour, rather than comparing this with slave labour. That some of them became disenchanted with the experiment and temporarily returned to slave labour was due fundamentally to the absence of one essential prerequisite, a surplus population of free labour that would have kept their workers intimidated and subjected to exploitation. Under these circumstances, neither the economic incentives contained in the sharecropping contract, nor the resort to frauds and occasional coercion succeeded in assuring a profitable and reliable labour force. The immigrants systematically restricted output in coffee and/or rebelled. It is this which led some planters to conclude that 'at the cost of any sacrifice the work of a slave is preferable to that of a free man'.[117] Others attempted to resolve their labour problems by introducing contract changes and more effective labour laws, yet with scarcely greater success. The permanent problem of labour productivity and discipline was only resolved in the 1880s when the state began to subsidise mass immigration and thus lay the foundations for an effective capitalist labour market.

NOTES

1. 'Bericht des schweiz. Generalkonsuls in Rio-Janeiro an den schweiz. Bundesrath über die Auswanderung nach Brasilien', *Schweizerisches Bundesblatt*, X. Jahrgang II, no. 34, 24 July 1858, pp. 183-188, quoting the speeches of an immigration agent named Mota on behalf of emigration to Brazil.
2. E. Viotti da Costa, 'Colônias de parceria na lavoura de café: primeiras experiências', in her *Da Monarquia à República: momentos decisivos*, São Paulo, 1977; W. Dean, *Rio Claro: a Brazilian Plantation System, 1820-1920*, Stanford, 1976; J.S. Witter, 'Um estabelecimento agrícola no estado de São Paulo nos meados do século XIX', *Revista de História*, no. 98, 1974; S. Buarque de Holanda, preface and notes to T. Davatz, *Memórias de um colono no Brasil*, São Paulo, 1941.
3. The transformation of Vergueiro's plantation Ibicaba in the first part of the nineteenth century is a good example of the evolution of São Paulo agriculture. Vergueiro had begun to replace sugar cane with coffee in 1828, but activities were limited by the small number of slaves he owned. In 1840, he made a first attempt to reorganise the plantation, gradually introducing 80 Portuguese families to replace his slaves. This experiment failed because of the political upheavals of 1842. Not discouraged, however, by 1847 he had founded an immigration company, Vergueiro e Companhia, and with the aid of a three-year loan from the provincial government, contracted a group of 64 German families. J.S. Witter, op.cit.; W. Dean, op.cit., p.88.
4. J. Vergueiro, *Memorial acerca de colonização e cultivo de café*, Campinas, 1874, p. 4. Cf. S.G. Kerst, *Uber brasilianische Zustände der Gegenwart, mit Bezug auf die deutsche Auswanderung nach Brasilien und das System der brasilianischen Pflanzer, den Mangel an afrikanischen Sklaven durch deutsche Proletarier zu ersetzen*, Berlin, 1853.
5. J.P. Carvalho de Moraes, *Relatório apresentado ao Ministério da Agricultura . . .*, Rio de Janeiro, 1870, p. 5. In view of the glowing terms in which São Paulo agriculture was presented to the immigrants, they presumably preferred the contract offering them a 50% share in net profit to one paying them a fixed piecerate per unit of coffee produced.
6. Le Comte Auguste von der Straten-Ponthoz, *Le budget du Brésil*, Paris, 1854, pp. 102-104. A very similar contract was signed between Senator F.A. de Souza Queiroz and a group of immigrants in 1852. The only differences were that in this case interest on outstanding debt was charged only after four years, the contract was for five years, and the planter agreed to let out land to the labourers on one of his plantations once the contract had been completed. E. Viotti da Costa, op.cit., pp. 175-177. The number of coffee trees assigned to each immigrant family depended in principle on the number, sex and age of its members. J.P. Carvalho de Moraes, op.cit., p. 7.
7. Some observers have maintained that under the first contract a significant number of immigrants managed to pay off their debts within the first years and move on to establish themselves on their own. J.P. Carvalho de Moraes, op.cit., p. 7; *Allgemeine Auswanderungs-*

Zeitung, Rudolstadt, 25 September 1857, no. 39; J.J. von Tschudi, *Viagem às províncias do Rio de Janeiro e São Paulo*, São Paulo, 1953, pp. 135, 149. The president of the province, Saraiva, reported in 1855 that of the initial 900 immigrants that had settled on Vergueiro's property in 1847, by 1855 about 670 were left. The rest had either fulfilled their contracts or indemnified the planter for the costs incurred on their behalf. By 1867, however, an emissary of the Prussian government, H. Haupt, noted that only under very favourable circumstances – such as fertile plots, a large family, the absence of diseases – could an immigrant family succeed in repaying its debt within a reasonable time. H. Haupt, in Sociedade Internacional de Imigração, *Relatório*, I, 1867, p. 39.

8. J.J. von Tschudi, op.cit., pp. 135, 166.

9. Ibid, p. 135; J.P. Carvalho de Moraes, op. cit., pp. 55-56.

10. J.P. Carvalho de Moraes, op.cit., p. 5; it was argued that this would allow labourers to tend a larger number of coffee trees, ibid., p. 59.

11. 'Relação das colônias existentes na província de São Paulo no ano de 1855', an anonymous manuscript dated 8 March 1856, located in the archive of the Instituto Histórico e Geográfico Brasileiro, Rio de Janeiro, lata 71/7. There were 117,731 slaves in the province in 1854. How many were engaged in coffee cultivation is not easy to say, but they surely constituted a majority. J.F. Camargo, *Crescimento da população no estado de São Paulo e seus aspectos econômicos*, São Paulo, 1952, II, p. 12.

12. J.P. Carvalho de Moraes, op.cit., p. 10.

13. E. Viotti da Costa, 'Política de terras no Brasil e nos Estados Unidos', in her *Da Monarquia à República*, op.cit. As she points out, the defenders of the law, mostly large landowners, were essentially concerned with the effects of this law in providing free labour to replace their slaves. Liberals opposed to the landowning interests rejected it, arguing instead for the donation of land to immigrants as an incentive to attract foreign settlers to civilise the country. Cf. J. de Souza Martins, *A imigração e a crise do Brasil agrário*, São Paulo, 1973, pp. 51-54; W. Dean, 'Latifundia and Land Policy in Nineteenth-Century Brazil', *Hispanic American Historical Review*, LI, no. 4, 1971.

14. Those who were attached to the plantations were a relatively unimportant group of *agregados*, a kind of retainer allowed the use of marginal estate land in return for occasional labour and other services. M.S. de Carvalho Franco, *Homens livres na ordem escravocrata*, São Paulo, 1969. Cf. Peter Eisenberg, 'O homem esquecido: o trabalhador livre nacional no século XIX. Sugestões para uma pesquisa', *Anais do Museu Paulista*, XXVIII, 1977-78.

15. J.J. von Tschudi, op.cit., p. 136. Tschudi was an envoy of the Swiss Federation, sent to Brazil in 1860 to inspect the conditions of Swiss sharecroppers. He was himself a landowner in his native country and reported sympathetically that the planters were well satisfied with the modifications of the contract since these offered them a solid guarantee for the capital invested.

16. *Allgemeine Auswanderungs-Zeitung*, Rudolstadt, 23 October 1857, no. 43, pp. 188-189. The Swiss consul-general David was exceptional in his sympathetic attitude towards the immigrants. Usually the consuls neglected their interests or worse. One of his predecessors was Charles Perret-Gentil, who was an immigration agent, as well as Senator Vergueiro's brother-in-law and business partner. Perret-Gentil's immediate predecessor had been dismissed after having misappropriated the funds of the Schweizer Hilfsverein in Rio de Janeiro. R.A. Natsch, *Die Haltung eidgenössischer und kantonaler Behörden in der Auswanderungsfrage, 1803-1874*, Zürich, 1966, pp. 171-72.

17. T. Davatz, op.cit., contains a detailed account of the events. See also the collection of documents reproducing reports by several observers in *Schweizerisches Bundesblatt*, X. Jahrgang II, no. 34, 24 July 1858, and XII. Jahrgang III, no. 61, 28 November 1860. The immigrants' discontent was aggravated by the fact that for a significant number their local governments had advanced transport costs free of interest and commission, for which they were being charged by Vergueiro.

18. T. Davatz, op.cit., pp. 105-106.

19. *Deutsche Auswanderer-Zeitung*, Bremen, 22 October 1857, no. 83, quoting a report of the Brazilian Imperial Colonisation Agency on the Ibicaba revolt. J.J. von Tschudi, 'Bericht des schweiz. ausserordentlichen Gesandten in Brasilien, Herrn v. Tschudi, an den Bundesrath über die dortigen Verhältnisse der Kolonisten, vom 6. Oktober 1860', *Schweizerisches Bundesblatt*, XII, Jahrgang III, no. 61, 28 November 1860. In his comments to the Swiss Federation, which Tschudi sent along with his report, he pointed out that in fact part of the blame for the difficulties belonged to the immigration agents who had not selected the labourers properly and had contracted many disabled people simply to earn the commission. He particularly blamed the immigrants themselves, who did not always show the necessary interest in their work. Tschudi had omitted these aspects in his report to the Imperial government, 'in order to emphasise those points which arise from the conditions in Brazil', *Schweizerisches Bundesblatt*, XII, Jahrgang III, no 61, 28 November 1860, p. 269. T. Davatz, op.cit., p. 100; M.J. Valdetaro, 'Colônias de S. Paulo', in Brazil, *Relatório da Repartição Geral das Terras Públicas apresentado em 31 de março de 1858*, Rio de Janeiro, 1858, p. 91.

20. *Allgemeine Auswanderungs-Zeitung*, Rudolstadt, no. 43, 23 October 1857, p. 189.

21. The theme is prominent, for example, in the previously cited works by J.S. Witter and S. Buarque de Holanda, as well as in W. Dean's *Rio Claro*.

22. R.A. Natsch, op.cit., makes this point quite clear.

23. A. Marshall, *Principles of Economics*, London, various editions.

24. S.N.S. Chueng, *The Theory of Share Tenancy*, Chicago, 1969.

25. T. Holloway, 'The coffee *colono* of São Paulo, Brazil: migration and mobility, 1880-1930', in K. Duncan and I. Rutledge (eds), *Land and Labour in Latin America*, Cambridge, 1977. It is unlikely that São Paulo planters did not know average coffee yields. They had the Paraíba Valley experience to go by. In any case, no entrepreneur can exactly predict future markets and prices.

26. J.D. Reid, Jr., 'Sharecropping and Agricultural Uncertainty', mimeo., 1973; cf. J.D. Reid, Jr., 'Sharecropping as an Understandable Market Response: the Post-Bellum South', *Journal of Economic History*, 33, 1973, which contains a number of sharecropping contracts strikingly similar to those used in São Paulo.

27. Juan Martínez-Alier, 'Peasants and Labourers in Southern Spain, Cuba, and Highland Peru', *Journal of Peasant Studies*, I, no. 2, 1974.

28. As José Vergueiro, op.cit., p. 5 pointed out in 1874: 'I understand colonisation can only progress if carried out in families. One estimates that each family is composed of an average of five members.'

29. J.P. Carvalho de Moraes, op.cit., appendix 5, p.2.

30. A. Ramos, *O café no Brasil e no estrangeiro*, Rio de Janeiro, 1923, p. 358.

31. The clause contained in the early contracts that half of any food crop surplus belonged to the planter, which was intended to discourage labourers from growing food in excess of their basic needs and in detriment to work in coffee cultivation, was soon abandoned as largely ineffective. J.P. Carvalho de Moraes, op.cit., p. 57.

32. C. Heusser, *Die Schweizer auf den Kolonien in St. Paulo in Brasilien*, Zürich, 1857, pp. 28-9. Among others, Kerst, op.cit., p. 70, considered the fact that the planters handled the selling of the crop a grave disadvantage. In view of the 'exorbitant deductions' made by the planters, he was convinced that the 'labourers would be better off if they received their share *in natura* and sold it themselves, instead of having the planter . . . as their intermediary . . . The buyers would then collect the product and save them any further endeavour. Of course, it could then occur that the labourers compete with the planter.'

The actual time taken by immigrants to amortise their debts is a controversial issue. Some authors have estimated that it took an average of nine years. See E. Viotti da Costa, 'Colônias de parceria', p. 174. W. Dean, *Rio Claro*, pp. 105-07, on the contrary, maintains that an average family could repay its debt within five years or less, and in addition accumulate some savings. Dean estimates that an average family of five members would tend about 3,000 coffee trees, yet most accounts indicate that this is an overestimate. See note 37 below.

33. There is evidence of other work stoppages. Tschudi, *Viagem às províncias* . . ., pp. 164-5, reports such a case for the plantation Boa Vista in Amparo in 1858. The labourers were apparently harvesting green berries along with the ripe ones to speed up the work. Reprimanded by the planter, who threatened pay deductions, they stopped the harvest. Another case is that of Portuguese immigrants on a plantation in the area of Rio Claro, reported in S. Machado Nunes, 'Colônias na província de S. Paulo', in Brazil, Ministério do Império, *Relatório . . . 1860*, p. 15. M.J. Valdetaro, op.cit., p. 94, reports that on the plantations Bery and Cauvatinga most Swiss and German labourers were lazy and careless in the fulfillment of their duties, to the point of abandoning work entirely. Cf. H. Haupt, op.cit., p. 39: 'The labourer in sharecropping ("obreiro colono por parceria") will moreover be a very bad worker, because he necessarily loses the love of work which gives him no result, nor independence. In this case, he will seek to avoid his duties and deceive the planter. Soon a feeling of enmity will grow up between them, and in many cases deplorable scenes will result.'

34. S. Machado Nunes, op.cit., pp. 2-3. Cf. also J.P. Carvalho de Moraes, op.cit., p. 86. Tschudi, *Viagem às províncias* . . ., p. 183, n. 69, reported that on the plantation São Lourenço some families tended no more than 500 to 700 coffee trees. When asked why, they replied that they were overburdened by their debts and thus not interested in cultivating coffee. Their food plots produced sufficient for them to live on, and they saw no reason to kill themselves working.

35. M.J. Valdetaro, op.cit., p.91

36. Ibid,; J.J. von Tschudi, *Schweizerisches Bundesblatt*, XII, Jahrgang III, no. 61, 28 November 1860.

37. Quantitative data on the number of coffee trees tended by immigrants, either on an individual or family basis, are scarce. According to Carvalho de Moraes, op.cit., pp. 70-71, it was initially expected that a labourer could tend up to 3,000 coffee trees in the case of mature groves between five and twelve years of age, provided he weeded them four to five times a year and had no additional obligations. In this case, additional labour for the harvest would be required. If the labourer, however, simultaneously grew subsistence crops, he could not tend more than 1,000 to 1,200. In actual fact, a labourer tended only between 800 and 1,000 trees, while women took on between 500 and 600 trees of five to twelve years of age, since 'in general labourers do not pay their debts with anything else but their income from coffee and for this reason they will try to tend as small a number of trees as they possibly can, in order to have more time for their own crops and to benefit from other activities. It is this difficulty with which the planters struggle . . . This is one of the reasons why in this province the work of a free labourer is considered less efficient than that of a slave.' Ibid., p. 86. Other scattered evidence supports the view that the labourers' productivity in coffee was indeed low. See E. Viotti da Costa, 'Colônias de parceria', pp. 173-4, and Tschudi, *Viagem às províncias* . . ., pp. 46-50, 183.

J.P. Carvalho de Moraes, op.cit., 'Mapa de São Jerónimo', published data on a plantation run under the sharecropping system which lists all the families that settled on the estate between 1852 and 1869, giving their size, composition (number of workers and consumers), income in that period, and date of arrival and departure. Of the total 141 families, 40% had paid their debts and made a profit by 1869, 4% had merely repaid their debts, and 56% continued indebted. On the basis of these data it is possible to calculate, if not the number of trees tended per labourer, at least the labourers' earnings per month for those families who had repaid their debts. The amount of outstanding debt is not given for those who continued indebted at the time. We can also calculate the consumer/worker ratio in order to determine the comparative productive capacity of the families. Since the greater the number of labourers in a family in proportion to the consumers, the greater would be its productive capacity, this should be reflected in the monthly earnings of its workers, provided that they worked at full capacity.

MONTHLY EARNINGS PER WORKER FROM COFFEE CULTIVATION
BY CONSUMER-WORKER RATIO OF THE FAMILY:
PLANTATION SÃO JERÓNIMO, 1852-1869

C/W ratio	1.0-1.4	1.5-1.9	2.0 and over
Mil. reis	6,508	7,714	8,111
Number of families	25	20	16

(We excluded three families from the total 64 with favourable results; one was a widow who married another immigrant and joined her accounts to his; the other two were single individuals.)

The earnings given refer to income from coffee cultivation, since it was with this that debts were paid. Harvests varied from year to year, but it may be assumed that over the years earnings roughly expressed the number of trees tended. As the table indicates, productive capacity of the families is inversely related to the number of trees tended by each of its workers as expressed in monthly earnings. This is a further indication that immigrants did not respond to the economic incentives contained in the sharecropping contract. As pointed out, these figures refer to those families who at some point managed to pay their debts and those who in addition could save. Thus, they do not tell us anything about the behaviour of those who were still indebted, although it may be presumed that even the debt-free families had been indebted for a considerable period of their permanence.

38. *Allgemeine Auswanderungs-Zeitung*, Rudolstadt, 2 October 1857, no. 40.

39. J.P. Carvalho de Moraes, op.cit., p. 21.

40. Ibid., p. 21. The initial debt incurred by the immigrants has sometimes been taken as an indication of the existence of debt peonage. Cf. W. Dean, *Rio Claro*, pp. 97-8, and E. Viotti da Costa, 'Colônias de parceria', p. 169. If we define debt peonage in its strict sense as being the intentional burdening of labourers with a debt to tie them to a property, this interpretation is inappropriate for the case of São Paulo. The primary aim planters pursued in charging immigrants with their transport costs and food advances was to recover their initial investment. Stability of labour on the plantation was a welcome by-product. Significantly, an 1852 contract stated both length of contract *and* amount of debt, in E. Viotti da Costa, 'Colônias de parceria', p. 177. Moreover, as some observers reported, it was not infrequent that immigrants moved from one plantation to another, taking their debt with them. C. Heusser, op.cit., p.1. In addition, the Vergueiro contract included a clause allowing for transfer of immigrants. Planters, it is true, were accused at times of treating immigrants like white slaves, as in H. Haupt, op.cit., p. 40. However, Tschudi who was quite critical of the São Paulo planters, was probably correct when he noted: 'I have not observed such a tendency on any of the plantations I visited; all planters declared that their greatest interest was in seeing their labourers free of debt' – surely not least because this would have meant that labourers worked in coffee cultivation. *Viagem às províncias* . . ., p. 186.

41. Tschudi, *Viagem às províncias* . . ., p. 143.

42. J.P. Carvalho de Moraes, op.cit., pp. 19-20.

43. Ibid., pp. 18-20.

44. Ibid., p. 12 and appendix 2 for an 1868 labour-leasing contract; cf. M.J. Valdetaro, op.cit., and S. Machado Nunes, op.cit., for descriptions of contract changes.

45. P. Denis, *Brazil*, London, 1911, pp. 216-17.

46. J.P. Carvalho de Moraes, op.cit., pp. 20-21.

47. Ord. Libro 4°, Titl 45, of the *Ordenações e Leis do Império de Portugal*, various editions.

48. Lei de locação de serviços of 11 October 1837, in *Coleção de Leis do Império do Brasil*, 1837, VIII.

49. Machado Nunes, op.cit., p.10, reports the case of one German and two Swiss labourers of the Laranjal plantation in the area of Campinas, arrested in São Paulo and condemned in accordance with the 1837 law to pay double their debt for having refused to fulfil their sharecropping contracts. However, he notes that in another case a similar sentence was overruled with the argument that the 1837 law did not apply to sharecropping. He concluded that 'with such a decision the planter will have to lose all his advances to the labourers, since persecuting them in civil court will have to produce this result.' Consequently, 'if immigration according to the sharecropping system is to increase in São Paulo . . . it will be indispensable to pass special legislation regulating the sharecropping contract and furnishing an easy means to resolve quickly the conflicts between labourers and planters.' Ibid., p. 20. Cf. Tschudi, *Schweizerisches Bundesblatt*, XII. Jahrgang III, no. 61, 28 November 1860, p. 260, for another case. Tschudi was highly critical of the practice of applying the 1837 law to sharecropping contracts, suggesting that this opinion is shared by a number of São Paulo jurists. Tschudi, 'Denkschrift an Seine Exzellenz den Senator João Vieira Cansansão de Sinimbu, Minister der Auswärtigen Angelegenheiten', ibid., p. 297. Finally, the police chief, Tavares Bastos, who was ordered to investigate the events on the Ibicaba plantation in 1857, suggested a number of legal reforms, in particular that the 1837 law be adapted to the sharecropping contract, 'Relatório de Tavares Bastos sobre colonização em. S. Paulo', in T. Davatz, op.cit., pp. 231-40.

50. As José Vergueiro, op.cit., p. 5, noted 'However, this same contract, as positive as it is, does not satisfy the *colono*, who detests being subjected to it; and since he does his daily job under constraint, he tries exclusively to make use of the plot of land [which he is assigned] because he has no hope of owning it someday.' Tschudi suggested a further reason for the immigrants' unresponsiveness to the new contract. With rising coffee prices in 1860, it was in the planters' interest to pay a piecerate rather than share of growing net profits. For the labourers it was the reverse; at a time of rising prices, labourers would have benefited more from a sharecropping than from a piecerate contract. Tschudi, *Viagem às províncias . . .*, p. 157. It must be kept in mind that the piecerate was established annually, while the share in net profit was defined by contract and could not be altered as long as the contract remained in force, usually until the debt had been repaid.

51. Tschudi, *Viagem às províncias*, pp. 176, 183. Both long-term productivity and length of life of coffee trees depend importantly on the number and quality of the weedings. At the time, the average number of weedings was five to six a year, including *coroação* before the harvest and *esparramação do cisco* after it. Although contracts by the 1860s usually stipulated the number of weedings, these could still be done with greater or lesser care. If weeds were eliminated by only lightly hoeing the ground, the soil would tend to harden and become impermeable to the rain, beside not allowing adequate ventilation. São Paulo, Secretaria da Agricultura, *Inquérito agrícola sobre o estado da lavoura cafeeira no estado de S. Paulo*, São Paulo, 1904, pp. 56-7.

52. J.P. Carvalho de Moraes, op.cit., p. 100, notes that a number of important planters introduced immigrant labour in the 1860s. See ibid., pp. 77-86 for a description of the different types of contract used on the plantations run with free labour in 1869. The separate payment of cultivation and harvest might have been introduced initially to counter the labourers' complaints about low profit from young coffee groves when tended under the sharecropping arrangement. Joaquim Bonifácio do Amaral, Visconde de Indaiatuba, 'Introdução do trabalho livre em Campinas', in *Monografia histórica do município de Campinas*, Rio de Janeiro, 1952.

53. Domingos Jaguaribe Filho, *Algumas palavras sobre a emigração*, São Paulo, 1877, pp. 34-5.

54. P. Denis, op.cit., p. 202.

55. See A. Ramos, op.cit., p. 104, on the labour-demand imbalance between cultivation and harvest, and its consequences.

56. For two contrasting positions on immigration systems, see the extended debate between Joaquim Bonifácio do Amaral and José Vergueiro, published in the *Gazeta de Campinas* between January and July of 1870. Vergueiro was totally disillusioned with privately sponsored immigration. In addition to the Ibicaba revolt, in the 1860s the Swiss Federation had sued Vergueiro and Co. for the restitution of passage money advanced by the cantons to the immigrants and eventually the company declared bankruptcy. R.A. Natsch, op.cit., pp. 207-208. In the early 1870s, the Angélica plantation, also belonging to Vergueiro, was sold off to creditors, the London and Brazilian Bank. The English managers, however, do not seem to have met with much more success. They are alleged to have been 'drunken incompetents' who brutalised the labourers and led the estate once again into bankruptcy. It is reported that in 1876, Angélica's administrator was murdered by some of his labourers. G.B. Marchesini, *Il Brasile e le sue colonie agricole*, Rome, 1877. Joaquim Bonifácio do Amaral, in contrast, travelled to Europe himself in 1870 to recruit labour personally. Even he, however, complained of the scant support he had received from other planters in this undertaking. *Gazeta de Campinas*, 24 July 1870. Cf. also J.B. do Amaral, op.cit.

57. P. de Turenne, 'L'immigration et la colonisation au Brésil', *Revue Britannique*, February 1879, p. 453. J.B. do Amaral, op.cit.

58. A Lalière, *Le café dans l'Etat de Saint Paul, Brésil*, Paris, 1909, appendix.

59. The law included the important proviso that until such children became 21 years old they could be used by their mother's owner. Though technically free, the *ingênuo*'s condition was almost indistinguishable in practice from that of a slave. Thus, the Rio Branco law would have little practical significance until 1892, when the first of those born under it would become 21.

60 The most thorough study is in Robert W. Slenes, 'The Demography and Economics of Brazilian Slavery: 1850-1888', Ph.D. dissertation, Stanford University, 1975, pp. 120-78.

61. Since the mule teams which the railroads replaced had sometimes required as much as 20% of the plantation labour force to operate, the number of slaves made available for field work was quite significant. E.Viotti da Costa, *Da senzala à colônia*, São Paulo, 1966, pp. 154-77.

62. The president of São Paulo noted in 1871 that the planters were no longer 'stubbornly sticking to tradition and routine' as they had previously done, but were now employing various foreign and domestic machines to replace 'their scarce and expensive labour'. São Paulo, *Relatório apresentado . . . pelo presidente da província . . . 5 de fevereiro de 1871*, p. 42. Cf. A. de Queiroz Teles, 'La culture du caféier à São Paulo, Brésil', *L'Economiste Français*, VI, no. 49, 7 December 1878, pp. 718-20 for a detailed description of the production process at the time.

63. J. Vergueiro, op.cit., p. 22.

64. *Informações sobre o estado da lavoura*, Rio de Janeiro, 1874, p. 149; the Dutch traveller Van Delden Läerne commented in 1883 that during the preceding ten years the planters had been able to increase significantly the productivity of their slaves who were now 'better treated, better fed and cared for', but who 'must work harder'. C. Van Delden Läerne, *Brazil and Java: Report on Coffee Culture*, London, 1885, p. 91.

65. In 1870 Carvalho de Moraes had suggested as an afterthought that the Imperial Government at least 'grant the planters who introduce free labour a subsidy of 30$000 to 40$000 for each child under the age of 14 years who comes in the company of his parents', op.cit., p.103.

66. O Lavrador, *Gazeta de Campinas*, 20 February 1870.

67. Congresso Agrícola, *Coleção de documentos*, Rio Janeiro, 1878, pp. 47-8. On the Congress, see Peter L. Eisenberg, 'A mentalidade dos fazendeiros no Congresso Agrícola de 1878', in J.R. Amaral Lapa (ed), *Modos de produção e realidade brasileira*, Petrópolis, 1980.

68. Decreto no. 2827, 15 March 1879, in *Coleção de Leis do Império do Brasil*, 1879.

69. The strike broke out on June 9th. The *Tribuna Liberal* of September 11 and 13, 1878, carried detailed reports on the strike by the Police Chief sent to investigate the events, as well as information from a special investigating commission.

70. They drew up a list of nine demands: that the planter fulfil his promise to provide new houses within six months; that foodstuffs be charged at current prices; that they receive the total amount of flour from the maize ground at the planter's mill; that they not be forced to abandon their food plots for new uncultivated land; that they be paid for the replanting of trees; that medical expenses be shared by the planter; that cloths for the harvest be provided free of charge; and that a school and a teacher be provided free. *Tribuna Liberal*, 11 September 1878.

71. J.B. do Amaral, op.cit; *Tribuna Liberal*, 11 and 13 September 1878. The previous year a group of German labourers which Amaral had recruited in the neighbouring province of Santa Catarina had gone on strike and some of the leaders had been sentenced to jail.

72. J.B. do Amaral, op.cit., pp. 248, 252.

73. C.F. Van Delden Läerne, pp. 213, 217. The question is also discussed extensively in Congresso Agrícola, op.cit.

74 Brazil, *Anais da Câmara*, 1884, V, pp. 541-3. In 1884, his brother, Martinho Prado, defending a bill introduced in the São Paulo assembly to subsidise the entire costs of the passage of immigrants, stated clearly the purpose of such a measure: 1) to replace the slaves with free labourers and prepare for immediate emancipation; 2) to allow immigrants to arrive free of debt; 3) to permit planters to employ free labour without having to advance money. São Paulo, *Anais da assembléia*, 1884, pp. 34, 224.

75. *Anais da Câmara*, 1884, V, p. 540.

76. *Anais da Câmara*, 1888, IV, p. 323.

77. *Anais do Senado,* 1887, annex to III, p. 6; another source estimated that between 1882 and 1888 a total of 103,571 immigrants had entered the province through official channels. Dr. Francisco de Paula Lázaro Gonçalves, *Relatório apresentado à Associação Promotora de Imigração em Minas*, Juiz de Fora, 1888, p. 85.

78. C.F. Van Delden Läerne, op.cit., p. 354.

79. A. Castro, *7 ensaios sobre a economia brasileira*, Rio de Janeiro, 1971, II, p. 78.

80. Decreto no. 213, 22 December 1890, in *Decretos do Governo Provisório.*

81. Decreto Federal no. 1162, 12 December 1890, 'Dos crimes contra a liberdade do trabalho', in ibid.

82. F. Mosconi, 'Le classi sociali nel Brasile e le loro funzioni', *La Riforma Sociale*, VII, 1897; F. Dafert, 'A falta de trabalhadores agrícolas em São Paulo', in Instituto Agronómico, *Relatório*, 1892, p. 206; F.P. Lázaro Gonçalves, op.cit., pp. 10-11, noted that freedmen demanded high salaries and were widely considered insubordinate and unreliable; L. Couty, *Etude de biologie industrielle sur le café*, Rio de Janeiro, 1883, pp. 120-8.

83. M. Hall, 'The Origins of Mass Immigration in Brazil, 1871-1914', Ph.D. dissertation, Columbia University, 1969, pp. 144-7.

84. The theme is a common one, for example, among the planters in the *Inquérito Agrícola*, op.cit.

85. Coffee has been an itinerant culture in Brazil. Simultaneously there existed a frontier region where coffee was just penetrating, a region where the groves were fully productive, and a declining region suffering from soil exhaustion. See, for example, A. Castro, op.cit., p. 61. Whereas until Abolition new land was cleared and new coffee planted mostly by slaves or free *empreiteiros*, from the 1890s onward immigrant labourers were also used for forming new plantations. This was done under a four-year contract which entitled the labourers to grow food crops at will in the coffee rows. At the end of the contract they were paid a certain sum per tree, and the contract was replaced by the usual mixed task and piecerate system. On new plantations, the virgin soil yielded much more abundant food crops. In addition, with intercalary growing, the labourers cultivated their crops while weeding the coffee trees, and food crops thus required little extra labour. Food growing on separate plots, however, as generally practised on the older plantations, implied doubling labour input without increase in profit for the worker. Consequently, labourers preferred new plantations and, if these were not available, they still preferred those plantations which allowed intercalary crops rather than assigning separate plots. J.P. Carvalho de Moraes, op.cit., pp. 68-9; P. Denis, op.cit., p. 213; A. Ramos, op.cit., pp.210-12; G. Maistrello, 'Fazendas de café – costumes (São Paulo)', in A. Ramos, op.cit., pp. 556-7, 572-3; *Inquérito Agrícola*, op.cit., p. 43; J. Brandão Sobrinho, *Apreciação da situação agrícola ... do 3° distrito agronômico do estado de S. Paulo ...*, São Paulo, 1903.

86. C.F. de Lacerda, *A crise do café: estudo das causas da crise do café e dos meios de combate-la*, São Paulo, 1903, p. 15.

87. C.F. de Lacerda, *Estudo da meiação, parceria, etc. e das suas vantagens*, São Paulo, 1905, p. 15. Only in the declining areas did planters return to sharecropping, granting the labourers, however, a larger share. F.P. Lázaro Gonçalves, op.cit., p. 42, describes the situation in the 1880s.

88. *Inquérito Agrícola*, op.cit., p. 46; S. Coletti, 'Lo stato di S. Paulo e l'emigrazione italiana', *Bollettino dell'Emigrazione*, no. 14, 1908, p. 41, states that around Campinas planters were unable to increase salaries, so they had to compete with other regions by allowing *colonos* to cultivate beans and maize in the coffee rows. He also notes that this was more advantageous to the *colonos* than 70-100 mil-reis more per 1,000 coffee trees tended.

While it was usually believed that these intercalary crops had an adverse effect on the coffee yields of trees over four years old, in practice planters' decisions in this respect depended much more on fluctuations of coffee prices and labour supply than on technical conditions. See A. Ramos, op.cit., p. 106.

89. F.P. Lázaro Gonçalves, op.cit., pp. 20, 28. For a typical *colonato* contract, see the *Bollettino Ufficiale della Camera Italiana di Commercio ed Arti*, São Paulo, II, no. 4, February 1903.

90. L. Couty, op.cit., pp. 166-7, comments on the planters' aversion to producing maize, rice, beans, and to raising animals in a country importing a large part of its foodstuffs from abroad. Cf. *Bollettino Ufficiale della Camera Italiana di Commercio ed Arti*, XV, July 1917, p. 10.

91. São Paulo, *Relatório apresentado ... pelo Presidente da província ... 17 de janeiro 1887*, pp. 120-22.

92. P. Denis, op.cit., p. 202.

93. E. Bonardelli, *Lo stato di S. Paulo del Brasile e l'emigrazione italiana*, Turin, 1916, pp. 71 ff.

94. As Denis noted, labourers were often more interested in the clauses in their contracts relating to food crops than in those determining wages, to the point of sometimes preferring lower wages but more favourable planting rights to higher wages. Op.cit., pp. 202-203.

95. G. Maistrello, op.cit., p. 564.

96. J.B. do Amaral, op.cit., p. 244; F.P. Lázaro Gonçalves, op.cit., p. 28; G. Maistrello, op.cit., p. 559. Planters clearly regarded labourers contracted in family units as cheaper. Thus Maistrello, op.cit., p. 562, noted in 1922 that 'the tending of the coffee groves is preferably done by families of *colonos* with annual contracts, but unfortunately they are not always sufficient and almost all the planters must resort to extra personnel. Needless to say, employing such labourers [*camaradas*] in weeding or harvesting increases quite noticeably the costs of the fazenda.' Planters usually employed single men as *avulsos* – labourers recruited exclusively for the harvest or special tasks such as pruning – or as *camaradas* for coffee processing and transporting. They were paid on a monthly wage basis, as young families generally were. Large families were hired as *colonos*. Carvalho de Moraes, op.cit., p. 66; *Bollettino Ufficiale della Camera Italiana di Commercio ed Arti*, February 1903, p. 73; cf. D. Jaguaribe Filho, op.cit., pp. 19, 32.

P. Denis, op.cit., pp.216, 318-23, noted that in comparison with Brazilian labour, Italian immigrants were not only more hardworking but also made their women work. Rather than a cultural trait, this is surely one further aspect of the different labour arrangements under which they were contracted. Immigrants being hired in family units, they were expected to make all able family members work, while Brazilians were usually employed as single wage labourers.

97. Maistrello, op.cit., pp. 558-9, calculated in 1922 the incomes of three immigrant families with different consumer/worker ratios, and showed that at the prevailing wage levels a family of four with only one worker was practically incapable of making ends meet. At the end of the year, the large family would have cleared 1,130$000, the family with 2 workers 620$000, and the small family only 140$000.

Family size	No. of workers	c/w ratio	trees per family	annual weeding	harvest earnings	extra work	annual expenses
10	6	1.66	16,000	2,400$	480$	600$	2,350$
5	2	2.5	7,000	1,050$	240$	460$	1,130$
4	1	4.0	3,000	450$	70$	120$	780$

With weeding wages calculated on the basis of large families, it is clear that families with a small number of workers, even exerting themselves proportionately more, would do comparatively worse. Earnings from extras derive from day wages received for tasks not stipulated in the contract.

98. The world price began to decline in the mid-1890s, partly because of increased supply, but planters went on planting new trees, protected by the falling exchange rate which kept prices in domestic currency high. By 1896, even the price in mil-reis began to decline sharply. However, coffee supply continued to climb as new bushes, planted under the stimulus of the high prices which had prevailed four or five years earlier, began to come into production. The amount of coffee grown in the 1900-1905 period was double that of 1890-5, and prices continued at a relatively low level, with some improvement after 1906. M. Hall, op.cit., p. 153.

99. Such as company stores, confiscation of produce, fraudulent measures, withholding of wages, and much else. Perhaps the most famous report on such matters, and one which helped provoke the 1902 Italian ban on subsidised emigration to São Paulo, is A. Rossi, 'Condizioni dei coloni italiani nello stato di San Paulo', *Bollettino dell' Emigrazione*, no. 7, 1902; part of the report dealing with complaints from immigrants is in P.S. Pinheiro and M. Hall (eds), *A classe operária no Brasil: documentos*, São Paulo, 1981, II, pp. 32-9.

100. These observers – largely, though not exclusively, diplomatic and consular personnel – were not opposed to emigration nor were they particularly hostile to São Paulo or Brazil. Nevertheless, they generally held that the conditions to which the immigrants were subjected on the plantations were extremely unfavourable, and many thought they were significantly worse than in other countries of large-scale Italian settlement. For sources and a more

detailed account, see M. Hall, op.cit., and M. Hall, 'Emigrazione italiana a San Paolo tra 1880 e 1920', *Quaderni Storici*, no. 25, 1974.

101. S. Coletti, op.cit., p. 375, writing in 1908, concluded however, that the fall in coffee prices had never prevented planters from earning a 'decent profit, if not as splendid a one as in the past', and he further noted that 'it was not so much the fall in the price of coffee as the acquisition of already prepared plantations at fabulous prices, and the sudden establishment of new ones with money borrowed at very high interest rates, which has led to the crisis'.

102. S. Coletti, op.cit., p. 53.

103. P. Denis, op.cit., p. 206. Denis considered the estimate – from a government source – to be high, but even on one well-run estate between 1895 and 1930, half the *colonos* stayed for less than 4.5 years, according to M.S. Beozzo Bassanezi, 'Fazenda Santa Gertrudes: uma abordagem quantitativa das relações de trabalho em uma propredade rural paulista', doctoral thesis, Faculdade de Filosofia, Ciências e Letras de Rio Claro, 1973, p. 153.

104. A. de Zettiry, 'I coloni italiani dello stato di S. Paulo', *La Rassegna Nazionale,* LXX, 1893, p. 78; A.L. Rozwadowski, 'San Paolo', in *Emigrazione e colonie: raccolta di rapporti dei rr. agenti diplomatici e consolari*, Rome, 1893, p. 177; *Emigrazione agricola al Brasile. Relazione della Commissione Italiana 1912*, Bologna, 1913, p. 232.

105. P. Denis, op.cit., p. 205.

106. P. Denis, op.cit., p. 205; A. Ramos, op.cit., p. 209. Strikes usually occurred at the onset of the harvest, since a delay at that point would immediately affect the quality of the coffee, thus making it the most strategically effective moment to try to put pressure on planters.

107. There are colourful accounts in *Fanfulla*, 5 October, 1900, 20 May, 6 July, 21 and 22 October, 1901.

108. From 1913, the annual reports of the Patronato Agrícola (published in the *Boletim do Departmento Estadual do Trabalho* and/or in the *Relatórios* of the Secretaria de Agricultura of São Paulo), usually give the numbers of rural strikes which have come to their attention, and sometimes a little additional information.

109. We have analysed these movements in M. Hall and V. Stolcke, *Greves de colonos na Primeira República*, CEDEC, II Seminário de Relações de Trabalho e Movimentos Sociais, São Paulo, 1979, mimeo; we hope to be able to present a fuller report in the near future.

110. The *colonato* itself has received considerable recent attention. Interpretations which differ substantially from our own (and from one another) are: V. Caldeira Brant, 'Do colono ao boia fria: transformações na agricultura e constituição do mercado de trabalho na Alta Sorocabana de Assis', *Estudos CEBRAP*, 19, 1977; J. de Souza Martins, *O cativeiro da terra*, São Paulo, 1979; C.R. Spindel, *Homens e máquinas na transição de uma economia cafeeira*, Rio de Janeiro, 1980; T.H. Holloway, *Immigrants on the Land: Coffee and Society in São Paulo, 1886-1934*, Chapel Hill, 1980.

111. Two of the most influential statements of the respective positions are A. Passos Guimaraes, *Quatro séculos de latifúndio, Sao Paulo*, 1964, and Caio Prado Jr., *A. revolução brasileira*, São Paulo, 1966.

112. The most vehement recent claim that the *colonato* was pre-capitalist is Jacob Gorender, 'Gênese e desenvolvimento do capitalismo no campo brasileiro', in Evaristo de Morais Filho et al., *Trabalhadores, sindicatos e política*, São Paulo, 1980. Gorender classifies the *colonato* as 'a latifundiary plantationist mode of production aided by dependent peasant forms'. In fact, Gorender posits a number of rather arbitrary and sometimes bizarre criteria for capitalism (such as contracts which do not extend beyond one month) and finds to his satisfaction that the *colonato* does not fulfil them. The high point of the article occurs on page 49 when, after explaining that the planters assigned pasture (*terreno de pastagem*) to the *colonos*, the author then helpfully enumerates the animals who grazed there: 'a horse, a cow, a calf, a pig, chickens'. While the image of grazing pigs and chickens is not without a

certain Disney-like charm, it does not enhance one's confidence in Gorender's agricultural expertise.

We hope it is evident by now that those features (such as the food plots) which some writers have used to classify the *colonato* as non-capitalist, seem to us to be best regarded as simply specific to the concrete conditions of São Paulo agriculture in the period. In any case, it would be difficult to deny that under the *colonato* free workers were obliged to sell their labour power in the market.

113. Francisco de Oliveira, 'Economia brasileira: crítica à razão dualist', *Estudos CEBRAP*, no. 2, 1972.

114. Sérgio Silva, 'Agricultura e capitalismo no Brasil', *Contexto*, no. 1, 1976, p. 31.

115. E. Viotti da Costa, 'Colônias de parceria', pp. 171-2. In her *Da senzala à çolonia*, Viotti argued somewhat differently, attributing the failure of sharecropping to the high cost of production of coffee on account of its very labour-intensive nature, low level of mechanisation, and low coffee prices, with consequent small profit margins, which discouraged both employers and labourers. W. Dean, in *Rio Claro*, while initially arguing that the sharecropping system was in fact more profitable than slave labour, attributes its failure to the fact that after the first years free labourers could only be kept working by the use of coercion or by offering them more favourable contract conditions. He suggests that the former would have required backing by the government which was unavailable, the latter the planters themselves were unwilling to grant, ideologically incapable as they were of dealing with a genuine proletariat on a purely contractual basis. He concludes that, rather than an increase in productivity or the observance of contractual relations, what planters expected of their labourers was subservience and loyalty.

116. In order to avoid invidious citations, we will only note that this was the argument in M. Hall, 'Origins of Mass Immigration'.

117. M.J. Valdetaro, op.cit., p. 93, quoting a planter whose labourers worked little and badly.

The Sharecropping Economy on the South African Highveld in the Early Twentieth Century

Timothy Keegan*

I. INTRODUCTION

The role of African peasants in capital accumulation in South African agriculture has only been fully recognised in quite recent years. According to the old dualist interpretation, contemporary underdevelopment and poverty in the African agrarian economy is the result of factors inherent in that economy: that it was backward, inefficient and unable to adapt, and therefore simply could not compete with the (white) progressive market economy. As their tribal self-sufficiency eroded, it was assumed, Africans entered the modern sector as unskilled labourers (see, for example, de Kiewiet [1957]; van der Horst, [1942]). However, largely as a result of the work of Colin Bundy among others, it has become clear that African agriculturalists in the nineteenth century responded positively and successfully to the colonial economy, its pressures and its opportunities, by producing agricultural surpluses for sale on the markets of southern Africa. The chronology of this response (and its eventual decline) was uneven in different areas, depending on the timing and intensity of colonial penetration, but it was a ubiquitous phenomenon, to be observed throughout the region [*Bundy*, 1979; *Arrighi*, 1970; *Palmer and Parsons*, 1977; *Ranger*, 1978].

Even in the white-settled arable highveld of South Africa, the primary producers feeding the growing urban centres in the early years of industrialisation were in large part black sharecropping tenants operating with differing degrees of independence from white landlord control. This article investigates the class forces which shaped this sharecropping economy and its significance in the eventual emergence of white capitalist agriculture. The geographical focus is on the arable highveld, the region in which the rising industrial heartland of South Africa – the Witwatersrand gold fields including the city of Johannesburg – was situated. It was also a region which was ecologically suited to grain production for the rapidly expanding urban markets which developed first at the Kimberley diamond fields in the 1870s, and more importantly on the Witwatersrand from the 1880s onwards.[1]

The relationship between sharecropping enterprises on the farms and primitive accumulation amongst white landlords is fundamental to an understanding of the nature of capitalist development in South African agriculture. South African historiography is replete with references to the irrationality of 'squat-

**School of Oriental and African Studies, University of London.* *I would like to thank the following for their critical comments and suggestions: Margaret Kinsman, Shula Marks, Terry Byres, Colin Bundy, Ted Matsetela, Tim Weiskel, Jeffrey Butler and Leroy Vail.*

ting' or 'semi-feudal' relationships. 'Uneconomical to the natives', de Kiewiet wrote in reference to black tenancy generally:

> the system was uneconomical to landowners as well. As in the reserves so also upon European land was there much wasteful agriculture and much overgrazing. With the growth of population and the rise of land values the demand came for a more profitable use of the land [*de Kiewiet:* 204].

This, for de Kiewiet, implied a 'cash economy' in which labour was hired for wages.

It is misleading, however, to regard sharecropping as a pre-capitalist relic, as an obstacle to the capitalisation of agriculture. This article, on the contrary, argues that sharecropping bore a functional relationship to the capitalisation of settler agriculture. Many of the less capitalised landowners probably would not have survived on the land if it had not been for the surplus production of their independent black tenants. It was partly on the basis of the appropriation of this surplus that many of the white landowners were able to establish themselves as capitalist farmers. This is not to underplay the crucial long-term role of the colonial state in promoting and subsidising settler agriculture at the expense of black producers. It is true that the transfer of capital from outside the agricultural sector – particularly through the state – was the vital precondition of sustained development in agricultural productivity. But many of the settlers might well have succumbed to the rapid penetration of finance capital into the countryside in the early years of industrial growth if it had not been for the expropriation of surplus from the black peasantry living on their land.

Eventually, however, on the highveld as elsewhere in the subcontinent, the commercial black peasantry declined as a factor in the colonial economy. Agricultural production was controlled more and more by white farmers who had access to state-directed marketing, transport and financial facilities, who owned the means of production as mechanisation replaced older methods, and who directly supervised the labour of their black tenants. Blacks were increasingly confined to a proletarian role in the industrialising economy. Economic development did not imply a liberalisation of the racially ascriptive order. In South Africa, as elsewhere, industrial capitalism battened on to, adapted and elaborated racial structures to its own ends [*Greenberg,* 1980; *Legassick,* 1974]. The black sharecroppers of the highveld had largely disappeared by the 1940s under the combined weight of the repressive white state and the capitalising white landlords [*Morris,* 1977]. As landlord capitalism matured, black peasants became an obstacle to its further development. Having served their purpose as producers of agricultural surpluses for their white landlords, black tenants were gradually squeezed into servile status as a labouring class, subject to repressive legislation designed to restrict their mobility and their bargaining power.

This argument concerning the instrumental role of sharecropping in capitalist development contradicts much literature on sharecropping, which is implicitly based on the assumption that capitalism is incompatible with labour that is not free from the means of production. This article is concerned with understanding the way in which capital can subsume and shape diverse forms of production, without directly taking over control of productive processes itself.[2]

Throughout most of the areas of intensive grain cultivation on the white-settled highveld, African communities had been broken up and dispersed during the wars of the Difaqane in the 1820s and 1830s prior to effective white colonisation. Therefore, white settlers on the highveld from the 1830s on moved into an area which had no continuous history of communal black settlement. This factor played a major role in explaining the nature of the relationships which emerged there between black and white. In the late nineteenth century many African families living on the white-owned farms on the highveld had few tribal affiliations, were Christian, and not infequently knew no life other than that under a white master (although the proximity of powerful African polities meant that they were never as vulnerable or as deculturated as, say, American slaves).[3]

The discovery of minerals in abundance and the consequent importation of large quantities of capital and a very considerable urban population in the final decades of the nineteenth century stimulated the rise of intensive cash-crop production which had profound consequences for the rural political economy of the highveld. Blacks generally adapted more successfully and rapidly than whites to the role of peasant commercial cultivators. This is hardly surprising: the trading towns and the Boer pastoralists of the interior had for long been largely dependent on African cultivators for supplies of grain in trade or in payment of rent to settler landlords [*Kimble,* 1978; *Germond,* 1967]. This dependence intensified with the emergence of large urban markets from the 1870s on. Many non-landowning Boers, including landowners' sons, turned to trade and transport riding. On many white-owned farms, black peasant enterprise became the basis of arable production for the expanding urban markets.

In particular, sharecropping emerged as a highly commercialised relationship. It spread rapidly in the 1880s and 1890s, particularly in areas with easy access to the Rand market. The railway line from the coast reached Johannesburg in 1892, enabling large quantities of foreign wheat and other cereals to be imported. The South African highveld, on the other hand, was pre-eminently suited for dry land maize cultivation. Maize provided the staple diet of the army of black migrants who filled the mining compounds and provided industrialising South Africa with its cheap proletarian labour. By the end of the century, the arable highveld was emerging as an area of specialised maize production, fuelling industrial growth based on the mining industry. And it was largely black peasants who provided the sustenance.

Prior to the opening up of internal urban markets, the Boers' relationship with external markets was based on the sale of animal products (wool especially) and products of the hunt (mainly skins). Arable production on white-owned farms was very limited, and generally confined to subsistence requirements. On farms where black tenant communities made it possible, the tenants might plough land for the landlord, and tend and reap the crop. This arrangement was only viable where the landlord's production was very limited and where the labour time required on the landlord's fields did not interfere with the productive activities of the tenant household.

Such arrangements were not suitable for intensive market production. There was a limit beyond which the landlord could not expect to intensify the extraction of labour service from his black tenants without resistance. Either the white landlord had to establish capitalist relations, which required access to capital and expropriated labour power beyond the means of most white settlers of the time. Or he had to batten on to and exploit the productive potential of independent black peasant farming.

This was especially so given the considerable competitive disadvantages of agricultural employers in the rapidly emerging capitalist labour market from the 1870s onward. Wages in towns, at the mines and at railway construction sites were relatively high in the early competitive phase of industrial development –30s or 40s a month, for example, on the diamond diggings. Although few Africans needed to sell their labour, there were a variety of reasons why short spells of migrancy were lucrative for young unmarried men. From the 1870s the seepage of the most serviceable young men from the farms into service elsewhere, leaving the elderly men, the women and children, as well as their stock, behind on the farms, predisposed white landlords to enter sharecropping arrangements [*de Kok,* 1904: 180; *Jacobs,* 1979: 214]. White farmers were unable to compete with urban and mining employment at a time of intensifying crop production, especially since tenants were rarely paid for services rendered.

In more distant areas which were at first remote from transport facilities – such as the richly watered eastern districts of the Orange Free State (where winter crops such as wheat thrived and which prior to colonisation had been settled by Sotho cultivators) – sharecropping did not reach a peak until the Anglo-Boer war. War devastation meant that the white landowners were in a particularly vulnerable position. It was under these circumstances, coupled with the construction of the railway line through the eastern districts which facilitated large-scale grain production, that commercially orientated sharecropping became the predominant relationship throughout the region [*Keegan,* 1981: Chapter IIIA].

The high period of sharecropping in the highveld grain districts was relatively short lived. The coincident capital accumulation in the landlord economy meant that the intensification of exploitation was more likely to take the form of expropriation of the sharecropping peasantry rather than increased surplus extraction on the basis of existing landlord–tenant relations, as a result of the great economies of scale possible in large-scale mechanised production using gang labour. But during its short ascendancy, through to about 1914, sharecropping was a dominant relation in many grain districts [*Natives Land Commission,* 1916: 2,15,22,28,36,68, etc.].

Still, any estimate of the geographical extent of sharecropping at any time, and the proportion of the commercial crop produced by black peasants as against white employers of labour, must be unsatisfactory and open to severe doubt. We are not here discussing the extent to which black cultivators had direct access to markets, since a large proportion of the African-produced crop reaching the market would have been channelled through the hands of white landlords. A very considerable proportion of maize or wheat sold by white farmers was not in fact produced by them, but expropriated from black tenants. This was a matter which not many self-respecting whites cared to admit.[4] There were powerful ideological incentives at work driving individual white farmers to exert greater control over productive activities on the land. The colonists were generally reluctant to admit the extent to which they were reliant on the enterprise of the colonised. Sharecropping was an ill-defined relationship and it was never formally recognised in law or in settler consciousness as a fully fledged system of production. If 'farming on the halves' existed, it was insisted, it was because certain of the white landowners were unwilling to shoulder the burden of civilisation. Consequently, the evidence on the extent of sharecropping is largely qualitative, providing little leverage for statistical analysis. Those who hazarded a guess at the time generally had a polemical purpose in mind. Alfred Barlow from the Heilbron district, who was amongst the few who publicly defended sharecropping, was possibly as well informed as any when he declared that if the government were to pass legislation 'stopping natives from ploughing for white men', the output of maize in South Africa 'would not be a third of what it is at present'.[5]

There were undoubtedly dotted throughout the highveld in the early years of this century 'progressive' white farmers who produced intensively and scientifically with substantial capital resources and who employed contracted wage labour. Amongst them were some of the more recent settlers who brought capital with them from Britain or from other colonies of settlement. Some of them were Boers who had acquired wealth out of land accumulation during times of low land values. However, the progressive farmer was still very much an exception in the early years of this century.[6] In between fully capitalist and sharecropping relations of production was a whole range of relationships encompassing varying degrees of organisational control exercised by the white farmers, and varying proportions in which the landlord and tenant respectively owned the means of production. Peasant crop production was still, as often as not, similar to production on landlord's fields in terms of the technology and division of labour. Thus even where sharecropping did not explicitly exist, the organisation of labour was not necessarily substantially different. The degree to which the landlord owned the means of production and supervised the labour process did not necessarily have any great effect on the productivity or organisation of labour in this situation.

The distinction between sharecropper and labour tenant was often blurred. Many a landlord condemned 'kaffir farming', while his lands were ploughed by his tenants using their own ploughs and oxen, for which they received in return a share of the income from the crop or ploughing lands of their own. This relationship was hardly a 'capitalist' one any more than sharecropping was, and depended as much on the surplus generated by the independent production of the peasant household as did sharecropping. It was particularly common for

African tenants' oxen to be included in the service of labour tenants and to be used on landlords' fields to supplement the latter's own oxen. Landlords' dependence on their tenants' means of production and productive skills was not confined solely to explicit sharecropping relations.

II. THE SHARECROPPING LANDLORDS AND CAPITAL ACCUMULATION

Sharecropping often resembled a 'putting out' system, in which the landlord provided the means of production and seed, or provided his tenants with credit or loans with which to acquire these necessaries. Such arrangements were to be found on company estates, such as the 22 farms of the Vereeniging Estates company, which not only earned considerable profits from their highly commercial sharecropping tenants, but in addition drew labour for their mining enterprises from their tenant families [*Trapido,* 1978: 36-37; *Keegan,* 1981: 205-10]. There was always a fair sprinkling of absentee-owned farms throughout the highveld, particularly during slump periods when speculators bought up land in anticipation of rises in prices, or, more likely, when land fell into the hands of financial or mercantile elements in payment of debt. Rent was drawn from the black tenants on such land in the form of a share of the crop or alternatively, in cash.

However, the great bulk of land in the arable districts of the highveld was occupied by resident white landowners (or lessees) whose sole source of income was from the land, either worked directly by themselves with the aid of black labour, or worked by surplus-producing black tenants, or some combination of the two. It would be fair to argue that those resident white landlords who were most dependent on the surplus production of their black tenants were generally the poorer and less capitalised of the settlers, who were therefore the most vulnerable to expropriation and foreclosure at a time when more and more private capital was seeking lucrative investment in the land against mortgage bonds, when mercantile credit was easily obtainable and the relative self-sufficiency of the settler family was eroding rapidly. This is not to say that all sharecropping landlords were vulnerable and undercapitalised. Sharecropping arrangements were potentially lucrative for most landlords at the early stage of cash crop production in the absence of widespread mechanisation, ready supplies of expropriated black labour or of sophisticated marketing facilities. The drive for greater productivity was not yet so compelling as to make capitalist production invariably preferable even to those whites who had sufficient capital to undertake it. But there was a clear relationship between sharecropping and shortage of capital amongst the white landowners.

The significance of the capital constraints blocking capitalist agriculture can best be understood by examining the productive processes involved. A one-share plough drawn by oxen can turn no more than about two or three acres of ground per day. According to the Bloemfontein newspaper, *The Friend*, by 1894, 90 per cent of the ploughs in the Free State were still one-share ploughs [*Jacobs*, 1979:191]. Climatic conditions on the South African highveld dictate that ploughing be completed in as short a period as possible, since rain comes in sharp, intense thundershowers, and the soil is wet enough for ploughing only for relatively short periods. Before the age of motor traction, farmers who wished to participate in the exchange economy as grain producers were hard pressed to get a competitive acreage under crops.

Thus some white farmers in the Free State by the late 1890s used more than twenty ploughs simultaneously [*Jacobs,* 1979: 191]. This might require the labour of up to sixty males (including children as leaders of the span), plus 120 or 160 head of oxen. Not surprisingly, given the labour and capital constraints experienced by white farmers, most were simply incapable of mobilising resources on this scale. Under these circumstances, it was a more viable proposition to draw off a surplus from a number of individual peasant households working with their own implements and oxen. Hence the partiality for sharecropping arrangements amongst so many white landlords. George Tylden, a farmer of Ladybrand district, defended sharecropping relationships in the following terms in 1908:

> Very few of us Boers . . . can afford to keep more than one span of oxen to plough with . . . it means too much capital locked up in animals which do not increase. This means that in our extremely short ploughing seasons we cannot bring a sufficient proportion of our land under cultivation . . . The farmer therefore gives one or more of these natives, who own perhaps three spans between them, a certain amount of land to plough. The boy finds the labour and often the seed, and gives the owner of the farm half the crop grown . . .[7]

Another Free State farmer wrote in 1913 that in many parts the rainfall was 'so small and far between that a farmer considers himself lucky if he manages to get in a few bags of grain.' In consequence, if blacks were prevented from ploughing on shares for their white landlords, the latter would ultimately be forced into bankruptcy. 'The rich man will perhaps be able to pull through, but the small farmer will suffer.'[8]

It should not be assumed that there was necessarily a division between sharecropping landlords on the one hand, and those on the other who directly controlled productive activities on their land. Both types of relationship generally co-existed on the same farm. There was a considerable degree of differentiation within the African peasantry, and while many tenants had the wherewithal to succeed as sharecroppers to their own as well as to their landlord's benefit, others were more dependent on the direct alienation of their labour power (as will be seen later). Thus, many a white farmer/landlord would use tenant labour to work the employer's implements and oxen, while supplementing this production with the surplus production of sharecropping families using their own implements and oxen. It was not unusual for sharecroppers to be called upon sometimes to provide labour in the landlord's fields, sometimes for a daily wage.

Sharecropping was especially indispensable to settlers who were newly established on the land. The highveld was becoming more densely populated in the years following the Anglo-Boer war, and farms were rapidly being subdivided. The amount of capital required to buy up and settle on relatively unimproved land was often out of proportion to the productive capacity of the land, given the level of technology available. Thus new settlers, many of them introduced by the reconstruction administration, commonly lapsed into sharecropping relationships. The larger Boer landowners of an earlier date had managed to maintain themselves on the basis of extensive pastoralism. The smaller landowners (or lessees) who were becoming more and more a feature of the arable districts had to survive on the basis of grain production – supplemented by dairy farming – or go under. Given high land values (pushed up by speculative buying and option hunting in anticipation of new mineral discoveries) and the need to enter debt in order to establish basic improvements on the land, it is not surprising that these newer settlers became dependent very quickly on black tenant enterprise. Although the Boers had always been dependent on the exchange nexus for inputs of commodities and capital without which their mode of production could not have been reproduced, the extent of the dependence of highveld landowners on commodity production was far greater by the beginning of the twentieth century than ever before. No producer could opt out of intensified commodity production. In the grain districts this increasingly meant intensive cash-crop production, given the declining size of farms. And this in turn meant sharecropping relationships for many, even most.

The significance of sharecropping in this context is illustrated by C.C. Chase of the farm Townley in the Heilbron district. In August 1913 he described the pattern of the 'many young, struggling and progressive farmers' like himself. He had been farming for two years, had little capital but 43 head of cattle. He could only afford to work one plough and one team of oxen. He held his farm on lease and hoped eventually to buy it. But with the value of land soaring, it was impossible for him to make enough money on the produce of but one team of oxen. Thus he had found it necessary to supplement his own efforts with those of two sharecropping families. In order to improve his finances, buy his farm and invest in stock and equipment for more intensive exploitation of the land, he wrote, he was dependent on the productive capacity of his sharecropping tenants.[9]

Francis Carroll, who farmed 700 acres near Kroonstad, was another example. He started with sixteen cows, one bull, eight oxen and a plough. At the age of 26, after a year or two on the land, he had increased his stock and declared himself satisfied with the progress he had made: 'I put it all down to ploughing on shares ... The half share system has been the salvation of the poor men like myself' [*Natives Land Commission,* 1916: 71].

G.J. van Riet of Thaba Nchu made the same observation as Chase with regard to the post-war settlers of that district:

> If you look round the country you will see some of our settlers who, were it not for the sowing on shares, could not stand today where they are now. The natives were planting and sowing under their supervision, and where settlers were not in a position to buy cattle those natives helped a lot. It is practised by nearly every farmer in the district and by the new settlers [*Natives Land Commission,* 1916: 15].

And a correspondent in *The Farmers' Weekly* wrote in 1911 that the settlers in the Thaba Nchu district 'could not get a quarter the crops they do now were it not for having squatters who have oxen . . . The settlers are not in a position to purchase much themselves as yet.'[10]

Finally, J.O. Oberholzer of Bethlehem told the *Natives Land Commission* in 1913 that he had five African tenant families on his one thousand acre farm, each with ten head of cattle, horses and ploughs. They cultivated six to eight hundred acres on half shares (120 to 160 acres per family!) and did all the 'loose work' on the farm, as well as supplying household servants who were paid a wage. There were many young beginners in his district who were in the same position, he said [*Natives Land Commission*, 1916: 55].

The evidence appears to contradict the belief widely held at the time by the guardians of racial supremacy, that independent black production was squeezing whites off the land, and that if sharecropping were prohibited the 'poor whites' would flock back to the farms as a small tenant class. The experience of the farmers quoted above, on the contrary, was that the productivity of the black sharecropping tenantry was keeping undercapitalised and often over-indebted white landlords afloat. If these African cultivators were to be uprooted from the land, the impoverished stratum of settler society would be greatly swollen rather than rehabilitated.[11] For many an impoverished white family survival on the land as lessees or 'bywoners' (white tenants at will who performed services for their landlords) was made possible by the black sharecroppers. Often white lessees or tenants served as go-betweens on absentee landlord farms, subsisting and paying rent from the produce of the black tenants. Their position was entrenched by laws prohibiting the formal renting of land by blacks, and by laws (designed to distribute potential labour) severely limiting the number of black families who might live on a farm unoccupied by a white man [*Keegan*, 1981: chapter II A].

Many white families became economically dependent on their altogether more prosperous black tenants and neighbours. As an instance: a landowner returned to his farm after the Anglo-Boer war with no working capital, having to sell off some four thousand acres of his ten thousand acre farm to pay off his debts. The Africans who had lived on his land returned from the town where they had been during the war. Within a year they had a herd of cattle and began ploughing on half shares. By 1912 the tenants had about 160 head of cattle against the landowner's one hundred. The Africans also had some five hundred head of small stock. They were reportedly far better off than the owner of the ground and declined to render service. The African-owned stock that year ate the fenced-off pasture veld, and the landowner had to wait for the tenants to reap their crops before he could feed his cattle on the arable fields.[12] The magistrate in Fouriesburg gave an instance in 1911 of a white landowning family in his district living entirely on the production of one African resident upon the farm who owned several spans of oxen and a couple of wagons. The African did all the ploughing and transport work. 'The owners do nothing', added the magistrate [*Blue Book on Native Affairs*, 1910:154].

It was not only the lack of working capital but the risks involved in investment that made sharecropping so attractive to these small landowners, whose viability was dependent on their ability to pay off their debts. One farmer wrote in *The Farmer's Weekly* in 1912 that, given the climatic conditions and the uncertainty of yields, the cost of production of cereal crops amounted on average to more than half the income from the crop. Under these circumstances it paid the landowner better to allow sharecroppers to bear the costs and the risks and get half the crop, than to do the work himself with hired labour and risk a loss on investment. 'Guarantee a farmer a fairly reasonable prospect of a crop every year as a result of his labour,' he wrote, 'and kaffir farming would be as dead as the dodo tomorrow.'[13]

In a situation in which agricultural technology was not very different in the peasant economy from the landlord economy, the direct subsumption of black labour under white capital did not have any great advantages in terms of productivity or profitability. Maize harvesting continued to be a labour-intensive activity quite late in the twentieth century, involving as it did simply the picking off of cobs (stalks being left on the land as winter feed). Where new technology brought competitive advantages – as in the case of threshing machines (generally owned by syndicates of whites who hired them out by the day in return for a proportion of the crop threshed) – it was compatible with the household organisation of labour. For the landowner who was unable or unwilling to capitalise his production, the best course was simply to leave the peasant in command and control of the means of production and to live off the surplus generated. This was especially so since the sharecropper was likely to devote more attention and care to a crop half of which was to be his own than to a crop which was to be of profit to the white farmer alone. A 1908 commission reported that whites who defended sharecropping asserted that 'some natives are as good or better farmers than some white men, and that unless the former has a direct interest in the crop, he will not exert his best endeavours towards its successful production' [*Natives Administration Commission,* 1909, 9: cf. *Rennie,* 1978].

There was another related reason why sharecropping was so congenial to landlord and tenant household alike. It protected the former against desertion and the latter against eviction, in an economy in which town employment offered more remunerative opportunities for younger family members than farm work. The landlord knew that even if the most productive workers went to town to earn the money wages obtainable there, the crop would still be tended and reaped, for it was in the peasant household's interest to see to it. And the peasant household knew that provided the crop was delivered, the landlord could not object too strenuously to family members oscillating between farm and town.

It would be misleading to interpret the sharecropping economy solely in terms of the capital constraints and other considerations confronting white landlords, however. It was a product of conflicting forces, one of which was the resistance of blacks to attempts to subvert their independence, coupled with the colonists' limited coercive powers. There was a constant agitation amongst capitalising white farmers against the independence and prosperity of black tenants on the farms who refused to alienate their labour regularly and intensively. Many a white landlord insisted that he was reluctantly bound to sharecropping arrangements because all attempts to transform them into more explicitly capitalist relations resulted in the desertion of his land by his tenants in preference for less servile arrangements on the farms of others. D.P. Liebenberg, a farmer in the Pretoria district, complained in 1910 that his white neighbour

> never did any work and the consequence was that I could not keep my natives. When I made them work from sunrise until sunset they would leave and go on to this other farm, where, of course, they had an easier time. I consider that it is a wrong principle to allow natives to lease land and give a portion of their crops to farmers [*Select Committee on Native Affairs,* 1910: 229].

In all the litany of agitation which came from white farmers, this was the most frequent. Given the ubiquity and importance of black peasant resistance, it is clear that the sharecroppers themselves need to be looked at more closely if the sharecropping relation is to be fully understood.

III. THE SHARECROPPERS AND THE SHARECROPPING RELATIONSHIP

By no means all sharecroppers were equally prosperous or productive. Nevertheless, sharecroppers who ploughed forty or sixty acres, produced three or four hundred bags of maize or more a year (as well as sorghum and – in the better watered areas – wheat), owned a few spans of oxen and perhaps a flock of sheep, as well as advanced implements, seem to have been sufficiently typical to have constituted a class of significance in the countryside [*Natives Land Commission,* 1916: 2, 10, 18, 28, 38, 39, 54, 56, 70, etc.: *Blue Book on Native Affairs,* 1910: 153, 178, etc.; *Select Committee on Native Affairs,* 1910: 136]. No doubt many sharecropping families led an altogether more precarious existence, mixing sharecropping with labour service and wage labour, and possibly drifting in and out of sharecropping relations as shifting family fortunes and circumstances permitted. The greater visibility and perhaps importance of the former should not lead us to underestimate the degree of differentiation amongst the sharecroppers. Nor should it be forgotten that sharecroppers and labour tenants were not always and everywhere distinct groups. Between the more successful sharecroppers and the propertyless farm labourers was a range of conditions combining servility with ownership of certain of the instruments of production indispensable for landlord survival. The extent of flux undoubtedly experienced by individuals and families from year to year should also not be underestimated.

From the share of the crop they received, the sharecroppers often had considerable surpluses for sale, which they disposed of to local storekeepers and merchants. The share of the crop received by the landlord was normally half, rising in some cases to two-thirds. The landowner would attempt to extract a larger than usual share where he had some control over the means of production: for example, if the tenant did not have enough oxen, or if he had considerable stock holdings and required large acreages for grazing. Landlord's costs might be deducted from the tenant's share, and perhaps also the threshing costs. The tenant would have control of all productive processes: ploughing, sowing, cultivating and reaping. The agreement would perhaps include the supply of a male adolescent throughout the year for work such as herding the landlord's stock, as well as one or two young girls or women for housework and washing. Some landlords paid a shilling a day for work performed by adult males from sharecropping families. Others expected labour without payment.[14] The tenant also had grazing rights for his animals. The landlord might have the right to maize stalks for the winter for his own stock.

Conflict was built into the sharecropping relationship. For example, the exact size of the crop which was to be distributed in predetermined shares was subject to manipulation. The African producer had plenty of opportunity to minimise the actual quantity which the landholder received. Generally, there was no supervision of the sharecropper's activities. Landlords were not unaware that they were being deceived, and often expressed their feelings on the matter. A correspondent wrote to *The Farmer's Weekly*: 'The native is supposed to give the owner one half or one third of the crop he produces, but after feeding his family and friends it is a very small share the owner gets.'[15] One source of dissension was the African practice of picking off the ripening cobs and consuming the green maize from about January through to the harvest in June or July. This practice was commonplace and served to diminish reliance on stored grain.[16]

The landlord tried to insure himself as far as possible against manipulation of another sort by insisting that he have first choice of the reaped and bagged crop. Thus the tenant could not neglect half the fields in the knowledge that that half constituted the landowner's share, nor could he separate off the poor quality grain after harvest and hand it over to the landlord.[17] The laws which forbade the ownership or leasing of land by Africans supported the landlord's rights with respect to the crop insofar as sharecroppers were not recognised as formal tenants. This physical control over the crop that the landlord exercised once it was in the bag no doubt also enabled him to exercise his own brand of fraud. The commissioner in the Basutoland district of Mafeteng remarked in his report for 1904 on the frequency of cases in which the white landowners in the Free State had refused to pay their Sotho tenants what they were entitled to. The frequency with which landowners evicted tenant families from the land on flimsy pretexts once they had ploughed and sown and even reaped was a source of much grievance.[18] W.A. King, Sub-Native Commissioner in the Pretoria district, said in 1910 that it was common for black tenants to be given notice once they had ploughed a large acreage: 'When the native claims compensation he is not only refused permission to remove buildings which he has brought but . . . he is told that he can go and take away the ground he has ploughed' [*Select Committee on Native Affairs,* 1910: 173].

Without much supervision and no policing mechanism other than the physical presence of the contracting parties at the harvest, the relationship was based on a precarious balance of interests. Despite the ubiquity of struggle over surplus, both parties needed the other. In the final analysis what made the system work was not coercion or force, but the peasant's own vigorous productive enterprise.[19] The viability and profitability of sharecropping for the landlord depended on the willingness of the black tenant to maximise his output.

The African sharecroppers, who possibly produced the bulk of maize on white-owned farms in the early years of the century, were frequently supplied with seed by their landlords. The intervention of the state in the provision of quality seed from government experimental farms to white farmers meant that the quality of grain produced on white-owned farms was likely to improve. Further, sharecroppers who sorted and bagged their grain badly were likely to get short shrift from their white partners. It may be that a good deal of grain marketed by Africans living in reserves was of an inferior quality, partly because of their lack of access to better quality and more consistent seed, and also because they often did not sort and bag their grain themselves but delivered it directly to the store. But as far as sharecropping was concerned, the mutual profitability of the relationship and the success of the maize export drive after 1907, which was to a large extent dependent on black peasant production (by whichever of the two parties – landlord or tenant – it was marketed), is evidence that the sharecroppers were producing a product fit for international markets [*Keegan,* 1981: chapter I D].

In short, the more successful of the sharecroppers were commercially oriented cash-crop producers. The more diligent and the wealthier the tenant, the more land he could plough; and clearly the landlord with limited resources of his own beyond the land wanted to see as much land cultivated by his tenants as possible. 'The ploughing on the halves depends largely upon the individual value of the worker,' said H. Reading, magistrate in Heilbron; 'If a boy is a good boy and has three teams of oxen and three ploughs the farmer reaps the benefit; his productive capacity is greater' [*Natives Land Commission,* 1916: 35]. New technology, new forms of family labour organisation and new attitudes toward productive activity were preconditions of successful sharecropping.

These conclusions are borne out by Emelia Pooe, an ex-sharecropper born in 1882 whose testimony was recorded in 1979 by Ted Matsetela. She stresses that considerable skill was a prerequisite to successful sharecropping. One had to 'plough properly (which was an onerous task), select good seeds, hoe the field perfectly and have the crop harvested and threshed timeously, especially when there was need to avert winter rains' [*Matsetela,* forthcoming]. It was not simply enough to have the requisite means of production – a span of oxen, a plough, yokes, chains and so on. Mrs Pooe stresses her husband's diligence as a matter of pride:

> Indeed, he worked very hard in his fields. He produced a lot from the soil. Hundreds of bags, half of which he gave to Theuns [Lindeque, the landlord] who in turn would proceed to sell them and get a lot of money from the labour to which he had never contributed anything.

Mrs Pooe remembers one year in particular:

> People in the neighbourhood ... other croppers, tenant labourers, farm-owners and *bywoners* [white tenants], black and white alike, talked about Naphtali's harvest. Many whites in fact remarked that Tol [another landlord] should be very pleased that he had engaged a cropper of Naphtali's ability. I remember Piet Smit, a white farmer ... brought his own maize for threshing. His maize stack was far smaller than ours. We beat him by far.

The way in which individual black households became locked into this commercialised and relatively large scale cash-crop production requires investigation. The testimony of Emilia Pooe, again, provides some illumination. When her father took to sharecropping in 1897, a major rinderpest epizootic had only recently subsided, as had a serious drought. Mrs Pooe tells us that many people had not ploughed. In order to get a ploughing team together, her father had to borrow young bullocks and heifers from neighbours [*Matsetela,* forthcoming]. So what might have led him to commercial sharecropping was the promise of seed and credit from the landlord and the prospect of recovering his stockholdings with the proceeds from his enterprise. Sharecroppers would presumably have had access to more ploughing land than otherwise; indeed, their acceptability to landlords would have depended on their willingness to maximise the crop. Under these circumstances, and given the monetisation of many items of productive and personal consumption, credit relations and tax demands which were already in some degree part and parcel of the peasant economy, economic survival as independent producers meant for many families a decision to throw in their lot with a white landlord who had both land and access to state aid and credit. In this way many black tenant households were inserted into the capitalist economy as sharecroppers producing surplus for the capitalising landlord class. Economic motivation and market relations must be seen in this material context for it was such factors as these which turned black cultivators into commercial sharecroppers, rather than any supposed innate, universal predisposition to respond to market opportunities for their own sake.

Africans' choices were changing by the 1890s. They were finding it more and more difficult to exist on the margins of the colonial economy. For one thing, land pressures were beginning to be felt. Those on the white-owned farms had to decide whether they were going to become ever more dependent on the alienation of their own labour, or succumb to the forces of commoditisation by transforming and extending their own productive activities. If they chose the latter course, the course of the sharecropper, they had to do so wholeheartedly, for their security of tenure depended on their capacity to satisfy their landlord's drive for surplus at a time of increasing commercial opportunity, rising land values and accelerating indebtedness. If they went in for sharecropping they had to have the skill and the capacity of commercial farmers, or risk failure.

This is not to say that Africans had not produced for sale prior to the mineral discoveries (they had on a large scale) or that a rent in the form of a share had never before been extracted from African tenants. But the nature and scale of the landlord–sharecropper relationship were qualitatively different by the 1890s and 1900s. It was not simply another form of surplus extraction such as Africans living on the white-owned farms had long since grown used to. The sharecroppers were entering a new realm of productive enterprise. The crucial new factor – the growth of a massive home market in the shape of the Witwatersrand – stimulated a new cash-crop economy and a new level of production. This in turn required a new breed of market oriented producers, amongst them the black sharecroppers. Sharecropping was an attractive option at this time of rapid commercialisation on the farms, when white farmers were intent on intensifying the extraction of surplus from their tenants.

So, despite the structural constraints which pushed households ever deeper into relations of surplus extraction, there was, nevertheless, often an element of choice involved in the way in which individual households succumbed. The decision to enter sharecropping relations implied an insertion into the colonial economy at a qualitatively different level than before.[20] Hence observers in the mid-1890s noted that many Africans were moving from other districts into the districts south of the Witwatersrand specially in order to enter sharecropping relations [*Jacobs,* 1979:200].

However, others on the farms resisted this voluntary insertion into the colonial economy as cash-crop producers. Many did not have the labour resources, the skill or the perseverance to make a success of commercial production. And there were those who by chance or design found themselves under landlords who preferred and had the capital for a more direct extraction of surplus value and saw no reason to alter their circumstances. All these remained labour tenants, whose increasing integration into the capitalising landlord economy was from the first predicated upon the rendering of labour service and the sale of labour power. Those who opted out of intensified commodity production were already on the path to proletarian status. Black tenants who were unable to meet the rising income requirements of their landlords by means of surplus from sharecropping were obliged to alienate their labour to an increasing degree. This emerging differentiation amongst the black tenants on the farms, and the class relations it entailed, will be discussed in the next section.

IV. SHARECROPPING AND THE HOUSEHOLD ECONOMY

In order to understand the nature of the peasant economy, it is necessary firstly to explore the organisation of labour in the black peasant household. This is no easy task, for the evidence is generally silent on such matters. White landlords were only concerned with heads of black household with whom they had contractual agreements. They rarely if ever expressed any interest in the forms of labour organisation which were employed in the production of crops, except for purely ideological purposes, such as in the constant references to the laziness of African men who used their wives and children as 'slaves' in the fields.

It would seem that the nuclear household was the basis of production.

Throughout most of the areas of intensive grain cultivation on the white-settled highveld, as we have seen, African societies had been disrupted prior to white colonisation, and white settlers broke up unwanted African groupings on white-owned farms in the process of asserting control over the land. Further, as land in African territories and unoccupied white-owned land became more difficult to acquire and grazing land scarcer, African families moved on to highveld farms. These families were probably small units, although kinship ties and grazing rights in African areas were sometimes maintained. The sharecroppers generally did not constitute large extended kin based communities.

The actual division of labour within the sharecropping family is difficult to reconstruct. Emelia Pooe stresses that work in the fields (unlike amongst the white bywoners) was a joint effort in which she and her children (when old enough) participated.[21] This is significant for this was a Christian family in which education was also a priority: Emelia's husband was a Sunday school teacher [*Matsetela,* forthcoming]. Women in sharecropping families did the hoeing and weeding and played a major role in the reaping and threshing. Quite young boys and girls also aided in these activities. Men were responsible for the heavy work. Indeed, the only activity in which men were necessary was ploughing. The widow Maria, who sharecropped with her two teenage children in the Hoopstad district, had to hire assistance to plough the fields, but otherwise managed to make a good living. The elderly Kgabale of the same farm did well from sharecropping with the aid of his own daughters, but depended on his two sons who came back from the Rand every year to do the ploughing [*Plaatje,* 1969: 78-9].

Indeed, there was a good deal of geographical mobility within peasant families. Young men and women of the most serviceable age for field work tended to spend a considerable part of the year in non-agricultural employment away from the farms. There was probably a disproportionate number of adolescents and of older members of peasant families employed in productive activities on the land for much of the year (during periods of low labour intensity in the productive cycle).

Despite the household basis of labour organisation, the black peasant enterprise was dependent on work parties drawn from neighbouring families and more distant kin at times of peak activity in the annual season. This was especially so during the reaping season, which was the high social point in the year. When the harvest on Zaaiplaas failed in 1895, Emelia Pooe and her mother paid an extended visit to her aunt in Rustenburg district in the Transvaal, assisting in the harvest and threshing of wheat. At the end of the season her aunt gave them ten bags of sorghum. The harvest was a great time for visiting relatives and sharing in the work load and festivities.

But it was not only members of the extended family who provided the sharecroppers with seasonal labour, but also unrelated black families living on the same or neighbouring farms. The harvest seemingly belonged to all, and all joined in the feasting that accompanied the work. The sharing of the crop amounted to a ceremony, according to Mrs Pooe: 'not only those who had worked daily watched over, but neighbours too who might have helped in the work parties. All took great pride in celebrating their neighbours' achievement' [*Matsetela,* forthcoming]. A farmer complained in January 1912, during the bumper winter cereal harvest of that year, of the system he called 'reaping with beer':

> A native gives notice that he is about to reap and will kill a sheep and supply unlimited Kaffir beer. It is astonishing to see the crowd that collects in no time, and the ease with which his crops come off. But the poor farmer when he gets up finds his boys missing, off to the beer drink. They turn up the next day, muddled and useless.[22]

The white farmer, on the other hand, was dependent on wage labour or tenant service for his seasonal labour requirements.

Mrs Pooe remembers that the labour tenants on the farms provided labour for the sharecroppers, especially during the hoeing, harvesting and threshing seasons. They participated in the work parties where beer and newly slaughtered sheep were served to the workers. Wage labour was exceptional, Mrs Pooe reports. Only if there were less than four people engaged in harvesting and threshing would the sharecroppers hire labourers. However, it was not unknown, and her husband hired labour when the crop was very large. So direct labour costs in the form of a wage bill were uncommon except under unusual circumstances, and this was an advantage to the peasant sector.

Nevertheless, it is not improbable that what Emelia Pooe is describing when she talks of aid given by the labour tenants to sharecroppers is an embryonic and disguised form of wage labour, hidden by the 'traditional' form which remuneration took. The extent and intensity of commodity production on the part of the sharecroppers would indicate that we are here encountering a labour form which was qualitatively different from anything that had gone before, despite the ostensibly co-operative and reciprocal nature of such labour organisation. The differentiation between the sharecropper with sixty acres and the capital to cultivate it and the labour tenant with a couple of acres and little more than his labour power to alienate suggests that the relations between them were asymmetrical to the point at which they began to seem like class relations.

On the other hand, this kind of differentiation among peasants is not at all clear cut. For individual members of sharecropping households might go out to labour for wages at different times and for different purposes.[23] At the same time, sharecropping families were generally expected to provide some degree of labour service to the resident landlord. Young boys, for example, were provided as herds and girls worked in the white homestead. Those who earned wages or provided labour service were probably the younger, dependent members of sharecropping families, whose labour was only intermittently required in the parental fields. Their labour away from the household economy was not necessarily indispensable to the reproduction of the household.

It would be quite misleading to suggest that differentiation within the black farm community was leading toward capitalist relations amongst tenants.

There were considerable barriers to capital formation amongst the croppers. Indeed, it seems that there was no general tendency toward increasing productivity above a certain level. A certain degree of investment in improved technology and methods was a precondition of sharecropping, to be sure. If the sharecropper was to produce sufficient surplus to satisfy his landlord's requirements, he had to invest in steel ploughs and good seed. But although there were, no doubt, instances where the croppers themselves used their surpluses for productive investment beyond the minimum necessary for sharecropping purposes, there were not many who owned implements such as sowers, harrows or threshers [*Blue Book on Native Affairs,* 1910: 53, 153-55]. Methods of cultivation similarly showed little indication of improvement over time. One constant complaint amongst whites was the destructive effect the black cultivator had on the soil. 'The natives plough a few acres of land until it is played out, then start another land, and so on, until nearly all the best land is spoilt with kweek or couch grass,' was a typical complaint,[24] although one which was as much a reflection of settler ideology as of empirical observation.

Insecurity of tenure in particular was a major factor in explaining why black cultivators did not rotate their crops or irrigate or manure their fields. The possibility always existed that the tenant family would be ejected, or that a conflict would arise which would make it difficult for the tenant to stay on. Many sharecroppers remained on the same land year after year. But during these years of rapid capitalisation and growing commercial opportunities for white landowners, tenure was uncertain at the best of times. This consideration explains the reluctance of African peasants to invest their often fairly substantial income in productive improvements.

Capital accumulation amongst the sharecroppers took the form of an expansion of stockholdings. The insecurity of tenure of black cultivators on white-owned land reinforced this tendency. African tenants could never be sure when they would have to move off in search of new land without any hope of compensation for improvements on the land. Stock at least could move of their own volition, unlike other forms of material wealth. Then again, Africans' lack of access to financial institutions made them dependent on stock as readily exchangeable commodities which reproduced themselves for the payment of debts or tax.

For the Africans stock had a far greater significance than as simple commodities. This they made clear in evidence before the *Natives Land Commission* in 1913. Adam Duiker, an African of Bethlehem district, said, 'I am fonder of stock than money, as we cannot live without stock, and that is our bank . . . What should I do with a thousand pounds if I converted the stock to money? We cannot go and buy motorcars' [*Natives Land Commission,* 1916: 58]. And Daniel Majake of Heilbron:

> We cannot sell our stock . . . for then we cannot live. We would sell if we could use the money to buy land or to lease land, but that is against the law . . . If we sell our stock we would have to use the money, and then would be without stock and without money after a while [*Natives Land Commission,* 1916: 39, 61].

This in turn meant that blacks were especially vulnerable to natural disasters such as rinderpest or drought. Animal disease was widespread and difficult to control given the nature of the state's resources. Thus a natural check was maintained on African herds and flocks and black accumulation of wealth, quite apart from the limitations placed on tenants' stock by landlords. War, epidemics and climatic vagaries meant that stockholdings fluctuated greatly. For example, the cattle population collapsed after the rinderpest outbreak of 1896-7, followed by the devastation accompanying the Anglo-Boer war of 1899-1902.

The rapid recovery of the stock population in the post-war decade presented the black tenants with new insecurities. A very common cause of tenant evictions was the unwillingness of landlords to cater for large stock numbers and the refusal of tenants to dispose of their flocks and herds. During the period of financial boom after 1909, for instance, when more and more land was being turned to maize cultivation and land was becoming a valuable resource for black and white alike, evictions of wealthy stockowning tenants reached alarming proportions [*Plaatje,* 1969; *Msimang,* n.d.; *Keegan,* 1981: chapter VE]. W.A. King, Sub-Native Commissioner in Pretoria, told in 1910 of an African tenant who owned some 150 sheep, 40 head of cattle, 12 or 15 horses and some goats, who could not find a single landlord in the Heidelberg district who would take him on despite the general shortage of labour [*Select Committee on Native Affairs,* 1910: 174-75]. And General Tobias Smuts, member of Parliament for Ermelo in the eastern highveld, said in 1914 that many Africans with large numbers of cattle had been turned off the farms and could find no land on which to settle: 'if a native has a considerable number he will have to sell some of his cattle ... if he wishes to live on a farm' [*Natives Land Commission,* 1916: 260]. The sustained accumulation of individual wealth was very difficult under these circumstances.[25]

It would appear that larger stockholdings tended to belong to older men [*Natives Land Commission,* 1916: *passim*]. It would be a reasonable assumption that the wealthiest sharecroppers were those with most family labour at their disposal. Older men would have recourse to the services of adult sons at peak labour times, even when, as was generally the case, such sons worked elsewhere for much of the year. Each new generation had to begin accumulating wealth from scratch; large stock accumulations were seldom passed on to one heir, for wealthy men generally had large families, all of whom had to be provided for. However, demographic differentiation based on the stage of the household in the cycle of generational reproduction was probably a minor factor in accounting for stratification amongst sharecropping families. Sharecropping was too shortlived on the highveld for any recurrent generational pattern to emerge. Indeed, given the uncertainties and vicissitudes of tenancy on the white-owned farms, it was more likely for an individual family to go through several cycles of impoverishment and reaccumulation in the course of a generation than for accumulation to be sustained through a full lifetime [*Relly and Nkadimeng,* 1981].

Apart from insecurity of tenure and the uncertainties attendant on stockownership as the major means of accumulation, another factor limiting the opportunities for extended reproduction was the black tenant's financial vulnerability and consequent tendency to indebtedness. The significance, extent and mechanisms of this factor are difficult to ascertain, given the silence of the evidence. But in other sharecropping social formations, most notably the post-bellum South of the United States, the 'lien system' was the very basis of exploitation and landlord-mercantile profit [*Daniel,* 1973]. As we saw earlier, landlord's credit was a common feature of sharecropping relationships on the highveld. But it is unclear whether the illiquidity of the peasant enterprise was such that debt peonage was ever a widespread or fundamental aspect of the relationship. This question relates closely to the question of stock ownership. For if it is true, as the impressionistic evidence quoted suggests, that Africans generally invested their income in stock and did not keep cash reserves, then a corollary would be a dependence on credit for the working and living expenses of the long growing season. This would in turn create opportunities for extortion, usury and crop mortgaging in advance of harvest.

That Africans were often in debt to storekeepers as well as to landlords is evident. Their vulnerability to financial and mercantile capital was noted in 1911 by the magistrate in Fouriesburg. He wrote that Africans were becoming 'far too fond' of borrowing from white men and, in many cases, had to pay very high rates of interest, such as 2s per pound sterling per month [*Blue Book on Native Affairs,* 1910: 192]. This usurious rate – 120 per cent per annum or 313 per cent compounded – is evidence of the weak financial position of Africans who were unable to borrow from banks or other financial institutions as they had no landed security to offer. They were therefore thrown at the mercy of the local trader, who then expropriated the peasant's crop after harvest at a price which was highly advantageous to himself.

If the African then found that he had sold a greater proportion of his crop than he could afford as a result of large debts or a poor harvest, he invariably had to buy back again at a much higher price later in the season than he had received. This predicament, which was common to black and white producers alike, was widely noted. The Bothaville magistrate tells us that Africans bought seed maize and other grain for sowing at from 15s to 20s per bag, yet they sold their maize at from 5s 6d to 7s 9d per bag and their sorghum at 12s [*Blue Book on Native Affairs,* 1910: 127-29, 192]. Several magistrates reported in 1910 that Africans kept only enough for their own subsistence after harvest. This financial vulnerability was noted by the magistrate in Winburg in 1911 when he wrote that 'unless the Native has a reserve of cash or foodstuffs (a remote contingency) he must either run into debt with his "partner", the farmer, or resort to questionable means of subsistence pending the marketable stage of his crops' [*Blue Book on Native Affairs,* 1910: 275].

Lacking adequate statistics, it is difficult to gauge what proportion of sharecroppers were actually affected by this indebtedness cycle. It would seem, however, that while a good many prosperous sharecroppers, like the Pooes, were able to maintain their relative independence from debt bondage, there were others who, like the cotton croppers of the American South, were sucked into a cycle of crop lien from which they were unable to extricate themselves short of the direct subsumption of their labour power under the control of productive capital. In these cases, the sale of labour power away from the farm by members of tenant households might have been a means of trying to extricate the household from financial bondage.

V. CONCLUSION: THE SUPPRESSION OF THE SHARECROPPERS

The reluctance or inability of sharecroppers to re-invest in improved technology and methods made them especially vulnerable. As land prices rose and the opportunity cost of sharecropping increased for the capitalising landlords, the croppers could not increase their output sufficiently to maintain the viability of sharecropping relationships. The limits of production were set by the amount of family labour available, their physical capacity for exertion and the amount of land they could cultivate. Under these circumstances, the sharecroppers were particularly vulnerable to any general flow of capital into the landlord economy, particularly under the auspices of the state. After all, many sharecropping landlords were themselves capitalising farmers. Landlord capitalism was itself a dynamic force, accumulating unevenly and developing in fits and starts. Even if the sharecropping landlords were not themselves necessarily capitalising as individuals, the surplus generated by their tenants was effectively a form of primitive accumulation, which all the time was creating the conditions for the victory of landlord capitalism and the levelling of peasant differentiation. Sharecropping thrived because and as long as settler farming was stifled through lack of capital. The survival of many of the sharecroppers depended on the lack of capital formation amongst a large proportion of the landlords. Embedded in the very structure of the sharecropping economy were elements of its own demise.

The capitalisation of settler agriculture has depended to a large degree on state intervention, which has taken the form of the transference of large quantities of capital to white farmers, especially through the auspices of the Land Bank, the provision of transport and cooperative facilities, government regulation of supply and price levels, practical and scientific aid and guidance by state departments.

The suppression of the sharecroppers was a process fraught with violence and coercion, of expropriation, forced stock sales and evictions. Most particularly, demands for labour service intensified as landlord capitalism developed and spread. S.M. Makgatho, President of the Transvaal Native Council, complained in 1914 that Africans throughout the area were being given ultimatums by the white landlords that, 'If you want to reside on my farm, I want a contract ... that you and your wife and children and your children's children will work for me.' Makgatho added that should the tenants refuse to work for twelve months a year, they were evicted [*Natives Land Commission*, 1916: 285]. The availability of alternative land was an important variable and as land became a more valuable resource and as population pressures began to be felt, so did the bargaining power of the white landowner increase. Severe limitations were increasingly imposed on African tenants' stock, as we have seen. This in turn made tenant families less independent and more reliant on rendering labour for their cash and credit needs.

The role of the state in the process of class formation was crucial. There always were forces strongly antagonistic to sharecropping enterprise in settler society. The idea that colonisers should shirk their responsibilities as the bearers of civilisation and economic progress to the extent of relying effetely on black production was inherently distasteful to the colonial mind. This was especially so since black peasant independence was regarded as being the reason why black labour was not forthcoming in the quantities and at the price required by capitalising landlords and other employers of labour. The legislature was always amenable to pressures from these 'progressive' quarters, as evidenced in the legislation going back to the nineteenth century designed to restrict to five the number of black families on any one white-owned farm in order to distribute potential labour supply. The major assault on sharecropping enterprise during the years under review came in the form of the Natives Land Act of 1913 which attempted to undermine all forms of independent tenancy and to enforce labour service in payment of rent, going so far as nominally to abolish sharecropping in the Orange Free State where the influence of absentee landlords and land companies was weak [*Keegan,* 1979].

The 1913 act was not immediately decisive, but it did provide the statutory framework within which future struggle and coercion could take place, as and when settler capitalism had matured sufficiently to require the transformation of class relations on the farms. The act remained the lynchpin of the settlers' drive toward proletarianisation of the black tenantry and the dispersal of those with large stockholdings and little labour utility. Tenants were also brought within the purview of masters and servants legislation, which brought the force of criminal law to bear on the contractual obligations of all tenants, and rendered them and their families vulnerable to the deprivations visited upon them by landlords desirous of servile labour [*Keegan*, forthcoming]. Furthermore, pass laws were employed to immobilise labour and prevent desertion and 'vagrancy', although never as effectively as the farmers desired, given the divergent interests of the urban and industrial employers.

Sharecropping was eventually replaced by more explicitly capitalist relations on the white-owned farms, in which cash wages and rations became the dominant form of compensation for labour, and labour was rendered throughout much of the year under close supervision and in non-household form. This transformation was very apparent during the 1920s, though tenancy was not replaced for two decades more. By the 1940s secondary industry was rapidly emerging as the dominant sector of the South African economy, and the agrarian hinterland of the highveld was finding it increasingly difficult to compete for labour with the burgeoning urban centres. As the lot of African tenants on the farms deteriorated, as their terms of labour became increasingly proletarian, more and more of the younger members of tenant families escaped to the towns. By the 1940s, this seepage had reached critical proportions, and, as M.L. Morris has argued, the electoral victory of the Afrikaner Nationalists in 1948 was partly predicated on a determination to solve the problem of agricultural labour mobilisation, allocation and control, as part of the emergent apartheid policy. By the 1960s labour tenancy had been in large part replaced by a tightly regimented contract-labour system [*Morris,* 1977].

This transition to non-tenant labour was facilitated by the mechanisation of production on the farms. The coming of the tractor in particular revolutionised labour processes and productivity. Dependence on tenant's means of production was a thing of the past once the tractor had become a commonplace on the highveld by the 1950s, replacing the ox as the primary means of traction. Mechanisation of the harvest as a result of the widespread introduction of the maize reaping machine also meant that farmers were no longer so dependent on a reserve of family labour on the farms on whom they could draw during the labour-intensive harvest season. The capital-intensive crop production on the contemporary highveld is a far cry from the peasant production of an earlier period.

But no amount of legislative moralising or ideological posturing could turn African peasants into proletarians while they remained indispensable for the survival of many white settlers on the land and for commodity production. The transformation of the countryside was gradual and uneven, and was the product of changes at the point of production rather than in law. In this analysis sharecropping is seen not as a pre-capitalist relic, but as a stage in the penetration of colonial agrarian formations by capital. It was a vital and integral stage in the capitalisation of settler agriculture. Rather than an archaism, it was a relationship created and structured by capital.

NOTES

1 The region under consideration was divided between the two Boer republics, the Orange Free State and the South African Republic (the Transvaal), in the nineteenth century. In the Anglo-Boer war of 1899-1902 they became British colonies, and were included as provinces in the newly formed Union of South Africa in 1910. The primary evidence cited has been reduced to a minimum for reasons of economy. More detailed documentation is presented in my thesis [*Keegan,* 1981].

2. Writers such as Hilferding, Bukharin, Luxemburg as well as the 'dependency' theorists and the French Marxist anthropologists analysed the interconnections that have occurred between capitalist and non-capitalist modes of production. But see, particularly, in the present context [*Friedman,* 1977; *Faure,* 1978; *Bernstein,* 1979 and *Wolpe,* 1974].

3. Many of the blacks on the farms at the end of the nineteenth century had been (or their parents had been) 'apprentices' or African children captured in battle or traded from African chiefs. These dependants, known as *oorlamsches,* were probably the most deculturated of the black farm inhabitants.

4. 'It is a very difficult thing to say what proportion of natives is living on farms on the share system in my district. Very few farmers will admit that that system has been carried on on their farms. All along we have had a shrewd suspicion that a very large proportion of the natives has been living on the farms under these conditions.' Magistrate, Winburg [*Natives Land Commission,* 1916: 1].

5. A.W. Barlow, Heilbron, to editor, *Farmers' Weekly,* 16 August 1911.

6. Even the most capitalised farmers usually tied their workers to the land by a method of wage payment in kind which took place at harvest time. The employer would designate an acre or two for the worker and his family, which would be worked with the landlord's own instruments of production. Tenant labour was much preferred to freely mobile wage labour, particularly given the much higher wage rates in towns.

7. Public Record Office, London, Colonial Office archives, CO 224/27, Lord Harris to Secretary of State, 21 September 1908.

8. 'Talis Qualis', Lower Vet River, OFS, to editor, *The Friend*, 29 October 1913.

9. South African Archives, Pretoria, Justice Department series, JUS 5/262/13, C.C. Chase to Prime Minister, 17 August 1913.

10. 'Free State Farming,' *Farmers' Weekly*, 27 December 1911.

11. The problems of white impoverishment and the emergence of a marginal white underclass was a burning issue of the time, and evoked deep social anxieties. Many Boers who had never been landowners lived as tenants on the land of others. With the increasing capitalisation of land and commercialisation of agriculture, it was becoming increasingly difficult to maintain an independent existence on the land as mobile pastoralists for a growing proportion of Boers. The fate of this growing 'residuum' of 'poor whites' was easily linked in the popular imagination with the emergence and consolidation of a class of relatively wealthy black sharecropping tenants. It was a matter of the gravest concern to whites generally that blacks seemed to be taking over the land from whites at a time of rapid economic growth and social change.

12. 'A Settler', to editor, *Farmers' Advocate*, July 1912, p.699.

13. 'No Compulsion', Thaba Nchu, to editor, *Farmers' Weekly*, 11 December 1912.

14. This practice was a source of considerable grievance amongst tenants, according to Rev. H.R. Ngcayiya, Ethiopian Church Minister of Klipspruit [*Select Committee on Native Affairs*, 1910: 422].

15. 'Progress', Kroonstad, to editor, *Farmers' Weekly*, 19 July 1911.

16. See R. Seggie, 'Squatting and Ploughing on Shares', *Farmers' Weekly*, 8 May 1912.

17. As a Wepener farmer put it, the African 'must not know what he is going to get before it is in the sack'. 'Wepener Congress', *The Friend*, 27 January 1909.

18. CO 417/394, Basutoland Reports, 1904, 21 November 1904. The provision of credit might also have enabled landlords to exact more produce in repayment than they were strictly entitled to. Little wonder that conflict was endemic in such relationships. The magistrate in Hoopstad estimated in 1911 that fully 80% of complaints received in regard to contracts on the farms arose from sharecropping arrangements [*Blue Book on Native Affairs*, 1910: 274].

19. The carrot rather than the stick was more likely to bear results. Some landowners gave a prize to the tenant who raised the largest crop in a year [*Plaatje*, 1969: 24].

20. Again, Emelia Pooe remembers vividly this transition and the soul searching that went into it.

21. The disadvantages of white sharecroppers for the landlords was their refusal to use family labour in the fields, and their dependence on black labour. They also refused to render labour to the landlord. The 'bywoners' played an inconsiderable role in arable production on the highveld, although many lingered on the land as a marginal element tied to the landed class by the ties of sentiment and kinship.

22. 'Our Weekly Causerie', *Farmers' Weekly*, 31 January 1912.

23. It was also common to supplement arable operations with transport riding. A.J. Bouwer, a Bethlehem farmer, said that 'they have a lot of wagons and are taking the work out of the

hands of the poor whites' [*Natives Land Commission*, 1916: 17].

24. 'Progress', Kroonstad, to editor, *Farmers' Weekly*, 19 July 1911.

25. Concentrations of stock were commonly dispersed among relations and friends on the farms. Many Africans working in towns invested in stock and grazed it on farms unbeknown to the white landowners, who assumed they belonged to the black residents. Much of the stock running on farms was not owned by the residents of the farms.

REFERENCES

Arrighi, G., 1970, 'A Study of the Proletarianisation of an African Peasantry in Rhodesia', *Journal of Development Studies*, Vol. 6, No. 3.

Bernstein, H., 1979, 'African Peasantries: A Theoretical Framework', *Journal of Peasant Studies*, Vol. 6, No. 4.

Blue Book on Native Affairs, 1910, U.G. 17-1911.

Bundy, C., 1979, *The Rise and Fall of the South African Peasantry*, London; Heinemann.

Daniel, P., 1973, *The Shadow of Slavery: Peonage in the South, 1901-1969*, New York: Oxford University Press.

de Kiewiet, C.W., 1957, *A History of South Africa, Social and Economic*, London: Oxford University Press.

de Kok, K.J., 1904, *Empires of the Veld*, Durban: J.C. Juta and Co.

Faure, Claude, 1978, *Agriculture et Capitalisme: Essai sur les Rapports de Production et Agriculture*, Paris: Editions Anthropos.

Friedman, H., 1977, 'Household Production and National Economy: Concepts for the Analysis of Agrarian Formations', *Journal of Peasant Studies*, Vol. 7, No. 1.

Germond, R.C., editor, 1967, *Chronicles of Basutoland*, Morija: Morija Sesuto Book Depot.

Greenberg, S., 1980, *Race and State in Capitalist Development: Comparative Perspectives*, New Haven and London: Yale University Press.

Jacobs, D.J., 1979, 'Die Ontwikkeling van Landbou in die Vrystaat, 1890-1910', D. Litt et Phil, University of South Africa.

Keegan, T.J., 1979, 'The Restructuring of Agrarian Class Relations in a Colonial Economy: the Orange River Colony, 1902-10', *Journal of Southern African Studies*, Vol. 5, No. 2.

——————, 1981, 'The Transformation of Agrarian Society and Economy in Industrialising South Africa: the Orange Free State Grain Belt in the Early Twentieth Century', Ph.D., University of London.

——————, forthcoming, 'The Sharecropping Economy, African Class Formation and the 1913 Natives Land Act in the Highveld Maize Belt of South Africa', *Industrialisation and Social Change in South Africa*, edited by S. Marks and R. Rathbone, London: Longman.

Kimble, J., 1978, 'Towards an Understanding of the Political Economy of Lesotho: the Origins of Commodity Production and Migrant Labour, 1830-1870', M.A. thesis, National University of Lesotho.

Legassick, M., 1974, 'South Africa: Capital Accumulation and Violence', *Economy and Society*, Vol. 3, No. 3.

Matsetela, T., forthcoming, 'Aspects of Sharecropping and Proletarianisation in the Northern

Orange Free State, 1890-1920: the Life Story of Emelia Pooe', *Industrialisation and Social Change in South Africa,* edited by S. Marks and R. Rathbone, London: Longman.

Morris, M., 1976, 'The Development of Capitalism in South African Agriculture: Class Struggle in the Countryside', *Economy and Society,* Vol. 5, No. 3.

_________, 1977, 'State Intervention and Agricultural Labour Supply Post-1948', *Farm Labour in South Africa,* edited by F. Wilson, A Kooy and D. Hendrie, Cape Town: David Philip.

Msimang, F.W., n.d., *The Natives Land Act, 1913: Specific Cases of Evictions and Hardships,* Johannesburg.

Natives Administration Commission, 1909, Orange River Colony, *Report of a Commission Appointed to Consider and Report upon and Make Suggestions for Modifying or Amending Laws with regard to Natives and Native Administration within the Orange River Colony,* published by the Union Government, Pretoria, 1911.

Natives Land Commission, 1916, Minutes of Evidence, U.G.22-1916.

Palmer, R., and Parsons, N., 1977, *The Roots of Rural Poverty in Central and Southern Africa,* London: Heinemann.

Plaatje, S., 1969, *Native Life in South Africa,* New York: Negro University Press. [Reprint of *ca.* 1916 edition.]

Ranger, T.O., 1978, 'Growing from the Roots: Reflections on Peasant Research in Central and Southern Africa', *Journal of Southern African Studies,* Vol. 5, No. 1.

Relly, G. and Nkadimeng, M., 1981, 'Kas Maine: the Story of a Black South African Agriculturalist', unpublished paper presented to History Workshop Conference, University of Witwatersrand, Johannesburg.

Rennie, J.K., 1978, 'White Farmers, Black Tenants and Landlord Legislation: Southren Rhodesia, 1890-1930', *Journal of Southern African Studies,* Vol. 5, No. 1.

Select Committee on native Affairs, 1910, Minutes of Evidence, S.C. 3-1910.

Trapido, S., 1978, 'Landlord and Tenant in a Colonial Economy: the Transvaal, 1880-1910', *Journal of Southern African Studies,* Vol. 5, No. 1.

Tylden, G., 1950, *The Rise of the Basuto,* Cape Town and Johannesburg: Maskew Miller.

van der Horst, S., 1942, *Native Labour in South Africa,* London: Oxford University Press.

Wilson, F., 1971, 'Farming, 1866-1966', in *Oxford History of South Africa,* Vol. II, edited by L.M. Thompson and M. Wilson, London: Oxford University Press.

Wolpe, H., 1974, 'Capitalism and Cheap Labour Power: From Segregation to Apartheid', *Economy and Society,* Vol. 1, No. 4.

Sharecroppers and Landlords in Bengal, 1930-50: The Dependency Web and its Implications

Adrienne Cooper*

INTRODUCTION. THE BACKGROUND AND THE QUESTIONS TO BE ADDRESSED

Sharecropping was an important form of agrarian relations in Bengal between 1930 and 1950. In 1947 Bengal, a province of British India, was partitioned into West Bengal (India) and East Pakistan (now Bangladesh). In both areas uneven and stunted agricultural development has been a primary concern, sharecropping being perceived as one of several mechanisms retarding growth. Each state has attempted to reform the land system inherited from British rule as a prerequisite for rural development. In Bangladesh few initiatives concerning sharecroppers have been taken; rather land reforms have reinforced the position of the landlords who are the mainstay of the rural elite [*Januzzi & Peach,* 1980: 10-12]. In West Bengal progressive reforms and campaigns have attempted to improve the economic and legal status of sharecroppers. To some extent successful, they have met with obstacles due to factors inherent in the sharecroppers' relationship of dependency on the landlords [*Khasnabis,* 1981: A 47]. In both states the number of sharecroppers and the area of land on share contracts continue to be significant features of rural society.[1]

Frequent and widespread peasant agitations are a shared heritage of both West Bengal and Bangladesh. In particular the *tebhaga andolan* of 1946-7 was an unprecedented movement of sharecroppers aiming to reduce land rents from one half to one third of the crop. Sharecroppers organised under the auspices of the Bengal Provincial Kisan Sabha (Peasants' Association), supported by other sections of the peasantry. This communist-led organisation was a branch of the All India Kisan Sabha which combined peasant and nationalist politics, working alongside the Indian National Congress and Muslim League.

Kisan (peasant) activists raised the *tebhaga* demand in all the districts of Bengal and, during the harvesting of the major *(aman)* winter rice crop, sharecroppers initiated struggles against their landlords. In some villages compromising cropsharing agreements were reached; elsewhere incidents led to further intensification of conflict. The peasants organised defence forces for protection against police and landlord harassment, and, in the process of resistance, established zones where *kisan* courts punished landlords and their agents. Where harvested paddy was already stored in the landlords' compounds, poor peasants launched raids to recover their crops. The movement extended to villages previously untouched by Kisan Sabha activities involving

*University of Sussex. I am very grateful to Terry Byres for his criticisms of earlier drafts and for his encouragement.

masses of sharecroppers in this agitation. Several (69) peasants were killed in firings inititated by the police or landlords and many hundreds were arrested.

The Bengal government responded to the *tebhaga* movement by publishing proposed legislation – the Bargadars Bill – in January 1947, conceding to some of the sharecroppers' demands. Seduced into believing that the agitations were successful, the Kisan Sabha leadership discouraged the sharecroppers from going on the offensive. At the same time the villages were declared to be 'disturbed', allowing the government to use more coercive powers to put down the movement. Armed police and military forces set up camps in the storm centres of *tebhaga* struggles to intensify repression. Poor peasants were arrested, their homes ransacked and fines imposed. By Independence Day, August 15th, 1947 the movement was crushed, with thousands of activists and participants in jail or underground [see *Tebhaga Songram,* 1973; *Sen, S.,* 1972; *GOB,* 1948]. But the echoes of the *tebhaga* movement reverberated. In following years sharecroppers' demands, including *tebhaga,* continued to feature in communist campaigns and agitations. Their struggles against landlords extended from the fields into the courts, and Bhagchas Boards were established in West Bengal to assist sharecroppers and landlords in negotiations [*Rasul,* 1969: 192-3. 210]. The experience of this movement and its implications for continuing the organisation of sharecroppers continue to be of relevance and are debated in academic and communist circles.[2] Whilst sharecroppers continue to be exploited by landlords, and the need for change remains, the history of the *tebhaga* movement can teach important lessons.

This article forms part of a wider study which situates sharecropping in its economic context, and *tebhaga* within a chronology of sharecroppers' struggles.[3] It aims to describe and analyse the sharecropping relationship, a 'web of dependency' incorporating both economic and non-economic structures of subordination, a complex relationship that cannot be captured in the formalistic mould of an economist's model. The secondary structures have frequently been ignored by economists, thus rendering their models defective and unfruitful if applied to actual situations. The landlords simply aimed at expropriating the surplus product of the sharecroppers. The dynamics of the dependency relationship functioned in order to maximise this exploitation. Secondary structures were essential for this process. At any historical conjuncture the dependency relationship assumed a specific form. In this article the economic and social relations that prevailed between sharecroppers and landlords in Bengal during the period 1930-50 are examined. Three broad questions are addressed: Why did the system survive and extend at a time when capitalist forces had such an impact on Bengali rural society? How did sharecropping operate? How was the dependency relationship continuously reinforced through secondary structures? Sharecroppers' agitations reached a peak in the *tebhaga* movement of 1946-7. The significance of their actions can be assessed only in the context of the intricate structures they challenged. An understanding of the complexities of the dependency relationship enables one to explain why purely legal measures enacted to protect sharecroppers have had limited success in south Asia in general and Bengal in particular.

SHARECROPPING IN HISTORICAL CONTEXT

Sharecropping in India has existed from 'ancient times'. Buddhist monastery and university lands were cultivated on a half-share contract, seeds being divided equally between the two parties. Hindu law books of the ancient period mention that non-cultivating landlords sub-let their land for a share of the produce. The Arthasatra of Kautilya, describing the land system at the time of the Maurya Empire in c.323 BC, explains that on Crown or government farms cultivators were employed on share contracts. Those who provided their own seed and bullocks received half shares of produce, those contributing only their labour received a quarter or fifth share. This evidence that sharecropping was 'as old as the country itself' was inevitably cited as defence of the system whenever there was any possiblity of change or legal reform [*GOB,* 1940a, 1:253; 11:136, 140, 170].

From this and similar evidence it was assumed that sharecropping existed down the centuries in Bengal. Early British administrators like H.T. Colebrooke found share contracts, and a comprehensive district survey in the 1870s mentions various forms of sharecropping in most areas, indicating that it had become integral to agrarian relations in Bengal [*GOB,* 1940a, 11:236; see *Hunter,* 1875-7]. In subsequent decades sharecropping became increasingly widespread, by 1940 covering at least 20 per cent of the total cultivated area of Bengal [*GOB,* 1940a, 11: 118-9]. Lack of statistical data frustrates any attempt to substantiate a strong impression that sharecropping extended considerably during this period of British rule.

It is now helpful to indicate some of the processes that contributed towards the expansion of sharecropping in Bengal. Since these were shaped by the land revenue system introduced by the British, a brief description is appropriate. The Permanent Settlement granted proprietary rights to *zamindars* in return for fixed and prompt revenue payments, guaranteeing a stable income hedged against risks. Responsibility for land improvement and management was handed over to the *zamindars* who were expected to develop their estates in a fashion similar to English gentlement farmers [see *Guha,* 1963; *Baden-Powell,* 1892]. The *zamindars* generally failed to live up to these expectations, discovering alternative means of stabilising and augmenting their incomes. They were allowed to transfer and divide rights, to sub-let to others. The vast areas of waste land, depopulated by the 1770 famine, were brought back into cultivation frequently by enterprising tenure holders. By the end of the nineteenth century net cropped area had doubled [*Chaudhury,* 1976: 309-314]. Intermediate tenures were created as a means of simplifying management of estates or raising income. Further subdivision and sub-letting of tenures created a large class of rent-receivers between the *zamindar* and the cultivator, all claiming their share of the agricultural surplus.

Clauses of the Permanent Settlement concerning the rents and rights of cultivators proved to be inadequate. Agrarian agitation pressurised the government to define through legislation the rights of different classes of cultivators.[4] However, these rights of *raiyats* (cultivators) to occupancy and 'fair' rents, became marketable assets, purchased by non-agriculturalists who sub-let to the actual cultivators. At the lowest rung of the ladder were sharecroppers, unprotected by law, with no rights or status [*Bhaduri,* 1976: 50-1; *Chaudhury,* 1975: 106; *GOB,* 1940a: 38-9].

Sharecroppers had very limited legal or customary rights to the land and were ignored by tenancy legislation except, of course, when there was potential for direct government gain. In an area of Jalpaiguri district, outside the realm of the Permanent Settlement, officials imposed a cess payment, a sort of rates, on sharecroppers and other cultivators, of eleven *annas* per acre (there were sixteen *annas* to a *rupee*) [*GOB,* 1919a: 49].[5] In this untypical case the sharecroppers were a source of income for the government.

The land revenue structure provided the framework for agrarian society. Accompanying the proliferation of sub-tenancies was the extension of sharecropping. Various patterns emerged which are summarised below.

Firstly *zamindars* often retained a segment of their land directly under their control to obtain produce for consumption by their large households and staff. These *khas, khamar,* or *nij jote* lands were cultivated by sharecroppers or agricultural labourers. Tenure holders emulated this practice, preferring to engage sharecroppers as they were themselves disinclined, for reasons of status, to be involved in cultivation. Sharecroppers on this land usually provided the inputs for cultivation. Middle-class tenure holders, for whom 'cheap rice was a necessity', had vested interests in maintaining sharecropping and were guaranteed to raise an outcry whenever there was any suggestion that sharecroppers should be legally protected, fearing that their livelihood was endangered [*GOB,* 1916a: 93; 1915: 168, Appendix, xlviii; 1925: 74; 1940a: 233, 253]. Accrual of rights, such as occupancy, was banned on *khas* land. Incentives, such as rising foodgrain prices, made produce-paying cultivators attractive and making land *khas* was one means to this end [*GOB,* 1916a: 94; 1925: 56; 1940d: 52; 1941: 24]. Sharecropping on *khas* land was one source of its extension.

Secondly, *zamindars* created favourable tenures to facilitate extension of cultivation of jungle and waste land. In areas outside the Permanent Settlement the government also offered tenures at low rates. In some districts these tenures were called *lots* or *jotes,* and the purchasers consequently referred to as *lotdars* or *jotedars*. In order to attract cultivators to break in the land implements, seed and animals were provided. The *giri* (landlord) in Jalpaiguri continued to supply all the inputs to sharecroppers in Jalpaiguri district long afterwards [*GOB,* 1919a: 74-5]. Tribal people were encouraged to do this work, being accustomed to growing crops under harsh conditions. Enticed by the promise of rights to their holdings, they were deprived of these ambitions and 'ousted from the land' when it had 'attained the desired point of tillage'. The landlords used to demand an exorbitant rent or employ other means to compel them to give up their holdings [*GOB,* 1908: 59]. Many of these cultivators remained on these holdings as sharecroppers [*GOB,* 1916a: 32; 1930a: 33; 1938:71].

Thirdly, sharecropping was an aspect of the process of differentiation among cultivators. Upwardly mobile peasants adopted the Brahmin (high caste Hindu) prohibition against ploughing which was a sign of high status, finding it more suitable for gentlemen to sub-let the land to sharecroppers than cultivate directly [*GOB*, 1916c: 15; 1919b: 19]. Sharecropping was utilised by 'the efficient cultivator who accumulates a little capital (and) tends to climb into the ranks of rent-receiver and money-lender where he takes as large a share as he can of the produce of the soil and puts nothing back' [*GOB*, 1928: 48].

Downwardly mobile cultivators lost their rights to land and became sharecroppers, often on the same holding. Land alienation through indebtedness was extremely common: cultivators unable to pay off their debts lost their land rights, which had been security for loans, to moneylenders. Well-off cultivators, benefiting from high prices in the 1920s took out loans which they were unable to repay when prices crashed during the Depression. This was another contributory cause in the increase in sharecropping during this period [see *GOB*, 1930b; and *Chaudhury*, 1975]. During the 1930s indebted cultivators continued to lose their land rights and become sharecroppers as the effects of the Depression permeated rural society. Some of these cultivators, although dispossessed, retained their animals and ploughs. Their moneylending landlords provided, at most, the seed.

The prices of land holdings fell during this period as part of the effect of the Depression. A major group of buyers were traders in foodgrains interested in 'rationalising [their] business vertically. This new kind of landlord is already a trader in jute or grain. It will pay him therefore to take his rent in jute or grain rather than cash and have to buy the raw materials for his business on the open market' [*GOB*, 1940a, VI: 46]. These landlords were inclined towards using share contracts, supplying seed, but rarely any other inputs for cultivation. Rising foodgrain prices during the Second World War culminated in the traumatic Bengal Famine of 1943. High prices and food shortages continued to feature throughout the 1940s. For those able to invest, vast profits could be made from marketing and hoarding raw materials, especially foodgrains. In this context share contracts were highly desirable.

Maximisation of income through evasion of tenancy legislation if necessary, and the convenience in specific circumstances of income in kind, were the major incentives for adopting sharecropping during the 1930s and 1940s. In the context of the colonial economy and world capitalism, sharecropping – apparently a pre-capitalist form of production – was integrated into the wider economy in different ways by various kinds of entrepreneurs. This history of the evolution of sharecropping forms the background for a description of the different share contracts found in Bengal. The circumstances in which sharecropping developed influenced the terms of contracts; the different economic relations between the sharecropper and landlord. The details of these relationships are described in the following section.

THE VALUE AND LOCAL FORMS OF SHARE CONTRACTS

Different forms of cropsharing were closely related to sharing of inputs, in turn influenced by the circumstances in which sharecropping had developed. Sharecropping is popularised as an *equal* relationship: whereby the cultivator and landlord provide equal inputs, take equal risks and receive equal share of the crops. The historical evidence suggests that the relationships were far more complex, often far from equitable, and the sharecropper extremely vulnerable to the demands of the landlord. Various types of rent agreements existed in Bengal, each cultivator and landlord coming to an individual contract at the start of the season.[6] The most common division mentioned by official and non-official sources was 50:50, half the crop going to the sharecropper and half to the landlord. *Barga* (share lease), *adhi* (half), and *bhag* (part, share), were terms used to describe this system, and sharecroppers called *bargadar, adhiar* or *bhagdar* respectively. One estimate in Rajshahi district was that 98 per cent of produce rents were *barga,* half-sharing, the remaining 2 per cent were *bhag khanjana* which were paid half in money and half in kind [*GOB,* 1916d: 94]. Unfortunately there are no figures for the prevalence of different share contracts in Bengal.

Taking a closer look at input sharing, some patterns emerge. Whoever provided the seed took an equivalent amount of paddy before the division of the crop took place, so that net output was shared [*GOB,* 1937; 121, etc]. However, in some places the landlord providing the seed was treated differently from the sharecropper. In Dinajpur district, if the landlord provided the seeds, rather than the sharecropper, a 50 per cent rate of interest was charged when the seeds were deducted before the division of the crop. There are no recorded instances of such interest being charged when the seed was provided by the sharecropper. In Birbhum district, if the landlord gave seeds they were deducted before the sharing of the crop. If the sharecropper gave seeds they were not deducted unless the land was of good quality, in which case the rent was high – two-thirds share of the crop. These instances illustrate an asymmetry in the share contracts. The sharecropper was expected to provide part of the working capital with less rewards than the landlord [*GOB,* 1941: 21; 1937: 121; *Bhaduri,* 1973a: 121].

In half-sharing contracts, the sharecropper provided the plough and other agricultural implements, and the animal power necessary for cultivation. The straw, or a portion of it, went to whoever provided manure [*GOB,* 1918: 54; 1937: 121; 1941: 21; 1942: 28]. In some contracts straw was a return for giving seeds [*GOB,* 1916: 14; 1937: 120; 1942: 28; 1923: 6-7].

There were share contracts where the proportional divisions were not equal. Shares could vary according to the quality of the land. Under the *takuthi* system in Malda district, the sharecropper paid a quarter of the produce as rent. Here the land was poor and jungly, and labour in short supply; and, therefore, the sharecropper was in a stronger position *vis-à-vis* the landlord than in other areas. The sharecroppers provided all the inputs and were able to refuse half-share contracts in favour of quarter-share [*GOB,* 1938:72].

Similarly rates of rent payable by sharecroppers in Birbhum district varied according to the quality of the land. The lower the fertility of the land, the greater the share for the cultivator: if the land was bad the sharecropper retained a half share, if ordinary land, he retained two fifths or nine twentieths, if good and well-irrigated he kept only one third. Similar variations were found in Murshidabad and Jessore districts, indicating that share contracts were not uniform [*GOB,* 1937: 120; 1934: 157; 1912: 83].

Cropsharing could be influenced by a variety of factors, including the circumstances in which sharecropping was adopted. Rack-renting forced rents higher than the usual half-share in parts of Bankura, to nine-sixteenths of gross produce payable as rent [*GOB,* 1926: 71].

Share contracts were affected by degrees of input sharing. Whilst the sharecropper who provided plough and animals received half of the crops, when the landlord supplied these inputs as well as the land, he (the landlord) received two thirds. In south-west Bengal this was known as the *krishani* system, the *krishan* being the sharecropper supplying only labour and keeping one third of the produce. This indicates that whilst land was worth half the crop, labour was worth one third and other inputs one sixth of the produce. Bhaduri's analysis of semi-feudalism is constructed from this pattern of sharecropping and, therefore, has an applicability limited to those areas of Bengal where this specific form of cropsharing occurred/occurs. It would have to be modified to be relevant to the other patterns of cropsharing, more common in many areas [*GOB,* 1937: 122; 1926: 70; 1910: 64; *Bhaduri,* 1973a: 120-37].

Half-produce sharing was probably the most common contract in Bengal, in the period from 1930 to 1950. The 50:50 sharing served an ideological purpose in sounding equitable, whilst in reality the sharecropper was often deprived of even a half share because of deductions for seed, and other inputs. Where the sharecropper was able to retain a larger share of the crop, this was because a half share on such poor land would have still been barely adequate for subsistence. In all cases the landlords maximised their surplus appropriation. It is clear that although sharecropping implied egalitarianism – sharing inputs, sharing output – it always concealed a high rate of exploitation.

Produce rents were invariably higher than cash rents, giving an 'abnormally large rate of profit' [GOB, 1916a: 85]. Sharecropping, being outside the realm of tenancy legislation, was a mechanism for rack-renting. In Nadia district half the produce was equivalent to Rs4 per *bigha* compared to cash rents of *raiyats* of Rs2-7-10 (there were approximately three *bighas* in an acre). In Pabna and Bogra districts, sharecroppers also paid on average twice as much as the lowest category of *raiyats* paying cash rents [*GOB,* 1928: 62, 171; 1930a: 45, 50]. The table below shows a 'fictitious' attempt by an officer to evaluate the cost of cultivating *aman,* the main winter rice crop, per acre. Rates for agricultural tasks are taken as the equivalent rates of pay for agricultural labourers, and are accounted in terms of man-days necessary for each task. The seed and produce were valued at prevailing prices during the Survey and Settlement in 1934-8.

Costs: Seed	Rs 0.14
Ploughing	Rs 3.12
Harvesting	Rs 1.8
Manure & Irrigation	Rs 2.0
Transplanting & Sowing	Rs 2.0
Reaping	Rs 2.0
Carting	Rs 1.0
Threshing	Rs 1.4
Total Costs	Rs14.6

(There were 16 *annas* to one *rupee*)

The average return was estimated as 17 *maunds* per acre, worth Rs 27-10, and straw worth Rs 3, the total value of the produce being Rs 30-10. If this was grown by a sharecropper, the seed would have been deducted before division and the by-products allocated to whoever had provided the other inputs. By this reckoning the landlord would receive at least produce worth Rs 13-6, in comparison to the prevailing average cash rents in the area of Rs 2-8 per acre. (Deducting Rs 3-14 for seed and straw, the remaining produce was worth Rs 26-12; the landlords share worth Rs 13-6.) The sharecropper would be left 'owing' 2 annas, if he paid himself and his family labour at the current market rate for agricultural labourers (the costs of cultivation excluding seed were Rs 13-8: the sharecropper's share worth Rs 13-6). Sharecroppers were also faced with secondary demands which are described in subsequent sections. In fact the sharecropper, the direct producer, was left with less than the half share, barely recompense for the tasks that contributed towards cultivation, not to mention the survival of the family household [*GOB,* 1941: 42].[7]

When jute was cultivated by sharecroppers, they did the laborious tasks of weeding, steeping and extraction. The landlord sometimes shared the increased labour costs or the sharecropper retained a greater share of the produce as payment for these onerous jobs. This indicates that there was some recognition that labour costs were to be rewarded [*GOB* 1923: 80; 1930a: 25; 1912: 83].

Share contracts were influenced by a number of factors. To see them merely as a product of a 'bargaining process' (as do Bell and Zusman [1976]) wherein shares 'reflect the agents respective bargaining strengths' suggests in its vocabulary a false equality between sharecroppers and landlords, and a degree of autonomy for sharecroppers that they did not possess. Rather, the share contracts were historically determined, reflecting the origins of the pattern of sharecropping and the unequal power relationship between sharecropper and landlord. The sharecropper retained a meagre share of the crop to ensure basic subsistence, while the landlord made a large profit. This scarcely reflected a 'consensus among both parties regarding that which is fair and just' as has been alleged [*Rudra,* 1981a: 5]. What justice was there in the share contract that maintained the landlord in affluence and the sharecropper just above destitution?

Agitations during the 1940s emphasise the extent to which sharecroppers were vulnerable to the landlord's power and unable significantly to alter the rental shares. Despite the widespread co-operation between sharecroppers, their unity in opposition to their landlords, in struggles to decrease the share of the crop payable by them to one third, in the years following the *tebhaga* movement the share contract reverted to one half. The power of the landlords to enforce such high rates of rent from sharecroppers was due to their ability to wield extra-economic powers as well as the threat of eviction. To this I will return below.

CRITICAL FEATURES IN THE DEPENDENCY RELATIONSHIP

There were particular features of the relationship between sharecropper and landlord which embodied the highly unequal relationship of dependency. These were expressed at critical points during the cyclical process of cultivation. That the demands and actions of sharecroppers during agitations centred on these, and that they constituted a burning sense of grievance among sharecroppers, suggests that any sense of 'fairness' or 'justice' inherent in the existing sharecropping relationship simply did not exist.

At certain stages of the annual agrarian cycle the sharecropper consulted the landlord, or the landlord demanded the right to make certain decisions regarding crop production. The sowing stage was crucial: 'The *jotedar* (landlord) decided what crop is grown' [*GOB*, 1941: 23]. Or the sharecropper paid for the privilege of deciding this [*GOB*, 1930a: 70-1]. Steps were taken to ensure a particular level of productivity: 'The landlord gave an order – you must grow fifteen *maunds* per *bigha*' [Interview Naru Kolkamar, sharecropper, Khanpur, Dinajpur, also *GOB*, 1919a: 75].

Harvesting the crop was done by the sharecropper, often at a time chosen by the landlord [*Jannayuddha* 1.12.43]. The paddy was carried to a place specified by the landlord and the threshing done on specially prepared ground, the *khamar*. This was usually in the landlord's compound, 'a sign of the master–servant relationship' [*GOB*, 1941: 23]. Here the crop was divided into shares according to the verbal contract. During the period between depositing the paddy and division, the crop was kept in the landlord's *gola*, a sort of wicker granary. Since the landlord had physical possession of the total produce during these stages of production, he controlled the situation and was able to make extra demands on the sharecropper. These I discuss in later sections.

The slogans of the sharecroppers' struggles related to threshing, storing and division as well as the basic demand to retain a greater share of the crop. They demanded that the threshing floors should be located at their own houses, or that a community threshing ground be established in a neutral place. The paddy should also be stored at the sharecroppers' homes. For while this practice of threshing crops at the landlord's compound continued, sharecroppers had no control over division and their dependent position was reinforced. One of the most significant features of the *tebhaga* movement was that sharecroppers harvested the crops collectively, took the paddy to their own or community threshing grounds, defying the landlords and tradition.

Where agitation intensified after the harvesting period, the landlords' *golas* (granaries) were major objects of attack by sharecroppers. 'Break the granaries' and 'take the paddy' were popular slogans as the sharecroppers took back what they considered to be their rightful shares of the crops [*Tebhaga Songram*, 1973; *GOB*, 1948; *Sen*, 1972].

A fundamental source of dissatisfaction and discontent was that the nature of the share contract was temporary and it was verbal. Bhaduri's description seems appropriate: 'Unregistered tenancy with the constant threat of eviction gives the landlord a highly personalised form of economic power over his tenants' [*Bhaduri*, 1973b: 11-12]. Even when the landlord did not utilise this power, the potential existed to reiterate the sharecropper's position of subordination every year when the contract had to be renewed. In reality, in many areas sharecroppers remained cultivating the same landholding for consecutive years, even for generations [*GOB*, 1919a 75; 1923: 80; 1940c: 75; 1958: 26]. One survey found that 21 per cent of sharecropping leases were annually renewed for fifteen years whereas 38 per cent held the share contract for the year of enquiry only [*GOB*, 1919b: 45]. Other cultivators, holding the same plot for twelve years consecutively, were eligible for occupancy rights according to the Bengal Tenancy Act. But sharecroppers, although perhaps farming the plot for an equal length of time, were excluded by legal and customary definition from any such rights. Other cultivators renting plots assumed continuation of cultivation unless they failed to pay the rent, whilst sharecroppers assumed a temporary agreement and had to renegotiate annually. This was the fundamental difference between share contracts and other forms of rent agreements in Bengal.

Cultivators on holdings paying rent in cash received receipts whereas sharecroppers rarely were given receipts for the shares taken by the landlords. This indicated the lower status of share contracts [*GOB*, 1916a: 13; 1958: 26].

Some economists argue that insecurity of tenure can be beneficial: the landlord can evict the tenant if the land is not being cultivated properly, and then employ another sharecropper in order to maximise production [*Newbery*, 1975: 272]. This narrow analysis ignores the alternative aspect of the relationship: the sharecropper's dependency on the landlord which was reiterated annually. Insecurity of tenure emphasised the landlord's control over an essential factor of production, the land. It enabled the landlord to pressurise the sharecroppers. In Khanpur village, for example: '*Barga* land wasn't available without loan, even if you had sufficient food, you were forced to take a loan' [Interview Dhabra Kolkhamar, sharecropper, Khanpur village].[8] I will discuss this aspect of the dependency relationship, the close connection between sharecropping and indebtedness, in a later section. It is mentioned here to illustrate the point that insecurity of tenure was a basic pillar of the landlord's power over the sharecroppers. The threat of eviction or failure to renew the share contract facilitated the extraction of a high rate of surplus and maximised dependency.

It is hardly surprising that a continuing theme of sharecroppers' movements in the 1940s in Bengal was the need to obtain protection against eviction. One alternative to this end was to gain tenancy rights similar to those of other tenants, such as the right of occupancy to a plot cultivated for twelve years [*Rasul*, 1969: 103]. Rent receipts were demanded as an intermediary step towards establishing tenancy rights, and the demand to stop eviction of sharecroppers was raised [*GOB*, 1948; Interview Ajit Rai, peasant-communist activist, Dinajpur].

Demands and actions of sharecroppers focused on some critical features of their dependency on landlords both because they constituted grievances in themselves and because they symbolised the greater power of the landlords. At critical stages of the production process sharecroppers were reminded of their dependency. They were scarcely equal partners with equal powers in this process.

SECONDARY STRUCTURES OF EXPLOITATION: BEGAR

The power of the landlord over the sharecropper, by virtue of his control over land, enabled him to demand use of the labour of the sharecropper's family. *Begar* was a form of corvée labour found with share contracts [e.g. *GOB,* 1936: 74-5]. The sharecropper worked for the landlord for a specified number of days without payment, receiving perhaps a meal. The work varied from one to forty-five days a year and was usually at peak cultivation periods such as sowing, ploughing, harvesting or jute steeping. It included different agricultural tasks such as digging irrigation channels, preparing the threshing ground, carrying produce to market. The landlord had an assumed access to the pool of the sharecropper's labour and could demand *begar* for emergencies at any time. Buni Oraon, a sharecropper in Oodlabari, Jalpaiguri district, had to prepare the threshing ground, make storehouses for the grain, dig water channels and work at other times, for his landlord [Interviews Buni Oraon, sharecropper; also Sadhu Charan Singh, sharecropper – *chowkidar* (watchman), peasant activist, Dinajpur; Lal Mohammed Munshi, poor peasant, Dinajpur; Santosh Mandal, activist, Burdwan; *GOB,* 1937: 122].

The extent of *begar* depended on whether the landlord lived in the locality and whether he had *khas* land for self-cultivation. The sharecropper was not expected to travel far to perform *begar.* These features are clear from the example of Khanpur village where the dominant landlord, Singho Bahini, also lived. He had a plot of land that was cultivated by agricultural labourers and *begar* from his sharecroppers. There was an approximate correlation between the number of days a tenant had to do *begar* and the size of the landholding. A sharecropper on 13 *bighas* of land did one day's ploughing a year *begar,* a cultivator with 26 *bighas* on share contract and 11 bighas on under-*raiyat* tenancy did three days' ploughing and one day harvesting a year as *begar,* for Singho Bahini [Interviews Khanpur]. A sharecropper cultivating plots of land for different landlords could be called upon to do *begar* for both [ibid].[9]

Begar was a distinct advantage for the landlord, guaranteeing labour at critical times in the year when demand was highest and labourers most expensive, and when correct timing was crucial for certain operations. *Begar* could be enforced at any time, as one sharecropper commented: 'Even if a man was eating, the landlord would take him away to work. If the man was a criminal, still he would not have been treated in this way' [Interview Nolikanta Barman, Khanpur]. The labour of the sharecroppers' families extended beyond purely economic tasks. Buni Oraon, mentioned above, worked for his landlord during social occasions such as marriages or funerals. Women at these times worked as servants, assisting food preparation. Children were also required to work in the landlord's house at these times when there was such need. As one person remembered:

> During the month of Chaitra, Goddess Bisalakshmi used to be worshipped for eight to ten days in the houses of the *Zamindars* and *Jotedars* (landlords). The relatives used to come from various districts.

> The task of us peasant boys was to supply water for their latrine, bring water for their bath, wash their clothes and attend their orders. If a boy refused to do it or used to be absent during this period of the week, he was brought to the *cutcherry* (landlord's office) and severely whipped . . . the extra food that was offered to the goddess was thrown in the river, never given to us. If anyone had thrown a net in the river, the *paiks* (landlord's guards) would have taken away the net [*Tebhaga Songram,* 1973: 115].

This vividly shows the nature of labour use by landlords.

There were many ways in which the labour of the family was exploited by the landlord, both paid or unpaid; use of women's labour took a specific form. In rural Bengal there was a sexual division of labour whereby women were prohibited from ploughing, and, according to social status, restricted in their range of agrucultural activities outside the home. Inside the household's courtyard women did much of the work necessary to convert the produce of the land into edible or marketable commodities. These areas of women's labour were appropriated by the landlords.

Paddy husking was done in this period by a simple husking machine operated by two women in a see-saw manner. It was time-consuming and laborious work. Landlords farmed out paddy to poor peasant women for husking, and paid them at low rates in cash or kind. A woman was paid between four *annas* and one *rupee* for husking one and a half *maunds* of paddy which produced one *maund* of rice. If paid in kind, for husking a *maund* of paddy, she was paid a tenth of a *maund* in paddy. (There were sixteen *annas* to a *rupee.* One *maund* is approximately 80 lbs.) Either way, paid in cash or kind, this was a poor return for considerable labour. Although poorly paid, women continued to seek out such work because the cash or grain income was significant in their subsistence budgets. Many sharecropping families relied on this additional income [Interviews, Khanpur village]. There were few rice mills in Bengal at this time, and a proportion of landlords were also traders. The availability of cheap women's labour for the essential tasks of paddy husking, and transforming paddy into *chira* (rice flakes), *mourri* (puffed rice) and other forms of preparations for consumption, was an added advantage of employing sharecroppers to cultivate the landlord-traders' holdings. Cheap labour added to the profits gained by the dependency of the sharecropping families.

Unwritten share contracts included these demands on family labour which benefited the landlord in a variety of ways including additional servants to tend visitors and additional labourers to plough fields. Slogans raised during sharecroppers' struggles in the 1940s included '*Begar Khata, Gondo Koro*' (stop *begar* unpaid work), indicating that this was yet another source of discontent and resentment [Interview Ajit Rai].

FURTHER EXACTIONS FROM SHARECROPPERS

The verbal and temporary share contract stipulated that a certain proportion of the crop was to be taken by the landlord, but there were other ways in which the landlord was able to extract more of the produce. There was a widespread practice in Bengal, the collection of *abwabs* by landlords. These exactions, in money or kind, were payable by both cash-paying as well as produce-paying cultivators, although some were specifically related to sharecropping. *Abwabs* date from Mughal times, and in the eighteenth century were a sort of tax extorted by *zamindars* for the upkeep of their staff and servants. *Abwabs* were made illegal but continued to be demanded by landlords and paid by their tenants.

Abwabs taken from sharecroppers were demanded at the time of the division of the crops. Some *abwabs* were deducted for agricultural operations connected with processing crops. These included *othan thachani,* for preparing the threshing ground; *gola bandani* for repairing the wicker bins for storing grains; *mahaldari* for the person who weighed the paddy for dividing the crops; *dosturi* to the person who guarded the paddy at the threshing floor, at night. *Abwabs* paid the person who swept the threshing floor, another who divided the crop and whoever kept account of division. The amount paid varied between one and three *cuttacks* for a single *abwab.* (A *cuttack* was approximately two ounces.) The total amount owing to the landlord for these *abwabs* was deducted from the sharecropper's share of the crop after division and there was no choice: the sharecropper had to pay, or risk not getting land for cultivation the subsequent year. The *abwabs* were taken on the threshing ground, at the landlord's compound, with the landlord's guards and servants ready to enforce these payments. The sharecropper had little choice but to pay up [Interviews, Baccha Munsi, activist, north Bengal; Kali Sarkar, peasant activist, Dinajpur; Kongsari Haldar, communist-peasant leader, 24 Parganas; *Das Gupta,* 1970: 18].

Paddy was taken by the landlord from the sharecropper's share of the crops to support the guards and *lathials,* (private forces armed with heavy sticks) of the landlords. These were more common in North Bengal and 24 Parganas where landlords employed large numbers of servants and sub-let extensive areas of land to sharecroppers. *Barkandaji, sipoy poesa* and *paharandani* were payments from sharecroppers for the landlord's guards and varied from Rs 3 to four *annas* if taken in cash to five *seers* to a *maund* if taken in kind. (There were forty *seers* to a *maund,* so this *abwab* could amount to one eighth of the produce) [Interviews Khanpur; Baccha Munsi, activist, Jalpaiguri; Kongsari Haldar, activist 24 Parganas, *Das Gupta,* 1970: 18].

In Jalpaiguri the landlords deducted *abwabs* to feed their animals, for the upkeep of elephants, horses, hens and ducks. This amounted to about one *cuttack* (approximately two ounces) each year [Interviews Baccha Munsi, Bimol Das Gupta, peasant and worker activist, Jalpaiguri]. *Abwabs* could be demanded on spurious grounds, for the marriage of the landlord's son or daughter, 'although ten years ago the son was married, still each year they had to pay' [Interviews Bhibuti Guha, communist and peasant activist Dinajpur; Buni Oron, sharecropper, Jalpaiguri; Noren Das, activist, Malda; Mullah Hasan Ali, peasant activist, Pabna]. Other large expenditures of the landlord such as on religious festivals or *jatras* (musical plays) were excuses for further *abwabs* [ibid; *Rasul,* 1969; 88]. Money or crops collected by landlords did not necessarily pay for the items they were intended for. The *zamindar* of Brikush collected Rs1,600 in *abwabs* to set up a school, and then invested this in his own business. New needs also developed: the *zamindar* of Raipur collected between Rs 2 and Rs 15 from each of his tenants in order to purchase a wireless [*Millat,* 16.5.47].

Salami or *nazrana* were payments, similar to *abwabs,* that were made by tenants to landlords when a tenancy was first taken up, or to guarantee the transfer of a holding to an heir. Although sharecroppers had temporary annual tenancies, in some places they had to pay *salami* to guarantee that the holding would be retained the following year, or given to an heir. In the 1940s in Khanpur one sharecropper had to pay Rs10 to retain ten *bighas* of land on share contract. (One *bigha* is approximately one third of an acre.) These exactions could also be demanded in kind; ten *seers* per *bigha* (one *seer* was approximately 2 lbs). Whilst these exactions were defined and limited for *raiyats* to certain rates, there were no legal constraints on these exactions from sharecroppers [Interviews Kali Sarkar, Khanpur].

Payment of *abwabs* was completely illegal, yet government officers repeatedly complained that the practice continued in the Bengali countryside. The Kisan Sabha (Peasants Association) campaigned for its abolition, but while the landlord controlled the division of the crop at his own threshing ground, and had power over his tenants, *abwabs* continued to be collected. They were a means of extracting even more from the sharecroppers, over and above the share of the crop. Along with deductions for loans, this could mean that the sharecropper walked away from the threshing floor virtually empty handed. Therefore, the actual share of the crop, specified in the contract, was merely a notional amount, since the landlord had other ways of augmenting his own income. *Abwabs* were a further means of surplus extraction, leaving the sharecropper with a small amount of paddy, hardly enough for the family's survival. This was an inherent characteristic of the sharecropper's circumstances and propelled the sharecroppers towards indebtedness. This aspect of the sharecropper's dependency is the subject of the next section.

INDEBTEDNESS – AN INTEGRAL FACET OF THE DEPENDENCY RELATIONSHIP

Sharecroppers were often forced into debt because their income from cultivation, after the landlord had deducted his share and more, was scarcely adequate for the family to survive on for the year. Since their landlords were also their creditors, debt bondage reinforced the dependency relationship of sharecropper and landlord. While in theory a sharecropper was free to change landlords, in practice debt tied the sharecropper to his landlord. Bhaduri's model of semi-feudalism in Bengali agrarian society relies on this feature of the rural ruling class, that they operated as both landlords and lenders [*Bhaduri,* 1973b: 11]. The model is somewhat static. In historical reality we witness both this phenomenon of sharecroppers being indebted to their landlords, and a downward spiral of pauperisation during the first half of the twentieth century in which indebtedness was a mechanism of downward mobility. In this process sharecropping was a stage between landholding and landlessness.

As indicated in an earlier section, downwardly mobile cultivators lost their rights to holdings, often becoming sharecroppers on those holdings. In fact, many sharecroppers combined sharecropping plots with self-cultivation on holdings to which they still retained rights, or with agricultural labouring. Few households were wholly dependent on sharecropping for their means of livelihood. To take one example, approximately one third of the families in Khanpur village cultivated landholdings on which they were *raiyats* or under-*raiyats,* as well as sharecropping other plots. One fifth of the families relied only on sharecropping plots and supplemented their income from labouring work such as paddy husking [Interviews Khanpur].

Ceasing to be cultivators on holdings to which they had rights was a complex and lengthy process. Once in the ranks of the sharecroppers, debts continued to be incurred, and gradually any remaining plots of land were alienated and the family relied increasingly on sharecropping or labouring for a livelihood. While having rights to holdings credit was easily available. For sharecroppers, their share contracts were not considered adequate security for loans, and their only source of credit was their landlord.

Debts fell into two categories: annual production and consumption loans; and emergency or life-cycle loans. The latter were cited repeatedly as the major cause of peasant indebtedness, since marriages, births and deaths involved lump sum loans for one occasion at high rates of compound interest. These were often impossible to repay for years.[10] Areas of debt that were specific to sharecropper families were paddy loans for consumption and advances in kind for seed and animals.

The *karja, dehri, dera* or *deria* system dates back to the nineteenth century and refers to paddy loans taken several months before the harvest when the sharecropping families had exhausted their food stocks. When the crop was harvested the landlord-creditor charged a 50 per cent rate of interest and deducted the total amount from the sharecropper's paddy at the threshing floor. The sharecropper suffered from this system in two ways. Firstly, the rate of interest was exorbitant, being 50 per cent for two to five months. Secondly, the sharecropper suffered from the annual price cycles. The paddy was borrowed at a time when rice and paddy prices were high, and charged at a cash value at the current prices. Repayment was made when prices were always low, at harvesting, but the cash equivalent of the loan, not the quantity of the loan, was deducted. For example, if a paddy loan was taken out when the price was Rs 20 per *maund,* and repaid when the price had dropped to Rs 10 per *maund,* because of the large supply of harvested paddy, the sharecropper owed Rs 20 worth of paddy, i.e. two *maunds.* With the 50 per cent rate of interest commonly charged, the sharecropper had to repay three *maunds* for the original loan of one *maund.* If the loan could not be repaid, it was carried forward 'from year to year for the purpose of keeping hold of the debtor' [*GOB,* 1930b, Lalbagh; *Sen,* 1972: 14-15; *Rasul,* 1969: 88; *Janajuddha, 8.12.43, 15.12.43; GOB,* 1940c: 18].

Being on the margin of subsistence, the sharecroppers were in need of loans for production. Seeds for sowing were frequently provided by the landlord and considered as an advance to cultivators to be repaid at harvest-time. Animals required for cultivation were loaned in various ways. A landlord could supply cattle and the sharecropper bore the costs of maintenance, the cattle continuing to belong to the landlord. In another system, the sharecropper looked after a cow for the landlord, feeding and caring for it. The first calf that was born would belong to the sharecropper but subsequent calves belonged to the landlord. In these ways the landlord ensured that the cultivator was provided with the means of production without actually bearing the costs of raising the livestock, and in the process even increasing his own stock of cattle. Advances in cash were also made by landlords to sharecroppers for the purchase of livestock. A poor cultivator in Khanpur borrowed Rs 100 each year to buy cattle essential for cultivation. Eight months later, when no longer needed, the cattle were sold and the loan repaid at 25 per cent per quarter. Although this was not very economical, the sharecropper was unable to afford to keep cattle all year round, and was therefore continually indebted [Interviews Khanpur; *GOB,* 1941: 21; B.L.C., 1930, 34, 1: 171].

Sharecropping and indebtedness were clearly closely intertwined. An investigation into fifteen villages in Dinajpur district found that of 504 families, 156 families took *dehri* loans, for subsistence. 852 *maunds* were borrowed and 1,280 *maunds* were due to be repaid. 25 per cent of those indebted were sharecroppers in debt to their own landlords, and all but three of the 99 creditors were landlords and rich peasants. One third of the debtors had to borrow again the following year [*GOB,* 1941: 24]. All the sharecroppers in Khanpur village, and many of those who were both sharecroppers and had other tenancies, took *deria* loans from the landlords who gave them land on share contract. Those with non-share holdings could also obtain cash loans at 25 per cent rate of interest which were more favourable [Interviews Khanpur].

Landlords preferred the *deria* system because there was a very high rate of return, and they could guarantee repayment, having the sharecropper's crop as collateral. The dependency of the sharecropper was reinforced. In Khanpur cultivators were forced to take these loans from the landlords, whether they needed them or not, in order to get land to sharecrop. In Khulna also:

> At times it is found that the *bargadar's* [sharecropper's] debt far exceeds the value of the produce he gets. About 25 per cent of the *bargadars* are in perpetual debt to the *jotedars* [landlords] and are therefore compelled to accept the unprofitable terms of the *barga* [share] settlement [*GOB,* 1925: 72].

In 24 Parganas again:

> what he (the sharecropper) receives after deduction of the landlord's half share and his previous debts to him will hardly keep body and soul together for three or four months. And he is at the door of the landlord again. It is difficult to extricate himself from the clutches of the landlords and *mahajans* (moneylenders) if he incurs any debt from them. In this manner the landlord can easily manipulate a large amount of cheap labour who are more submissive than assertive [*GOB,* 1936: 20-21].

These quotations reveal the generalised nature of the inter-relationship between sharecropping and indebtedness and the implication that in this way the landlords' control over sharecroppers was maximised. Whether the landlord was a *jotedar* with a large number of sharecroppers, or a rich cultivator sub-letting to one sharecropper, his power over the sharecropper through access to a form of property right, was complemented by the debt bond. It was in the landlord's interest that the sharecropper survived to work the land efficiently and through various loan systems the landlord managed to keep the sharecropper alive and equipped for production, with the minimum risk of default [*GOB,* 1919a: 56; 1925: 32; 1940a, 11: 236, 1940e: 17; 1941: 25].

The Kisan Sabha in Bengal recognised that the problem of indebtedness was central to the difficulties of the peasants and regularly organised campaigns to reduce the rate of interest, stop malpractices of moneylenders, pass legislation in favour of measures like a debt moratorium. Demands around this issue appealed to a wide section of cultivators who were periodically or perpetually in debt. That there were demands raised against the *deria* system during the *tebhaga* movement was not surprising; it was an integral part of sharecroppers' oppression. Agitation against *mahajans* invited the support of non-sharecroppers who were also indebted. In fact, sharecroppers were supported by other cultivators and by labourers due to an overlapping of interest such as their united antipathy to usurers.

SHARECROPPERS AND THE MARKET

Sharecroppers were integrated into the market in two ways: either as direct sellers or through their landlords as producers. All cultivators, even sharecroppers on the margin of subsistence, were forced to sell a part of their produce in order to obtain cash to buy basic necessities such as salt, cloth and kerosene, for life-cycle expenditures such as marriages, crisis purchases such as medicines or to repay cash debts. The sale of produce for these items were sales of 'distress surplus' rather than 'commercial surplus' grains. The produce did not represent an excess of production over consumption [cf *Narain,* 1961: 36-37].

There were two markets for the crops: one at harvesting time when prices were low, but when poor peasants were forced to sell because of the pressure to repay cash debts and pay for other items; another during the period prior to the next harvest when the same poor peasants were buyers or borrowers of grain to survive on, at high prices. The sharecropper was compelled to sell at harvest time, lacking holding power because of the lack of storage facilities and means of transport. Unable to wait until prices rose in the months after the harvest, the sharecropper was disadvantaged by price fluctuations, a victim of the market [*Bhaduri,* 1973: 123-4; and *Rudra,* 1981a: 7].

At the market place the sharecropper was at the mercy of the *zamindars* or *ijaradars* who had the rights to the land on which the market was held. A toll in cash or kind was rigorously exacted by the servants of the landowners for all the different items sold by anyone at the market. The sharecroppers had no choice but to pay or not to sell their produce [*GOB,* 1940b]. Like *abwabs* these tolls were illegal and were a means whereby the same class of landowners extracted surplus from poor peasants. The Kisan Sabha (Peasants Association) demanded the abolition of these tolls and organised struggles against them. Predominantly in North Bengal, agitations against tolls were extremely popular, uniting a cross-section of cultivators, including sharecroppers, against the landlords. In some markets peasants were able to reduce or abolish the tolls, or else alternative Kisan Sabha markets were established in nearby fields without tolls to undercut the original markets. The success of this movement in 1939-41 was significant as a precursor of the *tebhaga* agitation; the peasants were inspired through this experience of a limited achievement of their aims, challenging the power of the landlords, and the influence of the Kisan Sabha was extended [*Rasul,* 1969: 86-87; Interviews Baccha Munsi, activist Jalpaiguri; Sunil Sen, activist Dinajpur].

The other way in which sharecroppers were involved in the market was through their landlords. In the fluctuating economic climate of the 1930s and 1940s, it became increasingly advantageous for landlords and traders to combine operations. As foodgrain prices recovered and famine conditions threatened, it was particularly desirable to secure an income in produce from sharecroppers. As one official commented, 'Some of the bigger landlords deal in grain. When the two functions are combined the usual result is the progressive expropriation of the cultivating classes' [*GOB,* 1928: 24; 1940a VI 46]. Such alienation of land holdings from the actual cultivators implied increasing sharecropping, the original tenants becoming sharecroppers on their holdings.

Traders had traditionally secured their goods through the *dadan* system, whereby advances were offered to the cultivators in return for tying the sale of the produce to the trader at a low price. The traders toured the countryside in the months before the harvest providing cash advances when cultivators were in need, but they suffered when their produce was sold for low prices and the advances repaid [ibid: 62]. Sharecropping became a favoured alternative means of securing crops without bothering to actually buy from the cultivators. Another government official noted that landlords who 'are also the financiers of the countryside and market much of the exportable surplus of the district' also 'set more store on the grain which they get from *adhiars* (sharecroppers) than from other sources' [*GOB,* 1941: 17 and 1940c: 65]. In this way sharecroppers were integrated through their landlords into the market, and landlords maximised control and profits.

NON-ECONOMIC MECHANISMS OF THE SHARECROPPING RELATIONSHIP

The economic power of the landlords was manifested in several ways which have been described in preceding sections. The share contract in Bengal tied sharecropper to landlord in a relationship of dependency. The sharecropper was dependent on the landlord for access to land, credit and other means of production. This dependency facilitated further exploitation through mechanisms such as *begar,* and also made other exactions feasible which siphoned off the produce of the sharecropper's and his family's labour. The power of the landlord was cemented by social and political relations to which we now turn.

The extreme exploitation of sharecroppers was shrouded by an illusion of paternalism and protection: the landlords 'helped' the sharecroppers by providing them with loans when no one else would, and traditionally treated the sharecroppers as their children and dependents.[11] The Bengali language had three forms of verbs: the polite form is used formally and to address superiors, parents; the second form is used between equals in age and status; the third form is used by adults with children or to inferiors and servants. Landlords always used this third form of address with their sharecroppers, linguistically conveying their inferiority of status, and dependency. The philanthropic attitude of landlords towards sharecroppers concealed the reality of oppression. During the 1943 famine when some landlords refused to help the sharecroppers 'their real interests were exposed', the pretence of protection dissolved. One peasant and communist activist remarked that 'this was an important pre-condition of the *tebhaga* movement' since it proved to the sharecroppers that their interests were not served by the landlords [Interview Aubunir Lahiri, Dinajpur].

In this section some of the non-economic structures that operated in rural society are examined in the context of the landlord–sharecropper relationship. I would stress that neglecting these in any treatment of sharecropping renders an analysis inadequate. Non-economic mechanisms made possible and acceptable the surplus appropriation by the landlords and explains how sharecroppers succumbed to their own oppression. Concentrating on sharecropping merely as a formal rental contract between two neutral parties is a rather sterile approach. The sharecropper-landlord relationship existed in a complex political structure which tended to empower the landlord *vis-à-vis* the sharecropper. Social and religious hierarchies inclined to parallel economic structures, sanctioning the landlord's authority and validating the sharecropper's dependency.

Firstly, it is to be emphasised that landlords, as members of the dominant class, were supported by local administrative, legal and law-enforcing structures. The Bengal Self-Government Act of 1919 introduced local self-governing Union Boards providing machinery for tax collection to support local administration, to provide *chowkidars* (watchmen). Voting was limited according to landholding and landlords inevitably took advantage of this access to the administration networks, and got themselves elected as Union Board officials as stepping stones towards further political activities. In Dinajpur district 'most Union Board presidents would be found to belong to the *jotedar* (landlord) class from thirty to three hundred acres of land' [*GOB*, 1941: 16-17]. In Khanpur village, the brother of the major landlord was a Union Board president of nearby Patiram [Interviews Khanpur]. In the late 1930s peasant and communist candidates stood for these positions, and where there was support, were able to win seats and control of these units of local government. Many district leaders had stood for these posts, such as Kali Sarkar in Dinajpur or Nur Jalal of Jessore [Interviews with each].

The landlord class was eager to make use of educational possibilities foreseeing the potential for their sons: 'The young members of the family obtain university education, perhaps join the bar . . .' [*GOB*, 1941: 16-17]. Membership of the professions provided connections for these landlord families. As one settlement officer remarked: 'the advantages of the unhealthy artillery of the law courts is on the side of the landlords', rather than the sharecroppers [*GOB,* 1918:55; also 1915, Appx.: xiviii; 1930a: 15].

The widespread connections of the dominant class were obvious during sharecroppers' struggles when landlords rapidly mobilised the support of local administration, police and lawyers. Government officials took the side of the landlords in disputes, or at best tried to arrange compromise solutions over cropsharing. There were police parties camped in landlords' courtyards to defend them and their crops, and make arrests on the basis of the cases that the landlords promptly brought against the sharecroppers. The forces of law and order inevitably rallied to the side of the landlords [*GOB,* 1948; *Sen,* 1972; *Tebhaga Songram,* 1973].

Secondly, social relations in Bengal tended to reinforce economic hierarchy. Society was highly stratified; with the Hindu caste structure accompanied by Muslim segmentation, and tribal peoples considered inferior to both. Religious prohibitions and customs operated to reiterate superior and inferior status, and in the case of sharecropping, usually sanctioned economic subordination. The socio-religious situation varied from area to area. Frequently landlords were high-caste Hindus or Muslims of high status, whereas sharecroppers were low caste Hindus, outcaste, low status Muslims or tribals. Sharecropping was a logical course for the former since ploughing was considered by them to be demeaning and was prohibited for Brahmins (high-caste Hindus).

Patterns of historical expansion of sharecropping contributed to establish the recurrent phenomenom that sharecroppers were socially inferior. Landlords bringing land into cultivation often employed tribals and low-caste peasants who then became sharecroppers. Moneylenders or landlords who bought up holdings from impoverished cultivators tended not to be from agricultural castes, but from higher castes. Richer cultivators who bought out their neighbours may have come from the same social group but soon acquired the manners of superiors, such as refraining from ploughing (there are no figures for these phenomena although it was often remarked on). In eastern Bengal many landlords were Hindu whilst their tenants were Muslim [*GOB,* 1911: 48; 1915: 70, 87; 1941: 10-14; *Beteille,* 1974: 122].

In social interactions petty prohibitions symbolised the low status of sharecroppers as well as their economic impotence. They were not allowed to enter or pass by the landlord's house or *cutcherry* (office) wearing shoes, carrying an umbrella or riding a horse. Once inside the compound, the sharecropper had to sit on the ground, or a seat lower than that of the landlord [Interviews including Khoka Barman, Dinajpur, agricultural labourer; Salambhai, communist and peasant activist, East Bengal].

The social and economic authority of the landlord was enforced through physical coercion. Many landlords employed *lathials* (private forces armed with large sticks) to protect themselves and punish recalcitrant peasants. Tortures ranged from beating with shoes to murder. Extra large shoes were manufactured for the purpose of beating, which in the Indian context was degrading and humiliating as well as painful. One official remarked that:

> It is not severe beating, but beating with shoes, not starvation but the threat of forcible eviction, sometimes even the mere threat of refusing a stool in the *cutcherry* (landlord's office) when he comes to pay his rent, which cows the Mymensingh cultivator into submission [*GOB,* 1919b: 30; also Interviews ibid; Durga Sen, activist, Malda; Monikrishna Sen, activist Rangpur; Sunnyasi Sardar, activist, Khulna].

Elephants were used to demolish standing crops and the homes of insubordinate sharecroppers. In some instances 'an offender was dragged along the ground tied to the leg of an elephant' [Interviews P. Gupta, activist, Mymensingh; M. Sen; K. Barman]. Since landlords sometimes assumed legal power in the area of their estates, and had close contact with the police, 'when they murdered *kisans* (peasants), there was no retribution' [Interview Kongsari Haldar, activist, 24 Parganas]. According to custom, especially in eastern Bengal, *zamindars* had controlled rural administration and settled disputes at the *cutcherry* (office). This power could be abused; crimes were hushed up, for example [*GOB,* 1923: 82-3]. The traditions of *zamindari bichan* (trials) were taken over by other landlords, who by virtue of their control over land had authority over their tenants. This included moral questions, for example 'a couple having illicit relations were punished by being stripped naked and weighed down with stones in the sun by order of the landlord' [Interview Soylen Ghosh, activist, 24 Parganas].

The extent of these prohibitions and powers varied considerably from one area to another and the most extreme forms were found mainly in isolated places like the southern Sundarbans belt of 24 Parganas and Khulna districts, or more 'backward' places in North Bengal like Jalpaiguri and Dinajpur districts. These included areas where sharecropping had developed on large tracts of reclaimed land with tribal and low-caste labourers. Here social hierarchy was most polarised and sanctions against these abusive powers of the landlords were weaker. Where landlords were absentee, their *naibs* (managers) assumed the same powers over the sharecroppers [*Communist Party,* 1951:4].

The power of the landlords, reinforced by external networks and social hierarchy, was sanctified by attitudes of respect and esteem. In some places a pseudo-religious status was assimilated, 'the landlords were seen as kings, representatives of heaven, gods on earth' [Interview M.K. Sen; also Khetra Mohan, sharecropper, Dinajpur]. Therefore, the son of a landlord could justify the deference shown to his father, 'as showing honour . . . as any Hindu who is going by a temple will bow his head before the deity although the deity does not demand it' [Interview Satyendranath Chatterjee, Khanpur].

In fact, the landlord did have powers that achieved much that was out of the realms of possibility for the majority of sharecroppers; for example, intervention in legal or official proceedings. However, this access, through membership of the dominant class enabled the landlords to make excessive demands on sharecroppers while ideological validation ensured acceptance by the sharecroppers of their own subordination.

The landlord, by virtue of his position, had control over various aspects of the sharecroppers' lives. The sexual intimidation of women illustrates the powerlessness of the sharecroppers and the landlord's capacity to ignore any moral constraints on their behaviour. A woman activist expressed the vulnerability of women:

> like the mangoes of the *bargadars'* (sharecroppers) trees, like the bananas of his garden, like the gourds of his thatched roof, like the eggplant from his garden, his daughters and daughters-in-law were the *jotedar's* (landlord's) property. After reaching a marriageable age, whether the *bargadar's* (sharecropper's) daughter will be married or not, if the marriage will take place, depends on the whims of the *jotedar* (landlord). If the *jotedar* expresses his wish, the daughter or the wife of the *bargadar* would be sent to the *jotedar's* house [*Tebhaga Songram,* 1973: 107].

The landlord could exert considerable pressure on the family of a woman whom he wanted, either for sexual or economic purposes. In one case a landlord influenced caste elders and other village members to boycott the family socially and economically. Rumours were spread to discredit the family and economic resources were withheld from the family members, in order to pressurise the family to bring the young woman to his house. Only with the support of her family and sympathy of local communist cadres against this intensive pressure, was she able to resist. [Interview Krishna Malu, poor peasant, communist, Faridpur]. Sexual exploitation was connected to social status: it was more acceptable to harass or molest women of low caste, or tribal women whose moral codes were different to those of caste Hindus or Muslims. Social structures not only santioned inferiority but also sexual exploitation.

Sexual and economic exploitation were closely related. Women's labour was valued for certain tasks, such as paddy-husking. In some cases the landlord 'married' more wives to increase the labour power of his household. Polygamy was permitted by both major religions. These women 'could be used for cultivating the land during the day without pay for half a meal, and molested at night on the grounds of their (the landlord's) rights as husbands' [*Swadinata,* 6.7.46]. Women were particularly vulnerable to the articulation of 'feudalistic' power of the landlords; subjected to sexual harassment as well as economic exploitation.

During the sharecroppers' agitations women were active, not only through their brothers and husbands, but in their own right. Some participated in defiance of their families, or encouraged their families to become involved. In the agitation against market tolls in North Bengal, women had been active since they often did the marketing for their families, selling their produce [*Tebhaga Songram,* 1973: 107]. During the *tebhaga* movement in direct confrontations between peasants and the police and landlords, women took leading roles. Women led volunteer forces to harvest paddy collectively, parties of women surrounded police forces, disarmed them and held them captive.

Women took particular responsibilities such as organising warning systems in the villages to respond to police attacks, taking messages between activists, hiding cadres forced underground to avoid arrest, as well as being involved in general activities, alongside the men. But for women to be involved in these activities they had to challenge not only the shared constraint on all peasants to challenge established power structures, but also the specific social prohibitions on women of all communities. This ranged from strictest *purdah,* seclusion of women in the homes, to attitudes that women ought not to participate in politics. Given these constraints, it is not surprising that local women leaders were often tribal or Rajbansi (low caste) who traditionally had more mobility, women from families of communist and peasant activists who approved of their activities, or widows [*Tebhaga Songram,* 1973: 110-113; Sen, 1972:39]. Women were motivated by their double oppression as sharecroppers and women, and were militant fighters in sharecroppers' struggles. Their double oppression was evident when the small Liberated Areas were achieved. Within the peasant courts established at these times, complaints were made by women about their economic and sexual roles. Women demanded the right to earn and keep the money made through selling livestock they raised and produce of their gardens; others demanded an equal right to education, and some complained of wife-beating [*Tebhaga Songram,* 1973: 109-110].

Landlord's power over sharecroppers was manifested in many different ways, from overt coercion to social pressure. Nevertheless, sharecroppers were able to challenge this power and see the realities of their own oppression. These non-economic structures are extremely important to any analysis of sharecroppers' agitations. When sharecroppers challenged their landlords, they challenged a network of inter-connecting structures which reinforced their subordination. Any action was taken in that complex context.

SHARECROPPERS AND STRUGGLE

When a peasant activist was trying to argue the case for the abolition of the land system in Begal, a peasant answered, 'How is it possible? the landlords have been here and will be here, the system is eternal' [Interview, Moni Krishna Sen, peasant and communist leader, Rangpur]. This summed up an attitude of resignation to economic and ideological oppression. But this did not mean that the Bengali peasant was always a passive acceptor of fate. There was a long tradition of peasant agitation in Bengal, articulated in various forms, from agitation to *dacoity* (theft)[12]. Opposition to landlords could take a communal form in areas where there was a correlation between community identities and landholding structures. For example, a struggle by Muslim sharecroppers against Hindu landlords in Jessore in 1909 for a two-thirds share of the crop, was recorded as a communal struggle [*GOB*, 1912:83]. Social hierarchy cemented economic structures, and, inevitably, communal tensions could operate in economic conflicts such as these.

The Bengal Provincial Kisan Sabha (Peasant's Association) was the first all-Bengal formal organisation of peasants. In the 1930s and 1940s activists organised around a very wide range of issues and demands, including land reform, indebtedness, and political independence. At the village level eating customs and treatment of women were also questions for debate. These were issues that affected different categories of peasants, and vitally affected sharecroppers. Minor local successes gave confidence in the leadership and encouraged more extensive struggles. The *tebhaga* movement was the first to raise a specific demand for sharecroppers, in all the districts. It unexpectedly 'took off' partly because of the work previously done by peasant activists, which, although not aimed directly at sharecroppers, had affected different areas of their lives. The *tebhaga* demand was precisely appropriate and the movement extended to areas previously uninvolved in Kisan Sabha activities.

It has been suggested that the specific actions taken by sharecroppers during these struggles did not threaten 'the structure of authority or the system they were part of' [*Dhanagare,* 1976: 372]. But they assuredly did. The official framework of the struggle could be narrowly described as 'merely legal', but the actions taken such as setting up sharecroppers' threshing grounds, breaking into landlords' granaries, defying the landlords' servants and police, challenged powerful sanctions that existed in the structure of land control. Khoka Rai, a poor peasant of Dinajpur, snatched away the pen of a police official, openly defying the landlord's power. For him this was a significant and daring act that completely undermined the status quo [Interview, Khoka Rai]. There were a myriad of such details of individual actions and group activities which constituted threats to traditional authority and involved considerable risks.

Humiliation of landlords, their agents or the police that occurred during the agitations were inversions of 'normal' patterns of power and dependency. A landlord in Dinajpur was beaten in order to secure aquiescence to the sharecroppers' demands; in 24 Parganas the *naibs* (managers) of the landlord were brought to trial for previous misdeeds and fined; a police inspector in Jalpaiguri was forced to abandon his jeep and walk to peasant headquarters. These and other incidents showed that sharecroppers, in many ways, challenged authority structures. The capture of police parties, rescuing people who had been arrested, and disarming policemen, were further examples of defiance of the established support structures of the landlords [*Tebhaga Songram,* 1973: 119-120; *Peoples Age,* 26.1.47; Interview Samar Ganguly, peasant and communist leader, Jalpaiguri].

To portray sharecroppers' movements as 'partial struggles' aimed at obtaining limited economic gains for poor peasants is to present an unimaginative view of these events. Although actions in some areas were limited to achieving compromises between sharecroppers and landlords, in other areas the fabric of power relations was threatened, if only for a period of days or weeks. Liberated Areas, or 'struggle areas', were achieved, for example, in North Dinajpur, Narail in Jessore, Tamluk in Midanapore and in Southern 24 Parganas. In these places, over areas of between sixty and two hundred square miles, the landlords fled to the towns and peasant committees were constituted. For a number of weeks the police did not enter the villages, or, if they did, they were resisted by peasant volunteer forces. In these places, and several others, society was 'turned upside down' by the sharecroppers [*Peoples Age*, 5.1.47; 26.1.47; Interviews, Amal Sen, peasant and communist leader, Jessore; Krishna Binod Roy, Kisan Sabha leader and communist; also *Sen*, 1972; *GOB*, 1948; *Tebhaga Songram*, 1973].

The sharecroppers' struggles of the 1930s and 1940s were considered to have been unsuccessful since they failed to achieve legal reforms and provide long-term solutions to the sharecroppers' problems. In retrospect, clearly the adventurous actions and events during the struggles had temporary significance. But legislation enacted in West Bengal in subsequent decades was also unable effectively to protect the sharecroppers' interests, and in East Bengal (now Bangladesh) little has even been attempted to assist them. The abolition of the Permanent Settlement after Indepenence was a precondition for fundamental changes in the land system: but in East Bengal large holdings were retained by tenure holders to be cultivated with labourers or sharecroppers; whilst in West Bengal much land previously on share contract was brought under 'personal cultivation' and many sharecroppers were evicted [*Januzzi & Peach*, 1980: 10; *Bandopadhyay*, 1981: A41; *Rasul*, 1969: 199]. A large proportion of land continued to be cultivated by sharecroppers and in West Bengal legislation under Left Front governments aimed to provide sharecroppers with rights regarding sharing of crops, (retention of 75 per cent of the crop) and occupancy.

Operation Barga is a recent attempt to combine the efforts of officials and political activists to record sharecroppers in West Bengal. Beforehand approximately 0.2 million sharecroppers had been recorded, and between September 1978 and July 1979, 0.7 million more were recorded, at special camps. Despite this, about 50 per cent of sharecroppers remain unrecorded because the authority of landlords over their tenants remains a powerful force opposing such initiatives. Economic and non-economic coercion continues to prevent sharecroppers from becoming registered. Eighty sharecroppers were murdered between 1978 and 1981 when the campaign was on.

Sharecroppers still have extremely limited access to credit, both for subsistence and production expenses, and so continue to be economically dependent on their landlords. While practices such as *begar* continue unofficially, their exploitation is intensified. While landlords have better access to the law, police and politicians, even state-sponsored efforts can be frustrated. Operation Barga was challenged; landlords took a case to the High Court which ruled that the involvement of political workers was legally unsound and that officials should not be assisted to record sharecroppers, so crippling this campaign [*Khasnabis*, 1981: A 45-47; *Rudra*, 1981b: A65; Interviews, Khanpur].

The situation in rural areas of Bengal is, of course, different in the 1980s from what it was in the 1940s. But in many areas of West Bengal and Bangladesh sharecroppers continue to be impoverished and to be trapped in a dependency relationship remarkably similar to that of the earlier period. Sharecroppers continue to be relatively powerless, although the economic and political context has changed. The experience of the *tebhaga* movement shows that sharecroppers have not always been passive acceptors of this relationship.

Who is to say that, inevitably, they will continue to be so, either in West Bengal or Bangladesh?

NOTES

1. In West Bengal in 1980 it was estimated that one fifth of the land was cultivated by approximately two million sharecroppers. In Bangladesh the 1960 Agricultural Census found 16.2% of total cultivated area farmed by sharecroppers, but this was only declared sharecroppers and the actual extent was probably 30 to 40% of land. A survey in 1977 discovered that 35% of holdings on 22% of rural land were cultivated by tenants, of whom 83.6% were on share contracts [*Khasnabis,* 1981: A43; *Jannuzi & Peach,* 1980: 24-25].

2. For example, there has been a heated and bitter conflict concerning the role of communist leaders in the *tebhaga* movement [*Singh,* in *Tebgaha Songram,* 1973; *Umar,* 1974]. Or the *tebhaga* experience has been cited in debate on the role of different sections of the peasantry in agrarian revolution [*Alavi,* 1973].

3. The wider study is entitled 'Sharecropping and Sharecroppers' Struggles in Bengal, 1930-50', for submission as a D.Phil Thesis to the University of Sussex.

4. The Bengal Rent Act of 1859 defined the right of occupancy as twelve years continuous possession and the limits of customary rates of rent for certain classes of *raiyats* (cultivators). The Pabna agrarian riots of 1873 provoked major legislation – the 1885 Bengal Tenancy Act – aimed at improving landlord–tenant relationship. More extensive than previous legislation, this protected the *raiyats* if the superior tenure-holder sold up, facilitated transferability of holdings, tightened the law of occupancy, provided for sub-letting to under-*raiyats* and recognised the rights of *raiyats* as well as *zamindars*. This act was amended in 1928 and 1938 when holdings were declared easily transferable although the landlord had rights of pre-emption. Under-*raiyats* were given rights of occupancy and rent was limited to not more than one third of gross produce. Other clauses included granting to *raiyats* rights to trees on their holdings [*GOB*, 1940a, 1: 25-30].

5. The sources cited from the Government of Bengal cover the period 1900-50. This is due to the unfortunate lack of data for all the districts of Bengal during the 1930-50 period. Government Gazetteers were compiled in the first two decades of the twentieth century and not updated until after Independence. Survey and Settlement Reports, which contain the most detailed information on agrarian relations, were issued on one district at a time, from 1900 onwards. The sporadic nature of this information limits the scope of analysis.

6. Because of the complexity of agrarian relations in Bengal, I have ignored the different types of tenure-holders and refer to all the superior interest holders who cultivated land through sharecroppers as landlords.

7. It would appear that the high rate of profit from sharecropping was the major reason for its adoption, by comparison with, sub-letting. Risk sharing appears to be a less important consideration [cf. *Rao,* 1971; *Newbery,* 1977].

8. A survey was conducted in Khanpur village, Dinajpur district in 1977, by the author. A sample of 25 families, representing approximately different economic groups, of the 250 families that lived in the village in 1946-7, were interviewed. Data collected concerned economic, social and political areas of their lives around the time of the *tebhaga* movement. This information is referred to as the Khanpur Interviews.

9. Fragmentation of landholdings was a common feature of rural Bengal and the majority of families had more than one holding although the average size of these holdings was small. There are no figures that include plots taken in on share contract as well as other interests. The results of a survey undertaken by the Land Revenue Commission found that 37% of

families held only one *raiyati* or under-*raiyati* interest, 44% held between one and five holdings, 15% over five interests. The average size of a *raiyati* interest was 1.89 acres and .64 acres for an under-*raiyati* interest [*GOB*, 1940a, 11: 108, 116]. This suggests the extent of fragmentation.

10. Peasant indebtedness was a fundamental problem in Bengali agrarian society. In the aftermath of the 1929 Crash, rural credit 'dried up' and the government was forced to intervene by establishing Debt Settlement Boards to scale down debts with a view to their realisation. The Bengal Agricultural Debtors Act of 1935, however, excluded sharecroppers from access since they had no secure landholdings.

11. The *jotedar* (landlord) of Khanpur considered himself the 'head of society and fountain of justice', settling disputes, mediating and controlling many aspects of village life [Interview, Khanpur].

12. For a comprehensive reply to the accusation of passivity made against Indian peasants, see Kathleen Gough's extensive summary of forms of peasant protest [1976].

REFERENCES

Alavi, H., 1973, 'Peasants and Revolution', K. Gough and H. Sharma (eds.) *Imperialism and Revolution in South Asia*, New York: Monthly Review Press.

Baden-Powell, B.H., 1892, *The Land Systems of British India*, Oxford: The Clarendon Press.

Bandopadhyaya, N., 1981, 'Operation Barga and Land Reforms Structure in West Bengal: A Discursive Review', *Economic and Political Weekly*, vol XVI, nos 25-26.

Bell, C., & Zusman, P., 1976, 'A Bargaining Theoretical Approach to Cropsharing Constraints', *American Economic Review*, vol LVXI, no. 4.

Bengal Legislative Council Proceedings (BLC). 1930, Calcutta.

Beteille, A., 1974, *Studies in Agrarian Social Structure*, Oxford; Oxford University Press.

Bhaduri, A., 1973a, 'A Study in Agricultural Backwardness under Semi-Feudalism', *The Economic Journal*, vol. 83, March

__________, 1973b, 'An Analysis of Semi-Feudalism in East Indian Agriculture', *Frontier*, vol 16, nos 26-7.

__________, 1976, The Evolution of Land Relations in Eastern India under British Rule', *Indian Economic and Social History Review*, vol. XIII, no. 1.

Chaudhury, B.B., 1975, 'The Process of Depeasantisation in Bengal and Bihar, 1885-1947', *Indian History Review*, vol. II, no. 1.

__________, 1976, 'Agricultural Growth in Bengal and Bihar, 1770-1860: Growth of Cultivation since the Famine of 1770', *Bengal Past and Present*, vol. 95, part I.

Communist Party, 1951, *Save the Peasants of Kakdwip*, Calcutta; Kakdwip Defence Committee (Bengali).

Das Gupta, M., 1970, 'Old Memories', *Jalpaiguri Patrika*, 1.7.70, (Bengali).

Dhanagare, D., 1976, 'Peasant Protest and Politics – The Tebhaga Movement in Bengal (India)', *Journal of Peasant Studies*, vol. 3, no. 3.

Gough, K., 1976, 'Indian Peasant Uprisings', *Economic and Political Weekly*, vol. IX, nos. 32-34.

Government of Bengal (cited as GOB), 1908, Bankura District Gazetteer.

_________, 1910, Birbhum District Gazetteer.

_________, 1911, Rangpur District Gazetteer.

_________, 1912, Jessore District Gazetteer.

_________, 1915, Final Report on the Survey and Settlement Operations in the District of Bakerganj, 1900-8.

_________, 1916a, Final Report on the Survey and Settlement Operations in the District of Faridpur, 1904-14.

_________, 1916b, Final Report on the Survey and Settlement Operations in the Tippera Riparian Area, 1909-15.

_________, 1916c, Final Report on the Survey and Settlement Operations in the Pabna Riparian Areas, 1904-14.

_________, 1916d, Rajshahi District Gazetteer.

_________, 1918, Final Report on the Survey and Settlement Operations in the District of Midnapore, 1911-17.

_________, 1919a, Final Report on the Survey and Settlement Operations in the District of Jalpaiguri, 1906-16.

_________, 1919b, Final Report on the Survey and Settlement Operations in the District of Mymensingh, 1908-19.

_________, 1923, Pabna District Gazetteer.

_________, 1925, Final Report on the Survey and Settlement Operations in the District of Jessore, 1920-4.

_________, 1926, Final Report on the Survey and Settlement Operations in the District of Bankura, 1917-24.

_________, 1928, Final Report on the Survey and Settlement Operations in the District of Nadia, 1918-26.

_________, 1930a, Final Report on the Survey and Settlement Operations in the Districts of Pabna and Bogra, 1920-29.

_________, 1930b, Bengal Provincial Banking Enquiry Commission, Report, 1928-30.

_________, 1934, Final Report on the Survey and Settlement Operations in the District of Murshidabad, 1924-32.

_________, 1936, Final Report on the Survey and Settlement Operations in the District of 24 Parganas, 1924-33.

_________, 1937, Final Report on the Survey and Settlement Operations in the District of Birbhum, 1928-34.

_________, 1938, Final Report on the Survey and Settlement Operations in the District of Malda, 1928-35.

_________, 1940a, Report on the Land Revenue Commission, Bengal, vols. I-VI.

_________, 1940b, Department of Agriculture, File 7M-12/40.

_________, 1940c, Final Report on the Survey and Settlement Operations in the District of Rabgpur, 1931-8.

_________, 1940d, Final Report on the Survey and Settlement Operations in the District of Howrah, 1934-9.

_________, 1940e, Final Report on the Survey and Settlement Operations in the District of Burdwan, 1927-34.

———, 1941, Final Report on the Survey and Settlement Operations in the District of Dinajpur, 1934-40.

———, 1942, Final Report on the Survey and Settlement Operations in the District of Hooghly 1930-37.

———, 1948, Department of Land Revenue, File 6M-58/47.

Government of West Bengal (cited as GOB), 1958, Bargadars and their Problems.

Guha, R., 1963, *A Rule of Property for Bengal: An Essay on the idea of the Permanent Settlement,* Paris: Mouton & Co.

Hunter, W.W., 1875-77, *A Statistical Account of Bengal,* 20 vols.

Jannuzi, F.T., & Peach, J.T.,/1980 *The Agrarian Structure of Bangladesh: An Impediment to Development,* Colorado: Westview Press.

Jannyayuddha, 1942-45, Communist Party of Bengal, Calcutta, (Bengali).

Khasnabis, R., 1981, 'Operation Barga: Limits to Social Democratic Reformism; *Economic and Political Weekly,* vol. XVI, nos. 25-6.

Millat, 1946-7, Dacca.

Narain, D., 1961, *Distribution of the Marketable Surplus of Agricultural Produce by Size Level of Holding in India, 1950-1,* Bombay: Asia Publishing House.

Newbery, D.M.G., 1975, 'Tenurial Obstacles to Innovation', *Journal of Development Studies,* vol. 11, no. 4.

———, 1977, 'Risk-Sharing in Uncertain Markets', *Review of Economic Studies,* no. 94.

Rao, H., 1971, 'Uncertainty, Entrepreneurship and Crop-Sharing in India!, *Journal of Political Economy,* vol. 79, no. 3.

Rasul, A., 1969, *History of the Kisan Sabha,* Calcutta, (Bengali).

Rudra, A., 1981a, 'The Self-Contained Society', *Frontier,* vol. 13, no. 21.

———, 1981b, 'One Step Forward, Two Steps Back', *Economic and Political Weekly,* vol XVI, nos 25-26.

Sen, S., 1972, *Agrarian Struggle in Bengal,* Delhi: Peoples Publishing House.

Tebhaga Songram, 1973, Silver Jubilee Observance Committee, Calcutta, (Bengali).

Umar, Badruddin, 1974, *The Peasantry of Bengal Under The Permanent Settlement* (Bengali), Dacca: Subarna Prakashan.

The Major Mode of Surplus Labour Appropriation in the West Malaysian[1] Countryside: The Sharecropping System

Fatimah Halim*

Despite the fact that Malaysia boasted the highest growth rate of GNP in six selected Asian countries between 1960 and 1973 (at 3.9 per cent per annum) it seems likely that rural poverty increased absolutely over this period[2]. It is the purpose of this paper to examine one aspect of rural socio-economic structures which contributes to the problem of poverty in rural Malaysia. The central questions considered here deal with sharecropping as a mode of surplus appropriation. Our main emphasis will be on the major characteristics of this mode of surplus labour appropriation and on the various elements which serve to perpetuate this system. Identifying these elements helps us see the relations of the peasant with capital and how the struggle for effective possession and control between producers and capital takes place mediated through forms of household production.

Empirical data for this paper were collected from two villages, Merah and Manis, located approximately between 60 and 40 miles from the largest state-controlled irrigation scheme in Malaysia, the Muda Project.[3] To hold the racial factor constant, we chose homogenous villages where landlord, shopkeeper and poor peasant were all Malays. The study was carried out in 1977-8.

By looking only at the village level, this study concentrates on events at a very low point in the international structure of which Malaysia is a part.[4] This is not to deny the fact that the economic basis of village struggles is affected and partially sustained by production activities in industrial countries as well as by power holders in the capital city of Kuala Lumpur. And though the state is rapidly extending its tentacles in the sphere of agricultural production,[5] we will relate to the state only as an aside, for our primary aim is to explain the dominance and persistence of the sharecropping system as the major mode of surplus labour appropriation in the Malaysian countryside.

I. THE VILLAGE ECONOMY

Whether we examine the contemporary economy of Merah and Manis from the perspective of circulation, that is, the destination of products into use or exchange value circuits, or from that of labour power, we can classify both these villages as being based on simple commodity production and subsistence production. The socio-economic organisation for both commodity and subsistence production is rooted in the elementary family unit.

**Malaysia.*

(a) Small-scale Commodity Production

The objective of commodity producers is to produce specifically for exchange. The use values of these commodities lies in the fact that they can be sold in the market and the money obtained is then used to buy subsistence necessities such as rice, kerosene, cigarettes, bicycles, sugar and flour. In addition, some of the cash earned goes to the state in the form of land tax, sales tax etc.

In the villages studied the major commodity produced for export was rubber. All the rubber produced flowed into community circulation. A larger proportion of population was involved in rubber planting compared to padi (rice) cultivation. Other agricultural products such as fruit, poultry and other livestock also entered into the exchange circuit, more so in Manis than in Merah.

There was hardly any artisan production in either village. Iron had never really figured as a significant element in the agricultural pursuits of the Malay peasantry since most of it had to be imported.[6] Mat-making and basketry on the other hand had always been important, as such activities are closely interwoven with the subsistence economy. The products are used partly in padi production – baskets for storing padi, mat screens to deflect the flying grain during threshing – and also in the household for roofs, walls, as well as for seating.

With the decline of padi production, the receding jungle and the introduction of modern technological inputs, such crafts had declined together with the rearing of livestock such as buffalo which was formerly used for ploughing.

(b) Subsistence Rice Planting

Rice is the basic staple of Malaysians.[7] There is a great deal of flexibility in the composition of work groups and the manner in which rice production is organised. Which members of the family and how many are employed in the field at any one time fluctuated in accordance with the stage of the padi cycle (transplanting was traditionally and remains the job of women, while the preparatory work for planting is still mainly done by men), and other factors such as whether the husband can find work outside the subsistence sector. As the rate of activity in the commodity production sector depends very much on movements in the international market for rubber and as the supply of labour in the subsistence sector is itself dependent on the state of the commodity sector, productive activity in the subsistence sector is ultimately also dependent on the state of the world capitalist system.

We found that in our villages a high percentage of the rural poor did not produce rice (82 per cent of the poor in Merah).[8] On the other hand, around two thirds of the rich peasants and slightly over one third of the middle peasants were involved in rice cultivation. The history of both villages is dotted with cases of conversion of rice lands to rubber holdings. At the time of the fieldwork we found that six out of the seven cases of farmers switching to rubber because of padi failure came from the poor peasant class in Merah.

Most of the padi produced, as well as some of the fish, vegetables, tubers and fruits were consumed either by the family which actually worked the land and/or by the family owning the land. Amongst the poorer peasants who grew padi, more than two thirds had to purchase rice for their consumption. This was because only two out of this group owned adequate land for their subsistence. On the other hand, about 36 per cent of the middle peasants and half of the rich peasants involved in rice cultivation were self-sufficient.

With the exception of the largest landowner in Manis, all padi harvested was consumed by the producers. With the help of state-subsidised inputs – pesticide, fertiliser [*Sarji*, 1977] and credit [*SEACEN*, 1978] for the purchase of a tractor and wage labour – this man produced a surplus of 21 gunis[9] (which constituted two thirds of total production) which he sold in the market.

From the discussion above we get some idea of the oppressive nature of state intervention programmes. Firstly, in aiming for self-sufficiency in rice, the state provides facilities such as subsidies for fertilisers, pesticides and cheap credit for the purchase of agricultural equipment. These facilities may help the rich to become capitalist farmers as it is they above all others who own sufficient land to undertake production on a large scale. Moreover, subsidies are usually given on the basis of land titles. Poor tenants who need such subsidies are made even more dependent on landlords while those squatting illegally on state land do not benefit at all. Indications of negative distributional effects derived from our study are confirmed by the government itself: 'In as much as tenancy is common in Muda, landlords would appear to be a distinct group of beneficiaries of the development programs' [*Jegatheesan*, 1977: 16]. Secondly, the cost of price support for domestic rice is passed on to and recouped from consumers. The rich peasants will again be the main beneficiaries. Not only are they the only ones capable of producing in bulk for the market, but they also do not have to purchase rice for their own consumption. In contrast not only do the poor peasants produce very little for sale, they also have to purchase a higher proportion of their rice requirements. They therefore stand to gain little from state-subsidised agricultural schemes because they have to pay 19 per cent more for their rice as a result of the price support.[10] The poor peasants then tend to be disadvantaged at both the level of production and circulation.

II. THE SHARE-CROPPING SYSTEM AS THE MAJOR MODE OF SURPLUS LABOUR APPROPRIATION

Although exploitation in agriculture takes place in a context of generalised production of agricultural commodities (primarily for export), the dominant form of surplus labour appropriation is not direct, through the sale of labour power as a commodity. This is related to the fact that a substantial proportion of the households in both Merah and Manis (55 per cent in the former and 53 per cent in the latter) still owned the essential means of production, land. Despite this, 68 per cent of the farms in Merah and 72 per cent of those in Manis were either partly or wholly operated by sharecroppers.[11]

We can comprehend at least some of the reasons for this high rate of tenancy if we look at the ownership of land in the administrative district of Merah. Out of a total of 377,824 acres of cultivable land, 54 per cent were under forest reserve (that is, at the call of the state). A very large part of the remaining land had already been alienated to rubber estates (approximately 46 per cent of 81,143 acres), 65.7 per cent of which were under foreign control.

The land left over for village cultivation was therefore limited. Furthermore, at this localised level, ownership was highly concentrated. The Gini Index is extremely high at 0.7714 for Manis and 0.6567 for Merah. (Table 1 contains details on land ownership and class affiliation.) We thus find on the one hand a large group of peasants without any land to work on while on the other hand we find another group holding land which they cannot possibly cultivate with existing family labour. But this does not show us why sharecropping is the predominant mode of obtaining farm labour. We can only grasp this via an adequate historical treatment. We proceed to a brief historical review in the next section.

III. THE HISTORICAL BACKGROUND

The purpose of this section is to trace the development of monopoly rights over land and, conversely, the creation of a category of landless farmers. We will also relate the continuation of labour tribute as the dominant mode of surplus labour to the relatively weak position of village landlords as compared to owners of other capital assets such as plantations and trading enterprises.

The internal logic of the sharecropping system dictates that for such a system to take root a necessary condition is the recognition of proprietary rights over land, whether such rights are legally supported and institutionalised or simply enforceable through informal community pressures.

Prior to the establishment of British rule in Kedah [*Annual Report, Kedah*, 1909: 62] proprietary right to land was relative to usufruct and was not absolute in the modern sense of the word: it lapsed after a certain period of disuse or non-cultivation [*Maxwell*, 1884]. The weakness of the indigenous ruling class[12] and the open resource situation[13] made it impossible to incorporate all land under the absolute authority of the Sultan. But, subsequent to British colonisation formal registration of land for purposes of taxation [*Cowgill*, 1928] was pushed through, to the detriment of both small cultivators and big landowners. Where previously the peasant who could not tolerate the pressures of oppression still had the option of voting with his feet – he could freely open up jungle land elsewhere – he now became landless and dependent on the beneficence of the state to grant approval to cultivate and on village landowners for use of land on the most onerous terms.

In granting the title directly to the cultivator, the British hoped to eliminate any intermediary between itself and the small individual producers and so was able to appropriate the maximum tax for itself. The local chiefs' hold over agricultural surplus was removed with the abolition of corvée labour [*Maxwell*, 1884] and the 'rapai'.[14] For a start, then, the traditional rulers at the village level came to occupy a very low position in the hierarchy of potential accumulators. Any attempt by this group to penetrate the mercantile sector was frustrated by the presence of the Chinese merchant class and their agents stationed right at the source where agricultural surplus was produced

TABLE 1
THE CLASS CORRELATES OF LANDOWNERSHIP IN MERAH AND MANIS
(PERCENTAGE)

Class	Land Owned		Households	
Affiliation	Merah	Manis	Merah	Manis
Rich Peasants	43.1	49.4	8.7	9.6
Middle Peasants	38.5	28.1	29.3	14.9
Poor Peasants	17.4	21.8	57.3	62.2
Proletarians	1.0	0.7	4.7	13.3
Total (per cent)	100.0	100.0	100.0	100.0
	(617.4 acres)	(441.4 acres)	N= 150	N= 188

[*Suyama*, 1962]. Chinese merchant capital was given its particular niche in order to take the heat off competition with Europeans – planters and miners in production activities and traders in international trade.

In this way, opportunities for horizontal integration between landlord and merchant capital were also reduced. Landlords were thus deprived of external financing. Besides, financial institutions during the colonial period were largely controlled by three British banks whose lending activities were narrowly 'restricted ... to business which they consider acceptable and within their legitimate field – chiefly the finance of overseas and internal trade and other working capital requirements of larger firms' [*IBRD*, 1955: 642]. Conversely, the Malay Reservation Act[15] limited large-scale acquisition of land by merchant capital. This undermined their ability to finance projects for expanded capital reproduction. Consequently rural transformation was held in abeyance for more than a century.

With the abolition of forced labour, the indigenous ruling class was deprived of its primary source of labour power which had previously been used, amongst other things, to acquire large tracts of jungle land. Very short on cash, these landowners could not hope to compete with the capitalist sectors for wage labour to run their farms. Besides, most of them did not have much experience in agricultural pursuits having been too preoccupied with fighting to keep their positions while leaving most of the agricultural chores to slaves and debt-bondsmen. With lots of land and little inclination towards supervising farm-hands, it was only natural that these landlords would find the sharecropping system much more suited to their taste.[16]

IV. THE ESSENTIAL ELEMENTS OF THE SHARECROPPING SYSTEM

This section will concentrate on drawing out the essential features characterising the sharecropping system by a detailed examination of the system as practised in the villages studied, particularly in Merah.

The critical distinguishing feature of the sharecropping system is the unity of labour with the means of production, comprising both land and implements. Although about half of the farming households did not own any agricultural land at all access to land was attainable through the sharecropping system.[17] The landless were by no means the only candidates for sharecropping although they formed a substantial proportion of the tenant households: 63 per cent in Merah and 67 per cent in Manis. The remaining tenant households comprised mainly farmers owning such small plots that without resorting to sharecropping they would not be able to survive. Practically all tenants in the sharecropping system were poor peasants owning an average of less than one acre per family.

The problem of land fragmentation is exacerbated by the fact that in the Malaysian countryside at least half of the land is transferred to the next generation according to Islamic rules of inheritance[18] where besides the members of the immediate family other rightful heirs include siblings, nieces, nephews and a host of other relatives.[19] These rules, if implemented uniformly across class boundaries, will have some equalising effect on land distribution, especially when rich landlords tend to have more wives (the richest landlord in Manis had six wives). However, the rich are also prone to transfer their land to their children before death abiding by Malay Adat (Customary Law) where one is free to choose one's beneficiaries. This is done to avoid paying heavier death duties and to conserve their capital assets. On the other hand, the poor and ignorant, having nothing much to bequeath their children tend to allow the Muslim inheritance laws automatically to run their course after death.

Generally, it could be said that sharecroppers owned the tools of cultivation. In rubber tapping, these comprised essentially a limited number of cheaply available implements such as the tapping knife and bucket for collecting the latex. A tapper would also normally have cheap and easy access to supplies of acid coagulants, coagulating pans, and mangles if he wanted to process latex into sheets which could withstand delayed marketing. To put padi land into productive use, much more was required from the cropper. A set of implements would cost around $150, slightly above the monthly income of the poor peasant. Besides, with the introduction of high-yielding seeds, padi planters would now need more cash for purchasing chemical fertilisers, pesticides and tractor rentals. It could be for this reason that the share going to the padi cropper (two-thirds) was higher than that of the rubber tapper (half of the final produce).

Although it is required by law that tenancy agreements for padi[20] be in written form we have yet to come across one case where this is actually done. Even today when labour is relatively strong (six of our landlords had left their land idle for lack of hands), tenancy agreements are all verbal and informal. This is, perhaps, in accord with the necessity for maximum flexibility in an economy marked by extreme instability. The fact that in both villages a far greater proportion of rubber as opposed to padi land was worked by sharecroppers (68 per cent: 35 per cent and 74 per cent: 54 per cent in favour of rubber in Merah and Manis respectively) was probably dictated as much by the greater returns of rubber as by the greater degree of freedom associated with sharecropping in rubber. When price is declining or when the weather is unfavourable, or when a good offer for work turns up, the tapper can easily slip off for as short or as long a period as is desired. In contrast, one is committed to padi land at least for the period of preparation for planting and harvesting, altogether about four months: three months for preparation and about one month for harvesting.

Since labour is not separated from its means of production in the sharecropping system, labour is assumed to own its own product. Consequently, the product in which surplus labour is embodied has to be appropriated by the landlord under some kind of compulsion. The sharecropping system is, therefore, essentially a form of 'forced' labour.

Lacking access to the means of production, labour is forced to sell its power in a system of low remuneration. The price of labour within the sharecropping system has no determinate level. It is the outcome of the relations of servitude existing within the village. Marx [1959] and Lenin [1972] (see Cutler [1975]) both recognise this incomplete remuneration for labour power expended in the concept of *devalorisation of labour time* and posit it as the distinguishing feature of small peasant production operating within a commodity economy. This characteristic of the sharecropping system is even more pronounced when the system operates within the context of an economy geared to, and subdued by, a fast moving dominant sector. The price of labour in Merah was half that in the capitalist sector.

If the cropper's household seems to preserve its form as an autonomous peasant enterprise, this external feature has a radically different meaning from that of the small independent producer, for no major economic decision-making can now be undertaken in this household. The sharecropper's household is now bound by payments to the landlord and thereby is integrated into the household economy of the landlord. His total produce must now not only meet his family's own demands but also the rent to be paid to the landlord. In the most extreme form the cropper is reduced to a mere reserve for simple reproduction which effectively eliminates autonomy as a separate economic enterprise. Such households continued to function on this basis even when they sold a major share of their output in the market; even when their physical assets had acquired a cash value; and even when a labour market was begining to evolve.

Similarly, landlords were also rendered incapable of maximising profits from capital. In exchange for organising the rural electorate for the ruling party, UMNO (United Malayan National Organisation) landlords were often rewarded with licences for public and commercial transport and timber extraction, public work contracts, etc. (see Table 2). Normally, such licences would be traded with Chinese capitalists.[21] Due to their advanced age, insular life and inexperience, these first-generation landlords-turned-capitalists were often hesitant to grasp business opportunities that came their way. Instead, they preferred to hoard their money with the mobile National Post Office Savings, using it only to purchase luxury consumer items such as television sets and cars, and to send their children to private schools. However, with state help the transformation of landlord to capitalist may not be that distant in the future for changes activated by developments in the capitalist sectors of the economy

TABLE 2

TRANSFORMATION OF THE RICH PEASANTS IN MARAH AND MANIS

Business Interests / Rich Peasants		Taxis	Lorries and Buses	Contractors	Mining	Real Estate & Housing Development	Heavy machinery tractor, bulldozer, chain-saw	Political Affiliation*
Manis:	1	/				/		
	2	/					/	
	3		/	/	/			/
	4		/					
	5				/			
	6		/				/	/
	7		/				/	/
Merah:	8		/			/		
	9			/			/	/

Note: *At least at committee level of ruling Party Branch.

(both private and public) have brought such transformations into view.[22] For the moment, however, landlords in our villages were on the whole uninterested in maximising profits by employing wage workers; for the kind of wage they were able to offer, they could scarcely compete with the capitalist sector for a permanent work force.

The labour-means of production integration also implies that relative to the urban proletariat, the cropper is given a larger space of discretion in the timing and organisation of the production process. A major contradiction arises here because exploitation of the peasant keeps him poor and a poor peasant is to that extent less capable of paying for expensive inputs to boost productivity. Our study in Manis shows that a majority of the rich (87 per cent) were able to improve their yields through the application of non-human aids such as fertilisers, pesticides and mechanical energy for ploughing, whereas only 39 per cent of the poor peasants could afford these facilities. Moreover, among the poorer classes, there were still 18 per cent who worked in the fields with only their bare hands and some primitive tools such as the hoe. Landlords had also complained that, left to their own devices, croppers would not give top priority to tapping but would probably take up contract work whenever that was available. Here, then, exploitative relations of production could prove to be a severe constraint on the development of the forces of production.

To go beyond these limits would involve the landlord in adopting a system and form of labour exploitation that is consistent with trends in the market; although this system need not necessarily take the form of wage work, where labour possesses nothing except its labour power to be placed under the direct organisational control of the capitalist. Caught in the web of contradictions involving his relations not only with elements within the village but also his relations with external forces, the landlord could only operate to keep the sharecropping system ticking over.

V. STABILISING ELEMENTS IN THE SHARECROPPING SYSTEM

The fact that regardless of the deteriorating conditions in the countryside[23] the Malaysian peasantry has until very recently only given vent to their frustration on a massive scale on two occasions[24] would certainly prompt a casual observer to look for factors behind the apparent 'stability' and persistence of the sharecropping system. Confining our discussions primarily to the level of the village economy we have identified these factors as ideology and some access to land; temporary wage work and the inability of the capitalist sector to absorb rural labour as part of its permanent work force; consumption credit advanced by landlords and shopkeepers and other minor ways of improving the livelihood, including livestock rearing and fruit growing. These factors will now be considered.

(a) Peasant Ideology and Access to Land

Looked at from the position of the direct producers, one of the important reasons why the heavy burdens of the sharecropping system were borne by the cropper was precisely the patent preference for land and the way of life it typified. The importance that peasants attached to land, and their refusal to give up their access to it even if they had to concede their independent peasant status, has not been given its due recognition. Land in the rural area is not simply an economic means to an end. When asked why he did not sell off his land a landlord who not only had difficulty recruiting croppers but also had problems with the payment of land tax replied, 'sayang' which literally translated means 'love' or, in other words, 'did not have the heart to let the land go'. A young sharecropper told the author he personally knew five other young families who like himself had migrated to the towns but who had finally returned because they missed the wide space and 'kampong' life. A high wage was still not enough.

Moreover, the poor peasant seems genuinely confident that he can own a little plot within his lifetime. This was due to inadequate staffing of the early post-colonial administration; and even today the state is hesitant to bulldoze peasants who had illegally occupied state land unless a particular site was urgently required for development. Agriculturalists could still open up virgin jungle for cultivation and subsequently apply to the state for a permanent grant. Peasants know that land dispensation is an instrument of political patronage. They therefore learn to capitalise on their political clout by opening land en masse (according to one peasant 'beramai-ramai').

It is beyond question that in our villages a greater proportion of the land cultivated or settled by the villagers was obtained through squatting on state land and then subsequently applying for a grant. 44 per cent of the land held in Merah was acquired this way. Although 46 per cent of the illegal land in Merah was in the hands of the rich peasants as against 20 per cent for the poorer classes (including the proletarians), at least half of the poor peasants' land was acquired in this way.[25] This fact alone was responsible for firing the hope that it was still possible for the poor peasant to acquire land, although he had nothing but his bare hands. Perhaps the fact that four out of the 13 rich peasants in Merah managed to make it to the top through sheer hard work had helped to sustain the poor peasant in his suffering, at least psychologically if not materially. The possibility of pioneering new land and the low capitalisation in agriculture also served to depress prices of land such that at least 21 per cent of poor peasant land in Merah and 23 per cent in Manis was purchased. Although in mature communities such as Manis there is practically no more land to be colonised, the examples from Merah give some idea of the relevance of an open frontier in land when considering the transformation of rural relations.

(b) Wage Work

Colonial records are replete with instances of peasants abandoning their rice lands whenever crops were destroyed by drought, floods and pestilence [*Annual Report Dept. of Agriculture*, 1952: 6]. These problems of uncertainty were not alleviated for those peasants who had switched to rubber planting. And though low commodity prices did not affect the producers as badly, low rubber prices also led to frequent movements of peasants into towns and mines to seek work.

Prior to the establishment of a monopoly in land, it was not usual for the Malays to work for a wage.[26] As a result the labour markets in the mines, plantations, commerce and manufacturing sectors were controlled mainly by Chinese and Indian labour contractors: due to their initial headstart and cultural affinity with capital. Hence, when the Malay peasantry did venture out to seek work, and given the circumstances under which they were often propelled into the wage sector, it was not the least surprising that they would be most willing to work at anything for any wage. The latter was forced by circumstances but was also made possible by the fact that since their sojourn in the wage sector was considered to be only for a temporary period of time they would have left their families behind, thus eliminating the burden of having to feed and board them in the area of work. The rural migrant, as compared to other migrants who had fewer opportunities for gaining access to rural land, can therefore survive on a lower wage rate because his permanent location in the subsistence sector covers part of the expenditure of his family and makes it possible for his wage to be forced down below free hire level. It would, then, be ludicrous to consider him as part of the capitalist sector – as free labour – for his frequent but temporary forays into the wage sector are actually part and parcel of his role as tribute contributor to the landlord in the subsistence sector. Without wage work, it is doubtful if the rural structure based on extraction of labour tribute can survive for long.

Although the Malaysian capitalist sector was based massively on cheap alien immigrants in the past, it did also make use of rural migrants to solve temporary shortages. The openness of the economy to external influences makes a flexible wage bill especially attractive, more so for industries closely integrated into the world capitalist system. For instance, besides tying wage rates to the price of rubber, contract workers constituted one quarter of the labour force in the largest plantations [*Handbook of Rubber Statistics*, 1975]. After the Second World War, when the alien labour force had stabilised somewhat, it became even more dangerous to have widespread unemployment (which would inevitably happen if export prices were depressed), the consequences of which could threaten severely the foundations of the capitalist structure, as was proved by events prior to 1948.[27] With a migrant labour force, it would not be necessary to take in as many workers during peak periods, thereby reducing the number to be retrenched during troughs. Besides, migrant workers from rural areas help to depress the level of wages in the capitalist sectors. Indeed, the level of urban wages is so low now that many have to work outside normal working hours or hold more than one job to survive [*Halim*, 1981].

It is the contention here that until today the capitalist sector of the Malaysian economy has not expanded at a rate fast enough[28] to absorb rural workers as part of their *permanent labour force*. The latest national statistics show that the economically weaker sections of the manufacturing sector, for example, firms whose fixed assets fall below $50,000, employ a larger proportion of temporary workers – ranging from 11 to 13 per cent – as compared to enterprises with fixed assets of $100,000 and above [*The Census of Manufacturing Industries*, 1975: 19]. Temporary workers constitute less than 4 per cent of the latter's labour force. These larger companies are mainly foreign in origin (foreign companies own 47 per cent of the fixed assets in the manufacturing sector although they comprise only 6 per cent of total manufacturing companies in Malaysia). Generally speaking, the smaller (typically Malaysian) manufacturing company made use of temporary workers five times more frequently than the typically larger foreign company (the ratio of permanent to temporary workers is 18:1 and 90:1 respectively).

In the present situation, peasant workers seem to have become a more permanent feature of the Malaysian countryside. This particular trend is further encouraged by the government's policy of providing tax incentives to industries prepared to locate themselves in the countryside. Enterprises which avail themselves of this particular type of incentive are mainly the small Malaysian owned enterprises [*Abdullah*, 1979]. The weaker fraction of the capitalist sector in a dependent economy does indeed integrate 'nicely' with the subsistence sector which is also closely tied in a dependent manner to the world capitalist system. Since the rubber-tapping schedule falls between 6 a.m. and 12 noon and the padi regime leaves the peasant quite free between planting and harvesting, we can expect to find peasants on the road seeking employment or returning from employment to the land on which their families have been left to fulfil their contractual obligations.

In Merah, 36 per cent of the peasants had to take on additional work to support themselves during 'difficult times' – during bad weather or reduced rubber yields during the dry season. The comparative figure for Manis was 40 per cent. The difference could probably be attributed to Manis' fortuitous location nearer to the now highly capitalised agricultural area covered by the Muda Irrigation Scheme, the state-owned sugar plantation at Padang Terap and the mines about two miles away. In contrast, those finding additional work in Merah were confined mainly to work within the kampong economy such as 'kerja kampong' (rural odd jobs). 40 per cent of the jobs fell into this category in Merah whereas in Manis only 7 per cent of the jobs were of this description. A majority of the supplementary jobs (74 per cent) for Manis was located in sectors external to the kampong economy. In Merah, outside work formed less than one third of the total jobs found by the poor peasants. Supplementary work taken on by peasants ranged from the more familiar kampong work such as mat making and basketry, the collection of jungle products (which is fast dying out as witness the absence of such activities as commodity production in Merah), peddling, odd jobs, cooking, plucking coconuts to the more recently available job opportunities in the more highly capitalised farms of Muda, the tin mines and the sugar plantation.

The types and number of job opportunities available fluctuated unpredictably. For instance, during the good years, about 100 workers from Manis area were herded by labour contractors to work in the Muda area for the harvesting in January and August, lasting for a stretch of one to eight weeks. But at the time of our study, the peasants had to risk a small amount of capital to cover transport fares to go to the area to search for work. The 20 per cent of peasants who were willing to take up any work in and out of Manis was perhaps an indication of the inability of the wage sector to absorb the labour power available in the rural areas. Interviews with peasants indicate their willingness to leave the countryside if more permanent jobs could be secured elsewhere. For the moment peasant workers in our sample were all paid by the piece or daily rated.

(c) Consumption Credit

Prior to the widespread penetration of commodity production, people were more willing to part with the occasional surplus produced as one could scarcely find a market for it: everyone produced exactly the same things and it was too expensive to transport the products to population centres such as the tin mines, plantations and the towns. Moreover, one could never be assured of complete immunity from the unpredictable calamities brought by nature. These factors combined to produce an ethos of 'tolong-menolong' (mutual help) in the villages. Besides the sharing of fruits and vegetables this co-operation extended to labour power for lifting a house, preparation of fields, etc. . . . This ethos was much more prevalent in Merah than in Manis. We observed in Merah that nearly half of the poor peasants had erected their house on other people's land free of charge.

With the exhaustion of mutual goodwill accompanying the intrusion of commodity relations, a new source of credit came into existence, provided by the village shopkeepers. Their role as a source of consumption credit for the poor peasants was not unrelated to their position at the bottom of the heap of capital owners. Unlike the rice planter who is involved in a planting cycle with a long gestation period which can give rise to opportunities for the accumulation of debts, and opportunities for settling debts through advanced purchase of harvests, the rubber cultivator is relatively free from manipulation by merchant capital because he is able more often than not to obtain sufficient income for his daily reproduction. It could happen, though, that tappers could not tap for several days at a stretch e.g. during the rainy season. Some poor peasants had resorted to stealing rubber scraps from nearby estates while others had tried to dry the barks to keep the latex flowing straight into the cup below. Normally, if a tapper did not work for a week or so he would have to resort to the village shop for consumption credit. Unfortunately, those in most need of credit were also the least creditworthy. Often landlords would have to stand as guarantors for them. Due to a shortage of capital and the high risk involved (lending without collateral) such credit was normally extended for less than a week and confined only to essentials such as rice, flour and sugar. Even then several shops had foreclosed due to indiscriminate extension of credit. Such credit functioned more to gain a wider circle of business than for the appropriation of interest i.e. to deepen mercantile profits. It was agreed that as long as the peasant paid his debt on the first sale of rubber, he could still retain the goodwill of the shopkeeper. In practice, peasants indicated that it was 'not nice' to sell to a shop other than the one which helped one in times of need. Credit extended by shopkeepers was a widespread form of support used by peasants: 57 per cent and 40 per cent of the households in Manis and Merah respectively were dependent on this source of alternative support. This was definitely an underestimation for there was a tendency among peasants to hide the fact that they had to be thus indebted.

In times of special need – for example, weddings, funerals, illness, and the beginning of the school year – landlords were often approached for advances. In several cases in Merah, the horizontal integration of landlord and shopkeeper means that the landlord-shopkeeper can better monopolise and manipulate credit facilities to tie the cropper to him. The amount lent actually depended on the cropper's relationship with his landlord. On the average, landlords were known to lend out between $5 to $10 per request. The tight-fisted landlord would indeed find it difficult to get anyone to work for him.

(d) Livestock Rearing

As it was impossible to detail how much livestock contributed to the family's income we could only identify the ownership of livestock by each class, as presented in Table 3.

The first notable feature of this distribution is that the rich owned a greater proportion of the more valuable livestock – the buffalo and the cow. Conversely, the poor owned the smallest share of all categories of livestock with the exception of the duck (which can be cheaply acquired and maintained on zero capital). Compared to the poor peasants, the proletarians seem to own more livestock although these also tended to be of the cheaper variety such as the duck and the goat. Lacking the accumulation to purchase land, they had ploughed their savings into these more humble assets.

Ownership of livestock was also not evenly spread out within each class. For instance, of the 75 goats owned by the poor peasants in Merah, 54 were owned by just four families. One conclusion that can be drawn is that no two families adjust to their difficulties in the same way. Some tend to rely on supplementary work while others are prone to depend on livestock rearing. But, across class boundaries, we can say that the poor peasants who were poor because they did not own land were also less capable of ameliorating their poverty by alternative means. Even when these poor peasants were given chicks by the state, they had to give them away to those who could afford to feed them.

In Table 4, we present the various combinations of supports sought by the poor of Manis in their fight for survival in a remarkably unpredictable environment where their grasp on a livelihood could only be marginal and precarious. In Merah, however, it was impossible and perhaps even artificial for us to show all the combinations of resources marshalled by the poor peasants in times of distress. For instance, four peasants who derived some relief from remittances sent by the family members working and residing outside the kampong economy were also forced to find other sources of income through 'kerja kampong', selling home-produced commodities such as chickens, goats and bananas and living by credit from the village shop. The poor peasant beyond

TABLE 3

OWNERSHIP OF LIVESTOCK IN MERAH AND MANIS (PERCENTAGE)

Livestock / Class Affiliation	Buffalo		Cow		Goat		Duck	
	Merah	Manis	Merah	Manis	Merah	Manis	Merah	Manis
Rich Peasants	32.1	21.5	30.2	19.5	11.2	14.5	–	14.0
Concentration Index	(+23.4)	(+11.9)	(+21.5)	(+9.9)	(+2.5)	(+4.9)		(+4.4)
Middle Peasants	53.6	20.0	30.2	35.9	33.6	16.6	62.5	6.0
Concentration Index	(+24.3)	(+5.1)	(+0.9)	(+21.0)	(+14.2)	(+1.7)	(+33.2)	(–8.9)
Poor Peasants	14.3	49.2	32.6	35.2	49.3	52.3	37.5	72.0
Concentration Index	(–43.0)	(–13.0)	(–24.7)	(–27.0)	(–8.0)	(–9.9)	(–19.8)	(+9.8)
Proletarians	–	9.2	7.0	9.4	5.9	16.6	–	8.0
Concentration Index		(–4.1)	(+2.3)	(–3.9)	(+1.2)	(+3.3)		(+5.3)
Total: %	100.0	100.0	100.0	100.0	100.0	100.0	100.0	100.0
Total: No.	28	65	43	128	152	241	64	50

Note: The Concentration Index represents the difference between each class's proportion of ownership in a particular category of livestock and its proportion of the total households. A "–" sign denotes under-representation of a class within a particular category of livestock. Conversely, a "+" sign denotes over-representation.

TABLE 4

SUPPLEMENTARY SUPPORTS OF THE SHARECROPPING SYSTEM IN MANIS

Alternative Supports / Class Affiliation	Additional Work	Work & Credit from Shop	Work & Selling Livestock	Work, Selling Livestock & Credit from shop	Selling Livestock	Selling Livestock & Credit from shop	Credit from Shop	Savings	Savings & sell Livestock	Remittance from son	Credit from Landlord
Poor Peasants	15	9	4	18	4	31	8	2	3	1	2
Middle Peasants	2	–	–	–	–	–	–	–	–	–	–

question needed all these alternative lines of support because the benefits which accrued from each of these sources individually were small and were by themselves undependable. This variety in modes of adaptation is really symptomatic of the fluidity of the present rural structures with major supports changing quite frequently, depending not only on internal developments but also on developments external to the village economy.

CONCLUDING REMARKS

In discussing the relative stability of the sharecropping system we have noted that, besides factors internal to the village economy, the maintenance of the system stemmed from the integration of the peasantry with a weak local capitalist sector which was and still is subjugated to the world capitalist system of production. We also found that instead of being able to rely on props internal to the village economy to sustain him the poor peasant will have to resort more and more to the capitalist sector, not only for the sale of his labour but also for the purchase of production inputs.

Although no systematic attempts were made to describe the impact of the capitalist sector on the peasantry we did point out the general nature of their influence wherever appropriate. Our own observations would lead us to conclude that with further developments in productivity and widespread use of advanced technology in agricultural production, that is, when the mechanisms for surplus value appropriation are changed from absolute to those of relative surplus value appropriation, certain important consequences follow which we will now discuss.

Due to their weakness, local landlords cannot play the vanguard role of advancing agricultural productivity. The state is thus forced to intervene to organise and rationalise peasant production in order to stem the tide of rural migrants into the capitalist sector which is incapable of absorbing them and to ensure that control of the rural masses is retained in its own hands.

The introduction of technology, primarily by state agencies, serves not only to facilitate the subjugation of the peasantry to capital, but it has also led to deteriorating conditions for the majority of the peasants who are poor. Indeed, the average annual net income from the Muda Scheme farm households had declined from about $3,600 in 1975 to $2,430 in 1978/79 [*Jegatheesan*, 1980]. Increased productivity was inevitably followed by steep rises in prices of agricultural land, which is now beyond the reach of most peasants. As a result of the innumerable inputs subsidised by the state, rich landlords are now more willing and able to expand production on a commercial scale rather than rent out their land to croppers. As their position strengthened, they had succeeded in asking for cash rentals thus pushing all the risk of a bad harvest on the tenant. Mechanisation has further eroded the bargaining power of labour. Because the landlord no longer requires his labour for the whole of the production process, he also no longer has an obligation to the peasant to advance him small sums to cover him until the coming harvest. According to one Mada official, 'Longer term concerns over tenant farmer displacement and increasing inequality in farm size distribution as mechanisation enables large-scale operations also arise ... large-scale mechanisation will further impoverish farmers who are already poor ... it is evident that fuel shortages and rising fuel costs will be traumatic issues in heavily mechanised agriculture'. [*Jegatheesan*, 1980]. Caught in the midst of a rigid planting regime imposed by Mada, the poor peasant is forced to make use of fertilisers, fast-growing seeds and machines, all of which cost money. His debts are now large as compared to the consumption credit he used to take from the shopkeeper. As the whole risk of a bad harvest is now his undertaking, any unforeseen circumstances can wipe him out completely.

It is no wonder, then, that the recent demonstrations by padi farmers should be located in the Muda Scheme itself, the area where the state has poured in most money. After the demonstration, the state threatened that it would not hesitate to amend the law to impose heavier penalties on those found guilty of taking part in the demonstrations for the latter action could instil fear among would-be investors. The Kedah government report which stated that over half the households in this state (where the two major peasant demonstrations occurred) were living below the state-defined poverty line has been totally ignored [*Business Times*, 16/1/1980: 6].

The above discussion suggests that the sharecropping system will not be the major feature in peasant agriculture for very much longer. Where this system has given place to the cash rental system in areas of widespread, organised, high state input, as in the Muda Scheme, the poor peasantry have remained just as poor and oppressed. And unless a fundamental change in structure of land ownership and other means of production takes place, we can expect widespread unrest as the condition of the poor peasantry continues to deteriorate.

NOTES

1. Will henceforth be referred to as Malaysia.
2. Besides Malaysia, the other five countries investigated by the ILO were: Bangladesh, India, Pakistan, Sri Lanka and the Philippines [*International Labour Organisation*, 1977: 16 and 187].
3. The Muda Irrigation Scheme is the single largest irrigation project undertaken by the Malaysian government. In Malaysia, only half of the rice lands are under double cropping, mainly as a result of state-developed irrigation projects. Of the area under double cropping approximately 44% are located within the Muda Project which covers 260,000 acres (Calculated from figures given in *The Mid-Term Review of the Third Malaysia Plan 1976-1980*, 1979: 128-129). The total cost of developing the scheme will finally come up to $400 million ($1 Malaysian = U.S. $0.44). 59% of the present capital expenditure of $228 million consisted of a loan from the World Bank. In 1970 a semi-autonomous public agency, the Muda Agricultural Development Authority (Mada) was established to plan and implement the development programme [*Jegatheesan*, 1977].
4. The agricultural and mining sectors contributed approximately 30% to the GDP in 1978 [*Bank Negara Malaysia*, 1979: 7]. Malaysia is the world's largest exporter of rubber, oil palm, tropical hardwood, pepper and tin. In 1975, foreign ownership of share capital in limited companies came up to approximately 55% [*Bank Negara Malaysia*, 1979].

5. The presence of the state in peasant agriculture is expressed in primarily three ways:
   (i) Tight managerial control of all aspects of production and marketing in state-run land schemes such as those of the Federal Land Development Authority (FELDA). Here, peasants who have been bodily uprooted and placed in a totally new environment are organised under strict work regimes under the command of state officials. They are merely semi-proletarians although legally they are supposed to have a share in the land they work on. See Tungku Shamsul Bahrin and Perera, P.D.A. [1977] for details.

   (ii) Only slightly less control is exercised by the state in areas under irrigation authorities. But still, the control is substantial. We say this because these are primarily padi areas where control of water schedule is crucial to the production process although control here is exercised indirectly.

   (iii) In-situ development: where a host of state agencies such as Risda (Rubber Industry Smallholders Development Authority), MARDEC (The Malaysian Rubber Development Corporation), FAMA (Federal Agricultural Marketing Authority), Farmers' Organisation Authority were established to provide peasants with supportive service such as extension, credit, training, subsidies, research, marketing and processing. For details on these various state agencies and many more, see *The Fourth Malaysia Plan 1981-1985* [1981:265-269].

6. Crawfurd, J. [1820 Vol. III: 510] noted that 'among the causes which have contributed to retard their (Indian Island) ... progress in improvement, the scarcity of iron deserves a prominent place.'

7. Over the period 1956-61, it was estimated that 81% of Malaysia's starchy food calories were derived from rice [*Purvis*; 1966].

8. For a detailed discussion of classes in these two villages, see Halim [1980].

9. 1 guni (1 sack) of padi weighs approximately 30 gantang and when milled is equivalent to approximately 21 gantang of rice. 1 gantang = 1 imperial gallon.

10. To encourage domestic rice production the state had introduced protection which ultimately raised the market price of the nation's staple above the world level. 'This policy limits the government's loss but results in important distributive implications' [*Goldman*, 1975: 270].

11. This is not far from figures describing tenancy in other parts of Malaysia. Fisk's study of three villages found that no less than 80% of farms were run on a sharecrop basis [1961]. According to Hill, 'tenancy per se in the form of sharecropping is far more widespread than the Census (1960) would suggest' [1967: 103].

12. By ruling class is meant those of the 'kerabat class' which includes royalty and aristocrats; the 'orang keistimewaan' – common birth leaders who held high positions in the royal court and the government; and the 'orang besar-besar' of various ranks – constituting the primary officers of the state and other titled persons.

13. In the Malay States of the 18th Century, 'Land is so abundant in proportion to the population that they scarcely consider it as a subject of right any more than elements of air and water' [*Maxwell*, 1884: 117].

14. In addition to forced labour, the 'rapai' was collected from all cultivators and consisted of 100 lb. of rice regardless of land farmed [*Annual Report, Kedah*, 1909; *Crawfurd*, 1820 Vol.I: 43].

15. The Malay Reservation Act, passed in 1913, aimed at restricting dealings in land designated as Malay reserves thus controlling the sale and state alienation of such land to within the Malay population. See Wong [1975: Chapter 13] for details and further amendments to the original enactment.

16. It is not surprising to read that prior to the British imposition of legal proprietary rights over land, the sharecropping system was already in existence but only on the most fertile and densely populated lands which were under the absolute authority of the Sultan and which,

by no mere coincidence, were also nearest to centres of administration [*Maxwell*, 1884: 80; *Crawfurd*, 1820 Vol. III: 51].

17. Access to land can also be attained through the cash rental system but, outside the State-controlled irrigation schemes, this system was not common. For example, in the villages studied we found only four cases of cash rental ranging from $25 to $15 per season per 0.71 acres. The land, however, was located outside the villages but was owned by peasants in this study.

18. This is the rough estimation of the land officer.

19. A ridiculous situation is often reached where a part of a piece of land, for example, one seventh of a piece of land measuring 2 roods and 1.1 perch was divided amongst 11 heirs. A glance at the land records would show that ownership of such small fractions is common.

20. The Padi Cultivation (Control of Rent Security of Tenure) Act 1967 is the existing legislation aimed at controlling the level and system of rent (maximum level of rent to be one third of the total harvested yield of the rented land) in padi lands, thus affording some security of tenure to padi cultivators by controlling owner–cultivator relations between landlord and sharecropper.

21. After the 1969 'race riots' in Kuala Lumpur, the New Economic Policy was launched to expand the number of Malay capitalists by positively discriminating in favour of this group in all spheres of economic activity. It was estimated that in 1975, the Malay share of accumulated capital was around 10% [*New Straits Times*, 27.1.1980: 5 quoting the Chairman of UNMO Economic Bureau] while its share in the total population was 53%. These figures were supposed to compare unfavourably with figures on Chinese ownership and control. Although the Chinese formed 36% of the total population their share of equity capital in limited companies was 28% for the same period [*Bank Negara*, 1979:16]. However, it must also be mentioned here that:
    i) intra-race inequalities account far more for total inequality than inter-race inequality, [*Anand*, forthcoming];
    ii) that the glaring economic difference between the Chinese and the Malays is also largely a consequence of locational differences, i.e. in 1975 only 18% of the Malays but 50.8% of the Chinese are found in urban areas with a population of 10,000 and above [*Review of the Third Malaysia Plan*, 1978: 68], with the rural/urban mean income difference much larger at $228 as opposed to the Malay/Chinese mean income difference of $122;
    iii) foreigners own about 55% of share capital in limited companies [*Bank Negara*, 1979:16].

    One of the results of political intervention was the creation of a symbiotic relationship between Malay political elites with Chinese capitalists. For instance, although almost 100% of logging licences issued by state governments had gone to Malays [*The Star*, 11.2.80:3] and about 62% of the padi wholesale licences were issued to Malays [*New Straits Times*, 27.1.80:5], very few of them were actually carrying out the padi or logging operations. For instance in the case of padi wholesale business, only 24% of the Malay licence holders 'are active in the trade'. These Malays had sold their licences to non-Malays.

22. This is most clearly exemplified in the Muda scheme where, besides 'the emergence of a class of farmer-contractors' one also finds the intrusion of urban-based capitalists into farming and 'an increasing number of gentlemen farmers comprised of Malay bureaucratic elites who would invest in rice lands, many of which are under Malay reservation areas' [*Ismail*, 1980: 8, 16].

23. According to a Mada economist 'with rising cost of production the economic status of padi farmers on the whole has deteriorated and the proportion of peasants living under the poverty line is expected to increase from year to year' [*Ismail*, 1980:15]. A Kedah government report states that 'the farmers outside Muda are poverty stricken by relative and absolute poverty' [*Business Times*, 16.1.80: 6]. The per capita GDP in Kedah–Perlis had in fact declined from 67% of the national average in 1970 to 56% in 1978.

24. For accounts of the Baling demonstrations by rubber smallholders, see *Far Eastern Economic Review (FEER)* [10.1.1975: 29-31]. A balanced account of the demonstrations by rice farmers can be obtained by referring to *Asia Week* [8.2.1980, 15.2.1980] *FEER* [22.2.1980], *The Star* [27.1.1980] *New Straits Times* [27.1.1980], *Business Times* [6.2.1980] and *Watan* [8.2.1980].

25. In Merah, the price for 0.71 acre of rubber land was approximately $1,000 when it carried a grant. A similar piece of land without a grant would cost $300. It was not preposterous that there was this trade in illegally occupied land; there was as yet no case where land opened illegally had been taken over by the state, at least not in Merah, which is located in the more remote part of Kedah.

26. This could stem from the fact that with an open resource situation 'every man could earn a wage living by working for himself and none need earn a wage' [*Gullick*, 1951: 55]. Biddulph commented that 'For many years, there has been a reluctance on the part of the Malays to resort to a wage economy. They have preferred to live in the village and only to work to the extent which they considered necessary in order to live' and that means, mainly as migrant workers [1953: 30].

27. Widespread labour unrest culminated in the declaration of Emergency rule which lasted from 1948 to 1960. A comprehensive overview of this period can be culled from Short [1975]. For information on strikes and labour opposition see Stenson [1977] and Palmer [1964].

28. From the 1970s onwards when import-substitution industrialisation had reached its logical limits growth had to depend on export demand. To gear industries to the competition of international markets, firms had to rationalise. As a result more industries will have to be capital intensive. Recent studies have shown that local manufacturers have not been able to fill this role. It is, therefore, the foreign multinationals which have contributed to our phenomenal industrial expansion [*Chee*, 1975]. The recent high rates of manufacturing exports are artificial to the extent that they only reflect the temporary boom of the electronic industry while much of the value increases in other manufacturing exports between 1972-74 involved price increases and very little in the way of volume increase [*Bell*, 1977]. With a predicted downturn in the growth potential of advanced countries and an increase in large firms (with elasticity of employment with respect to output) we can continue to expect a lower rate of absorption of rural workers.

29. We must also keep in mind the countervailing tendencies against mass resistance, e.g. the frantic efforts of peasants who dissipate their energy roaming the countryside to look for a living; and the repressive machinery of the state under the Internal Security Act.

REFERENCES

Abdullah, M., 1979, 'Investment Incentives Legislation and National Growth' reported in *The Business Times*, 23 May.

Anand, S., (forthcoming), *Inequality and Poverty in Malaysia: Measurement and Decomposition*, New York: Oxford University Press.

Annual Report, Department of Agriculture, 1953.

__________, *Kedah*, 1909-1910.

Asiaweek, various issues, Hong Kong.

Bank Negara Malaysia, 1979, *Money and Banking in Malaysia*, Kuala Lumpur.

Bell, P.W., 1977, *Industrial Growth and Economic Development in Malaysia* Discussion Paper, Series No. 2, Penang: Universiti Sains Malaysia.

Biddulph, J.P., 1953, *Annual Report of the Ministry of Labour*, Kuala Lumpur.

Business Times, various issues, Kuala Lumpur.

Census of Manufacturing Industries, 1975, Kuala Lumpur; Department of Statistics.

Chee, P.L., 1975, 'The Role of Small Industry in the Malaysian Economy', Ph.D Thesis, Universiti Malaya

Cowgill, J.V., 1928, 'System of Land Tenure in the Federated Malay States', *Malayan Agricultural Journal*, Vol. xvi, No. 5.

Crawfurd, J., 1820, *History of the Indian Archepelago*, Vol. I, III, Edinburgh: Archibald Constable & Co.

Cutler, A., 1975, 'The Concept of Ground Rent and Capitalism in Agriculture', *Critique of Anthropology*, No. 4/5

Far Eastern Economic Review, various issues, Hong Kong.

Fisk, E.K., 1961, 'Productivity and Income from Rubber in an Established Malay Reservation', *Kajian Ekonomi*, 6

Fourth Malaysia Plan 1981-1985, Kuala Lumpur.

Goldman, R.H., 1975, 'Staple Food Self-sufficiency and the Distributive Impact of Malaysian Rice Policy', *Food Research Institute Studies*, Stanford University, 14 No.3.

Gullick, J.M., 1951, 'The Negri Sembilan Economy of the 1890's', *Journal of the Malayan Branch of the Royal Asiatic Society*, Vol. xxiv.

Halim, F., 1980, 'Differentiation of the Peasantry: A Study of the Rural Communities in West Malaysia', *Journal of Contemporary Asia*, Vol. x, No. 4.

__________, 1981, 'Rural Labour Force and Industrial Conflict in West Malaysia' to be published in *Journal of Contemporary Asia*, Vol. xi, No. 4.

Handbook of Rubber Statistics, 1975, Department of Statistics, Kuala Lumpur.

Hill, R.D., 1967, 'Agricultural Land Tenure in West Malaysia', *Kajian Ekonomi*, 12.

International Bank for Reconstruction and Development (IBRD), 1955, *The Economic Development of Malaya*, Singapore.

International Labour Organisation (ILO), 1977, *Poverty and Landlessness in Rural Asia*, Geneva.

Ismail, Z., 1980, *'Economic and Social Aspects of Padi Production'*, Seminar on Economics, Development and the Consumer, Penang.

Jegatheesan, S., 1977, *The Green Revolution and the Muda Irrigation Scheme*, Muda Agricultural Development Authority (MADA), Kedah, Malaysia.

Lenin, V.I., 1972, *The Development of Capitalism in Russia*, Collected Works, Vol. III. Moscow: Progress Publishers.

Maxwell, W.E., 1884, 'Law and Customs of the Malays with reference to the Tenure of Land'. *Journal of the Malayan Branch of the Royal Asiatic Society*, xiii.

Marx, K., 1959, *Capital*, Vol. III, Moscow: Progress Publishers.

Mid-Term Review of the Third Malaysia Plan 1976-1980, Kuala Lumpur.

New Straits Times, various issues, Kuala Lumpur.

Parmer, J.N., 1964, 'Chinese Estate Workers' Strike in Malaya in March 1937', Cowan, C.D., (ed.) *The Economic Development of South East Asia*, London: G. Allen & Unwin.

Purvis, M., 1966, 'Evaluation and Use of Under-developed Agricultural Statistics: The Food Economy of Malaysia', Ph.D Thesis, Cornell University.

Sarji, A., 1977, *Farmers' Cooperatives*, Kuala Lumpur; Farmers' Organisation Authority.

Short, A., 1975, *The Communist Insurrection in Malaya 1948-60*, London; Muller.

South East Asian Bank Countries Research and Training Centre, 1978, *Agricultural Credit in South East Asian Banks Countries*, Petaling Jaya.

Star, various issues, Kuala Lumpur.

Stenson, M.R., 1977, *Industrial Conflict in Malaysia*, Kuala Lumpur; Oxford University Press.

Suyama, T., 1962, 'Pang Society: The Economy of Chinese Immigrants', Tregonning, K.G. (Ed.), *Papers on Malayan History*, Singapore.

Tungku Shamsul Bahrin & Perera, P.D.A., 1977, *Felda: 21 Years of Land Development*, Malasia: Federal Land Development Authority.

Watan, various issues, Kuala Lumpur.

Wong, D., 1975, *Tenure and Land Dealings in the Malay States*, University of Singapore Press.

INDEX

(Note: sharecropping is abbreviated to s/c throughout except for its own entry.)